THERMODYNAMIC PROPERTIES OF THE ELEMENTS

Tabulated values of the heat capacity, heat content, entropy, and free energy function of the solid, liquid, and gas states of the first 92 elements are given for the temperature range 298° to 3000°K. Auxiliary data include temperatures and heats of transition, melting, and vaporization and vapor pressures. Literature sources are listed. The published values have been analyzed and are supplemented by estimates when experimental data are lacking. With the aim of providing the basic data for the elements needed in the calculation of the thermodynamic properties of chemical compounds, the tables were compiled by D. R. Stull and G. C. Sinke at the Thermal Laboratory of the Dow Chemical Co. These up to date tables especially fill the need for data in the increasingly important high temperature region.

Number 18 of the Advances in Chemistry Series
Edited by the Staff of *Industrial and Engineering Chemistry*

Published November 1956 by
AMERICAN CHEMICAL SOCIETY
1155 Sixteenth St., N.W.
Washington 6, D. C.

Copyright 1956 by
AMERICAN CHEMICAL SOCIETY

All Rights Reserved

Preface

Preparation of consistent tabulations of thermodynamic data is a difficult task because of the complex interrelations of the data. One new set of data can require changes in numerous related values. For inorganic thermodynamic compilations, the logical starting point is a reliable tabulation of data for the elements. If data for compounds are to be compared, they must be based on the same elemental data. Once the thermodynamic data for the elements have been fixed, then equilibria involving the elements and compounds can be treated to fix the stability of the compounds. When the heats of sublimation and ionization of the elements are available, Born-Haber cycle calculations can be carried out for ionic compounds to check the reliability of data for the compounds.

The present tables are thus an important step in the preparation of complete thermodynamic tables for inorganic compounds at high temperatures. Because of recognition of the importance of data for the elements, there has been a great upsurge in experimental work on the elements during the last decade. The availability of large calculating machines has made the calculation of gaseous thermodynamic data from spectroscopic data almost routine. In particular, there has been a great deal of work on the determination of vapor pressures and heats of sublimation of the elements. In view of these recent developments, the present compilation is greatly improved over similar previous work, and is the first complete compilation for the elements at closely spaced temperature intervals.

The authors have chosen 298.15° K. instead of the conventional 0° K. as the standard reference temperature for preparation of free energy function tables. 0° K. is the most logical reference temperature for calculating thermodynamic functions from spectroscopic data. However, most heats of formation of condensed phases are given at 25° C. or 298.15° K. and data are often lacking for converting to 0° K. Whenever data are available for calculating thermodynamic functions using the 0° K. reference temperature, data are also available for converting to the 298.15° K., but the reverse is not always true. Thus, for high temperature thermodynamic calculations, 298.15° K. is a more convenient reference temperature. Inasmuch as it will be necessary to combine the $(F° - H°_{298})/T$ functions given here for the elements with $(F° - H°_0)/T$ functions given elsewhere for compounds, the authors have provided the $(H°_{298} - H°_0)$ values that will allow conversion from one reference temperature to the other.

It is perhaps regrettable that the authors have not elected to attempt the difficult task of assigning uncertainties to the data, particularly the heats of sublimation. Often, differences between thermodynamic quantities are known with much higher accuracy than the absolute values are known. Thus, it is frequently necessary to retain many figures beyond the last significant figure to retain the accuracy of the relative values. However, when no indication of the absolute accuracy is given, the person using the tables can be deceived by the number of figures presented. It is important to know the limitations of the calculations that one is making. Some of the heats of sublimation given in the tabulations are uncertain enough to cause uncertainties in the calculated vapor pressures by as much as a factor of 2, and in some instances by a factor of 10. It is difficult to assign uncertainties because the uncertainty is usually not due to random error in the determinations but often to unknown systematic errors. The compiler must, from his experience with the technique used, his appraisal of the experimenters, and possible checks with theoretical or empirical rules, try to guess the odds by which the values he listed may be off by a given amount. Although difficult, it is important to try to do this.

Because of the recent change in the temperature scale, as well as changes in the best values for the fundamental constants, thermodynamic tabulations from different sources are not quite consistent. These differences are usually negligible from the practical point of view, but they can be annoying when thermodynamic calculations are being checked for arithmetic errors, because different ways of carrying out the calculations will give slightly different answers. The change due to the change in the temperature scale and the resulting change in R is 1 in 27,000, and should be kept in mind when the values in this tabulation are combined with values from earlier tabulations based on the old temperature scale. Also, the values used for h, k, and N have fluctuated in recent years and the values used in this compilation are probably not the ones that will be generally accepted in the future. Here again, the corrections are nuisance corrections rather than significant ones, but it appears likely that the data tabulated in these tables for gaseous elements will have to be recalculated for complete consistency with future tabulations when general agreement has been reached on the values of h, k, and N, as well as the ratio of the physical and chemical atomic weight scales. Fortunately, the modern calculating machines make this chore relatively easy.

A word should be said about the use of these tables in evaluating vapor pressure data or in calculating vapor pressures from the heats of sublimation. Because of the difficulty in obtaining accurate temperature coefficients, the calculation of heats of sublimation or vaporization from the temperature coefficient of the vapor pressure is often not reliable. When entropies are known and free energy functions are available, the preferred method of treating the data is to calculate the heat of vaporization or sublimation from each vapor pressure by means of the relationship, $\Delta H^\circ_{298} = T\,[\Delta F^\circ/T - (\Delta F^\circ - \Delta H^\circ_{298})/T]$, where the function $(\Delta F^\circ - \Delta H^\circ_{298})/T$ is tabulated in the tables and $\Delta F^\circ/T$ is obtained from the equation $\Delta F^\circ/T = -RT \ln P$. If the data have no serious temperature-dependent errors, the values of ΔH°_{298} derived from data at different temperatures will show no trend with temperature. If the data are subject to error, ΔH°_{298} will show a trend with temperature. However, a reasonably good value can still be obtained from the average ΔH°_{298}, whereas the temperature coefficient of the vapor pressure would yield a heat greatly in error.

Because the heats of sublimation and vapor pressures are related through the free energy functions, it is important that they be used consistently. Heats of sublimation derived through the use of free energy functions in other tabulations should not be used with the tables given here. Comparison of the data in this compilation with those in other compilations will show differences in the tabulated values, even though the same original data were treated, because of different methods of preparation of the free energy functions. In spite of the differences between tabulated heats of sublimation and free energy functions in different tables, the original data can be reproduced from either set if the heat of sublimation is used together with the free energy function that was used to obtain it.

Likewise, the heat capacity values tabulated in the present compilation may appear different from those of other compilations, even when the original data are the same. This is due to the fact that the original measurements are usually heat content measurements at high temperatures and the accuracy of the heat content measurements is not sufficient to allow the temperature dependence to be fixed explicitly. Different people assume different functions to represent the temperature variation of the heat content or heat capacity. For example, some prefer to take an average constant heat capacity to represent data for a limited liquid range. Others will assume a linear variation with temperature with some relationship between the two coefficients of the heat capacity equation.

Clearly, the user of any thermodynamic tables must become familiar with the tables and the interrelationships of the data if he plans to make extensive use of the values. Moreover, he must not use them blindly. The actual numbers tabulated for the different thermodynamic functions are not so significant as the final equilibrium constants that are to be calculated from them. These tables are designed to yield equilibrium constants of as high an accuracy as can be obtained from the available data. Thus, the uncertainty of a given heat of sublimation may be considerably smaller in regard to its use for calculation of vapor pressures than in regard to its use for heat balance calculations.

The above considerations point out the importance of having all thermodynamic tables prepared in a consistent way, preferably by a single group. New data are being obtained at a rapid rate and it is important to have some permanent staff of experienced people providing continuous revisions, either through use of loose-leaf additions or through lists of revised values. The National Bureau of Standards has made a start in this direction with the publication of Series I and II of their thermodynamic compilations in Circular 500. Since the publication of Circular 500, however, the NBS group appears to have lost its momentum; work on Series III, the high temperature compilation, seems to have come to a halt except for some tabulations published in the *Journal of Research of the National Bureau of Standards*. The job of obtaining a complete tabulation of all available data is such an enormous one that no single group could hope to do it adequately. It is to be hoped that many groups will contribute by tabulating data in their fields of interest so that the first stage of a complete compilation can be achieved. Then it might be possible for the National Bureau of Standards to keep these tables up to date, but even this would require much more adequate staffing and support.

In many instances, estimates were necessary to carry out the calculations. Even for the elements a surprising amount of experimental data are necessary to put the tables on a firmer basis. It is hoped that research workers will take note of these gaps and endeavor to fill them when they have the equipment and materials on hand. Thermodynamics can be an extremely powerful tool, but its edge is severely blunted when the fundamental starting data are lacking.

Berkeley, California LEO BREWER
August, 1956

CONTENTS

Introduction	3
Physical Constants and Terminology	6
Sources and Discussion of Data	10
Tabulated Values of Thermodynamic Properties	36
Bibliography	227
Index	234

Introduction

RECENT YEARS have seen a considerable widening of the scope of chemical technology. Elements previously neglected or unavailable have become important as fission products in atomic reactors. The search for materials possessing the unusual properties needed for applications in atomic energy, aircraft, electronics, and many other industries has raised laboratory curiosities to quantity production status. Higher and higher temperatures are being used in routine chemical operations, with the prospect of high temperature process heat from atomic energy increasing the attractiveness of this field. Along with the broadened horizon of technology has come a growing recognition of chemical thermodynamics as a useful research and engineering tool. When accurate data are available, a screening program based on thermodynamic calculations frequently points out the most favorable approach to the production of the elements and their compounds. Even estimates can sometimes indicate the most promising of the available routes or the most suitable material for a particular purpose.

Examples of thermodynamic calculations are discussed by Kelley (*181*, *183*, *184*) (italic numbers in parentheses refer to the bibliography on page 227) in his pioneer bulletins on practical applications of thermodynamics. Other examples are detailed by Brewer and coworkers (*36*). More recently, Margrave (*222*) has presented the advantages of the free energy function in thermodynamic calculations. In the hydrocarbon field, representative papers by Rossini and coworkers (*103*, *189*, *339*) demonstrate the value of thermodynamics for the petroleum industry.

In order to make thermodynamic data for chemical compounds consistent and directly comparable, values for the heats and free energies of formation must all refer to a single set of data for each element. It was to provide such a set of data for the temperature range from 298° to 3000° K. that this project was undertaken. The choice of a reference state for any particular element is somewhat arbitrary, but we believe the most practical choice is that of the condensed state up to the temperature at which the vapor pressure of the element reaches one atmosphere and the ideal gas state above this temperature. We have therefore elected to use the crystalline solid from room temperature to the melting point at one atmosphere, the liquid from the melting point to the normal boiling point, and the most representative ideal gaseous species in the temperature range from the normal boiling point to 3000° K. In two cases, arsenic and phosphorus, the vapor pressure of the solid reaches one atmosphere at a temperature below the melting point and the liquid is not used as a reference state for these elements.

Tables for this defined reference state, including the heat capacity, the heat content relative to 298.15° K., the absolute entropy, and the free energy function at even 100° intervals from 298.15° to 3000° K. have been assembled for the first 92 elements. These tables are arranged alphabetically beginning on page 36. The choice of 298.15° K. as the reference temperature is made because the low temperature heat capacities of many elements and compounds are not known. Most of the thermodynamic data now reported in the literature refer to 25° C., which, when combined with the recent international agreement on 273.15° K. for the ice point (*319*) gives a reference temperature of 298.15° K. The figure 298° K. quoted in the tables and text should be understood to be the reference temperature, 298.15° K. For those who prefer to use 0° K. as the reference temperature, we have included, for cases in which it is known, the heat content at 298.15° K. relative to 0° K.

Changes in phase in the tables are indicated by lines drawn across the tables in the appropriate temperature interval, while the nature of the phase change is easily determined from the description of the reference state at the top of each table. The tables are based on 1 gram atomic weight of the element, except for hydrogen, oxygen, nitrogen, and the halogens, for which the more familiar diatomic form is used.

In addition to these reference state tables, we have tabulated the thermodynamic properties of all but a few of the ideal gaseous species over the entire range from 298.15° K. to 3000° K. These values can be readily calculated from molecular constant and spectroscopic data by methods described in standard texts (*153*, *225*, *271*). Pertinent data were mainly taken from the compilations of Moore (*241*), Herzberg (*152*), and Landolt-Bornstein Tabellen (*208*). Estimates were made for a few molecules for which spectroscopic data were not available.

At temperatures below the normal boiling point of an element, the heat and free energy of formation of these gaseous species refer to the process

$$xE \text{ (condensed)} \rightarrow E_x \text{ (gas)}$$

The equilibrium constant of this reaction is equal to the equilibrium pressure of the gaseous species E_x over the condensed state. The logarithm of the equilibrium constant of formation, given in the last column of these tables, is, therefore, identical with the logarithm of the partial pressure in atmospheres of the gaseous species E_x in the saturated vapor of the element E. The total vapor pressure of the element E is obtained by adding together the partial pressures of the various species which make up the vapor. In the frequent cases in which only the monatomic form is present, the vapor pressure is directly determined from the logarithm of the equilibrium constant of formation of the monatomic gas.

Above the normal boiling point, the heat and free energy of formation refer to the process

$$\frac{x}{y} E_y \text{ (gas)} \rightarrow E_x \text{ (gas)}$$

in which E_y is the gas species selected as the reference state and E_x is any other form of interest. The equilibrium constant in this case can be used to calculate the amount of each form present in the vapor, provided the saturation pressure of the element is not exceeded. The vapor density can be readily calculated from these data by considering the equilibria among the various gaseous species.

Upon examination of the tables, it will become evident that the gaseous reference state selected may not be the most stable state over the entire temperature range in which it is used. For example, our selection of ideal diatomic gas for the reference state of phosphorus is based on the fact that this form predominates in most of the temperature range from 704° to 3000° K. For a short range just above 704° K., however, the equilibrium constant of formation of the tetratomic form indicates that this form is more stable than the diatomic form. Similarly, monatomic chlorine is more stable than the reference state of diatomic chlorine at temperatures above 2100° K. Since the transition from one gaseous species to another takes place over a considerable temperature range, it is not possible to select a single species for these elements which is always the most stable. In order to keep an unambiguous expression for the equilibrium constant, however, it is necessary to use a single form for the reference state. When considering the equilibrium point of a reaction involving the reference state of an element, therefore, it may be necessary also to take into account the equilibrium between the reference state and other elemental gas species.

Sources and discussion of the data presented for each element begin on page 10. The elements are arranged alphabetically according to their names. In many cases, the tables for the condensed states represent an assemblage of information from previous exhaustive compilations, particularly those of Kelley (*180*, *182*, *185*, *186*), Rossini and coworkers (*274*), and Brewer (*35*). We have searched the literature through 1955 in order to bring these compilations up to date. A considerable mass of new thermodynamic data is now available and it is believed that in many cases the present tables are an improvement. For some of the gaseous species, earlier calculations by Brewer (*35*) and by Katz and Margrave (*173*) agree in general with our values. Since our data are given at smaller temperature intervals, interpolation should be simplified. Several investigators have made their experimental information available to us in advance of publication and we express our gratitude to them. Comments and suggestions by readers of an earlier limited edition have been very helpful.

In spite of the diligent efforts of many able scientists, there are still many gaps in our knowledge of the thermodynamic properties of the elements. In cases where other workers have made reasonable estimates needed to fill these gaps we have used them, while in numerous instances we have made our own estimates. It is difficult to assess the reliability of these estimates and we have operated on the principle that an "educated guess" may be of some value. When experimental data are available we will be among the first to abandon our estimates.

We wish to thank the many coworkers who assisted in the assembling, computation, and printing of this report.

D. R. STULL AND G. C. SINKE

The Dow Chemical Co.
Midland, Michigan
July 1956

Physical Constants and Terminology

ATOMIC WEIGHTS

The values used are the 1953 international atomic weights published by Wichers (*345*) except for the 12 elements revised on recommendation of the Commission on Atomic Weights of the International Union of Pure and Applied Chemistry (*101*).

Element	Symbol	Atomic Number	Atomic Weight[a]
Actinium	Ac	89	227.
Aluminum	Al	13	26.98
Antimony	Sb	51	121.76
Argon	A	18	39.944
Arsenic	As	33	74.91
Astatine	At	85	(210)
Barium	Ba	56	137.36
Beryllium	Be	4	9.013
Bismuth	Bi	83	209.00
Boron	B	5	10.82
Bromine	Br	35	79.916
Cadmium	Cd	48	112.41
Calcium	Ca	20	40.08
Carbon	C	6	12.011
Cerium	Ce	58	140.13
Cesium	Cs	55	132.91
Chlorine	Cl	17	35.457
Chromium	Cr	24	52.01
Cobalt	Co	27	58.94
Copper	Cu	29	63.54
Dysprosium	Dy	66	162.51
Erbium	Er	68	167.27
Europium	Eu	63	152.0
Fluorine	F	9	19.00
Francium	Fr	87	(223)
Gadolinium	Gd	64	157.26
Gallium	Ga	31	69.72
Germanium	Ge	32	72.60
Gold	Au	79	197.0
Hafnium	Hf	72	178.50
Helium	He	2	4.003
Holmium	Ho	67	164.94
Hydrogen	H	1	1.0080
Indium	In	49	114.82
Iodine	I	53	126.91
Iridium	Ir	77	192.2
Iron	Fe	26	55.85
Krypton	Kr	36	83.80
Lanthanum	La	57	138.92
Lead	Pb	82	207.21

THERMODYNAMIC PROPERTIES OF THE ELEMENTS

Element	Symbol	Atomic Number	Atomic Weight[a]
Lithium	Li	3	6.940
Lutetium	Lu	71	174.99
Magnesium	Mg	12	24.32
Manganese	Mn	25	54.94
Mercury	Hg	80	200.61
Molybdenum	Mo	42	95.95
Neodymium	Nd	60	144.27
Neon	Ne	10	20.183
Nickel	Ni	28	58.71
Niobium	Nb	41	92.91
Nitrogen	N	7	14.008
Osmium	Os	76	190.2
Oxygen	O	8	16 (defined)
Palladium	Pd	46	106.4
Phosphorus	P	15	30.975
Platinum	Pt	78	195.09
Polonium	Po	84	210.
Potassium	K	19	39.100
Praseodymium	Pr	59	140.92
Promethium	Pm	61	(145)
Protactinium	Pa	91	231
Radium	Ra	88	226.05
Radon	Rn	86	222
Rhenium	Re	75	186.22
Rhodium	Rh	45	102.91
Rubidium	Rb	37	85.48
Ruthenium	Ru	44	101.1
Samarium	Sm	62	150.35
Scandium	Sc	21	44.96
Selenium	Se	34	78.96
Silicon	Si	14	28.09
Silver	Ag	47	107.880
Sodium	Na	11	22.991
Strontium	Sr	38	87.63
Sulfur	S	16	32.066[b]
Tantalum	Ta	73	180.95
Technetium	Tc	43	(99)
Tellurium	Te	52	127.61
Terbium	Tb	65	158.93
Thallium	Tl	81	204.39
Thorium	Th	90	232.05
Thulium	Tm	69	168.94
Tin	Sn	50	118.70
Titanium	Ti	22	47.90
Tungsten	W	74	183.86
Uranium	U	92	238.07
Vanadium	V	23	50.95
Xenon	Xe	54	131.30
Ytterbium	Yb	70	173.04
Yttrium	Y	39	88.92
Zinc	Zn	30	65.38
Zirconium	Zr	40	91.22

[a] A value given in parentheses denotes the mass number of the isotope of longest known half life.

[b] Because of natural variations in relative abundance of the sulfur isotopes, its atomic weight has a range of ±0.003.

PHYSICAL CONSTANTS

Rossini and coworkers (*272*) give values for the necessary physical constants. The new value of the ice point has been used (*319*), necessitating changes in some of the derived constants. Although some calculations were made using older constants, the difference does not affect the thermodynamic functions in the second decimal place.

Name and Symbol	*Value and Units*
Velocity of light, c	2.997902×10^{10} cm./sec.
Planck constant, h	6.62377×10^{-27} erg sec./molecule
Avogadro constant, N	6.02380×10^{23} molecules/mole
Faraday constant, F	96,493.1 coulombs/equivalent
Absolute temperature of ice point T (0° C.)	273.15 ° K.
Pressure-volume product for 1 mole of gas at 0° C. and zero pressure $(PV)_{T=0°\text{C.}}^{P=0}$	2271.16 joules/mole
Electronic charge $e = F/N$	1.601864×10^{-19} coulomb
Gas constant $R = \dfrac{(PV)_{T=0°\text{C.}}^{P=0}}{T\,(0°\text{C.})}$	8.31469 joules/deg. mole 1.98726 cal./deg. mole
Boltzmann constant $k = R/N$	1.38031×10^{-16} erg/deg. molecule
Constant relating wave number and energy $Z = Nhc$	11.96171 joule cm./mole 2.858917 cal. cm./mole
Standard atmosphere (atm.)	1,013,250 dynes/cm.^{2}
Thermochemical calorie	4.1840 (exact) joules 4.18331 int. joules 41.2929 cm.^{3} atm.

TERMINOLOGY

g.f.w. = Gram formula weight
$H_{298.15} - H_0$ = Enthalpy at 298.15° K. relative to 0° K. in cal./g.f.w.
C_p = Heat capacity at constant pressure in cal./deg./g.f.w.
$H_T - H_{298.15}$ = Enthalpy or heat content at temperature T° K. relative to 298.15° K. in cal./g.f.w.
S_T = Absolute entropy at temperature T° K. in cal./deg./g.f.w.
$\dfrac{F - H_{298.15}}{T}$ = Free energy function in cal./deg./g.f.w. = $\dfrac{(H_T - H_{298.15})}{T} - S_T$
ΔH_f = Heat of formation from reference state in cal./g.f.w.
ΔF_f = Free energy of formation from reference state in cal./g.f.w.
$\text{Log}_{10} K_p$ = Logarithm to the base 10 of the equilibrium constant of formation from reference state.
M. P. = Melting point in ° K. at 1 atmosphere pressure.
B. P. = Boiling point in ° K. at 1 atmosphere pressure.
ΔH_m = Heat of melting in cal./g.f.w. at the melting point.
ΔH_v = Heat of vaporization in cal./g.f.w. at 1 atmosphere total pressure.

THERMODYNAMIC PROPERTIES OF THE ELEMENTS

S. P. = Sublimation point in ° K. at 1 atmosphere pressure.
T. P. = Transition point in ° K. at 1 atmosphere pressure.
ΔH_s = Heat of sublimation in cal./g.f.w. at S. P.
ΔH_t = Heat of transition in cal./g.f.w. at T. P.
T_c = Critical temperature in ° K.
P_c = Critical pressure in atmospheres.
T = Absolute temperature in ° K.
K = Kelvin scale of temperatures where 273.15° K. represents the ice point.
e. u. = entropy unit = cal./deg. mole.

Circular superscript, °, denotes the thermodynamic standard reference state of unit activity.

Sources and Discussion of the Data

ACTINIUM

Foster (*115*) has made a preliminary report that indicates the melting point is about 1470° K. and the normal boiling point is about 3600° K. The rest of the data are all estimated and are intended to serve only until measured data are available.

ALUMINUM

Giauque and Meads (*123*) have measured the low temperature heat capacity from 15° to 302° K. and calculate at 298° K. an entropy of 6.769 ± 0.02 e. u. and an enthalpy of 1094 cal./gram atom. The heat capacity and heat content data for the solid state are taken from the work of Kelley (*185*). The melting point appears well established at 932° ± 1° K. (*185, 206, 274*). A value of 2550 cal./gram atom has been selected as the heat of melting on the basis of Kelley's heat content data, since the recent determinations of Oelsen, Oelsen, and Thiel (*254*) and Wittig (*348*) are not in good agreement. The liquid heat capacity value given by Kelley (*185*) has been extrapolated to the boiling point.

Huff, Gordon, and Morrell (*163*) have calculated the thermodynamic properties of the ideal monatomic gas using spectroscopic data given by Moore (*241*). Vapor pressure measurements have been made by Brewer and Searcy (*39*), Baur and Brunner (*23*), and Farkas (*108*). Giving Brewer and Searcy's data the most weight, we derive a heat of sublimation at 298° K. of 77,500 cal./gram atom, a boiling point of 2720° K., and a heat of vaporization at the normal boiling point of 70,200 cal./gram atom.

ANTIMONY

The low temperature heat capacity has been measured by Anderson (*9*) and by DeSorbo (*83*), who calculates the entropy at 298° K. to be 10.92 ± 0.05 e. u. From these data we calculate the enthalpy at 298° K. to be 1410 cal./gram atom. Kelley (*185*) lists the heat capacity of the solid above 298° K. and the liquid as well as the melting point of 903° K., with an associated heat of melting of 4740 cal./gram atom. These values agree well with those of Kubaschewski and coworkers (*206*), and the heat of melting is about the average of the heat measured by Oelsen, Oelsen, and Thiel (*254*) and by Wittig (*347*). The thermodynamic functions of the monatomic gas were calculated from the spectroscopic values given by Meggers and Humphreys (*233*). The thermodynamic functions of the diatomic gas are based on the data given by Kelley (*185, 186*) and were extended to 3000° K.

In addition to the vapor pressure data listed by Kelley (*180*) and Brewer (*35*), measurements have been made by Nesmeyanov and Iofa (*247*) and by Richards (*267*). Considering all these data, the best fit is obtained by combining the thermodynamic functions of the monatomic and diatomic species with the following values: (1) an entropy of 83.65 e. u. at 298° K. and a heat of sublimation at 298° K. of 49,000 cal./mole for the gaseous species Sb_4; (2) a heat of sublimation at 298° K. of 56,400 cal./mole for the gaseous species Sb_2; and (3) a heat of sublimation at 298° K. of 62,700 cal./mole for the monatomic gas. The last value is consistent with the value of 69,000 cal./mole given by Gaydon (*118*) for the dissociation energy of the diatomic gas. These values lead to a total vapor pressure of all species of one atmosphere at 1910° K., in good agreement with the measured value of von Leitgebel (*211*). The heat of vaporization of 1 gram atomic weight at 1910° K. to the equilibrium vapor is 16,230 cal. The reference state selected is the condensed state below 1910° and the ideal diatomic gas state above 1910° K. Note that the reference state table is based on 1 gram atomic weight (121.76 grams) for all phases.

ARGON

Clusius and Frank (61) find 83.78° K. for the melting point with 280.8 cal./gram atom for the heat of melting as well as 87.29° K. for the normal boiling point and 1558 cal./gram atom for the associated heat of vaporization. These vapor pressure data are substantiated by the more recent work of Clark, Din, Robb, Michels, Wassenaar, and Zwietering (57). Thermodynamic properties of the ideal gas have been calculated at the National Bureau of Standards (295). Kobe and Lynn (193) select 151° K. for the critical temperature and 48.0 atmospheres for the critical pressure.

ARSENIC

From the heat capacity measurements of Anderson (7) from 57° to 291° K., Kelley (186) calculates the entropy at 298° K. to be 8.4 ± 0.2 e. u., and we calculate the enthalpy at 298° K. to be 1226 cal./gram atom. Kelley (185) gives the heat capacity of the solid from 298° K. to the melting point of 1090° K. where he estimates the heat of melting (182) as 6620 cal./gram atom. Our data indicate a vapor pressure of about 28 atmospheres at this temperature. Thermodynamic functions for the monatomic gas were calculated using the spectroscopic data listed by Moore (241). Kelley (185, 186) lists heat content and entropy for the diatomic gas, while Gaydon (118) adopts 90,800 cal./mole for the dissociation energy of the diatomic gas at 0° K.

Vapor pressure data, as reviewed by Brewer and Kane (37), can be represented by assuming the tetratomic molecule to be the gaseous species, with a heat of sublimation at 298° K. of 34,500 cal./mole, an entropy at 298° K. of 75.00 e. u., and a reasonable estimate of the heat capacity. According to this view, there is no appreciable concentration of the diatomic species in the vapor at saturation pressure below 1000° K., which sets a lower limit of about 48,000 cal./mole for the heat of sublimation at 298° K. for the diatomic gas. Comparison with the bond energies of P_4 and Sb_4 gives support to this value.

The data of Preuner and Brockmöller (262) lead to an unreasonably low figure and are somewhat discredited by comparison of their values for phosphorus, antimony, and sulfur with those of other workers. The heat of sublimation at 298° K. to the monatomic species was calculated to be 69,400 cal./mole using Gaydon's value (118) for the dissociation energy of the diatomic gas. This treatment leads to a total pressure of one atmosphere at 886° K. and a heat of sublimation of 7630 cal./gram atom. For the reference state we have selected the solid below 886° K. and the ideal diatomic gas above the sublimation point. Note that the reference state table is based on 1 gram atomic weight (74.91 grams) for all phases.

ASTATINE

These data are entirely estimated by comparison with the other halogens and are intended to serve only until measured data become available.

BARIUM

Kelley (181) estimates the entropy at 298° K. as 16.0 e. u. while Latimer (210) estimates 15.1 e. u. We adopt 15.5 e. u. Kubaschewski (205) has reported solid and liquid heat capacity data, a heat of transition of 140 ± 80 cal./gram atom at 643° K., and a heat of melting of 1830 ± 70 cal./gram atom at 983° K. The heat capacity of the solid beta phase and the liquid phase appear to be extraordinarily high, and when combined with vapor pressure data lead to a very unusual Trouton's constant of about 14. Consequently, we have estimated a lower heat capacity of the condensed phases by comparison with calcium.

Thermodynamic functions of the ideal monatomic gas were calculated using spectroscopic data from Bacher and Goudsmit (19). Vapor pressures have been measured by

Hartmann and Schneider (146) and by Rudberg and Lempert (278). We consider the data of the former workers to be more reliable. Their data give a heat of sublimation at 298° K. of 41,740 cal./gram atom, a normal boiling point of 1910° K., and a heat of vaporization at the normal boiling point of 36,070 cal./gram atom.

BERYLLIUM

Hill and Smith (156) have measured the heat capacity from 4° to 300° K. Their results lead to an entropy of 2.28 e.u. at 298° K. and an enthalpy of 468 cal./gram atom. The recent measurements of solid heat capacity of Ginnings, Douglas, and Ball (127) have been adopted and extrapolated to the melting point. The melting point accepted by several sources is 1556° ± 1° K. (206, 207, 274), while the recent review of Kubaschewski and coworkers (206) gives 2800 ± 500 cal./gram atom for the heat of melting. In the absence of any liquid heat capacity data, we have used the value of 7.50 cal./degree/gram atom estimated by Kelley (187).

Thermodynamic functions for the ideal monatomic gas have been calculated using the energy levels given by Moore (241). The vapor pressure data of Gulbransen and Andrew (138) and of Holden, Speiser, and Johnston (160) are in good agreement, while the results of Schuman and Garrett (288) are too low and the values given by Baur and Brunner (23) have a wrong temperature dependence. We calculate a heat of sublimation at 298° K. of 77,900 cal./gram atom, a normal boiling point of 2750° K., and a heat of vaporization of 70,400 cal./gram atom at the normal boiling point.

BISMUTH

Low temperature heat capacity measurements by Anderson (10), Bronson and MacHattie (42), Keesom and van den Ende (176), and Armstrong and Grayson-Smith (16) were used to calculate an entropy and enthalpy at 298° K. of 13.58 e. u. and 1536 cal./gram atom, respectively. From many sources, Kelley (185) derives an equation for the solid heat capacity above 298° K. Kubaschewski and coworkers (206) select 544.5° K. as the melting point and 2600 ± 50 cal./gram atom for the heat of melting. Data on the liquid heat capacity are discordant and the average value for liquid metals of 7.50 cal./degree/gram atom has been used.

Thermodynamic properties of the ideal monatomic gas were calculated using the energy levels listed in Landolt-Bornstein Tabellen (208). Kelley (185, 186) gives data for the diatomic gas. The dissociation energy given by Gaydon (118), 39,200 cal./mole, indicates the saturated vapor must be largely monatomic. Of the vapor pressure measurements, those of O'Donnell (252) are about an average of the low pressure region while in the normal boiling point range the determination of von Leitgebel (211) is considered most reliable. When corrected for the actual composition of the gas, the results of O'Donnell and von Leitgebel are in excellent agreement and lead to a value of 47,500 cal./gram atom for the heat of sublimation at 298° K. of the monatomic species. Combining this value with the dissociation energy given by Gaydon yields a heat of sublimation at 298° K. of the diatomic form of 55,300 cal./mole. From these data we calculate a normal boiling point of 1832° K. and a heat of vaporization to equilibrium gas at 1832° K. of 36,200 cal./gram atom.

BORON

Johnston, Hersh, and Kerr (168) have measured the heat capacity of the crystalline form from 13° to 305° K., and calculate the entropy at 298° K. to be 1.403 ± 0.005 e. u. and the enthalpy at 298° K. to be 292 cal./gram atom. In the absence of definite information, we have estimated that the solid heat capacity will reach a value of 7.5 cal./degree/gram atom at the melting point and have extrapolated the low temperature measurements in a reasonable manner to obtain this value. Cueilleron (77) has measured the melting point of the crystalline variety and reports a range of 2273° to 2348° K., which we have rounded to 2300° K. Elements with a hexagonal close-packed structure have an average entropy of melting of 2.3 e. u. Using this estimate gives 5300 cal./gram atom for the heat of melting. We estimate the liquid heat capacity to be 7.5 cal./degree/gram atom. Huff, Gordon, and Morrell (163) have calculated the thermodynamic properties of the ideal monatomic gas using the spectroscopic data given by Moore (241). The vapor pressure has been measured by Myers (243) from which we calculate the heat of sublimation at 298° K. to be 141,000 cal./gram atom, a normal boiling point of 4200° K., and a heat of vaporization of 128,800 cal./gram atom at 4200° K.

BROMINE

McDonald (227) has measured the melting point to be 265.95° K., while Rossini and coworkers (274) list 2520 cal./mole (2 gram atomic weights) for the heat of melting. McDonald (227) has also measured the heat of vaporization in the temperature range from 298° to 308° K., from which we derive 7450 cal./mole for the heat of vaporization at 298° K. Evans, Munson, and Wagman (106) have calculated the thermodynamic properties of the ideal diatomic and monatomic gases, while Gaydon (118) gives 45,440 cal./mole for the heat of dissociation at 0° K. Combining the statistical entropy of diatomic bromine gas at 298° K. with McDonald's heat of vaporization and the vapor pressure data selected by Stull (322) gives the entropy of liquid bromine at 298° K. as 36.25 e. u. This value is lower than that of 36.7 e. u. given by Kelley (186) based on low temperature heat capacity data. Since the liquid heat capacity and heat of melting are based on very old measurements, we consider the entropy derived from spectroscopic data to be the more reliable.

The normal boiling point has been selected by Stull (322) to be 331.4° K., although the recent measurements of Fischer and Bingle (112) give a somewhat higher value. The heat of vaporization at the normal boiling point is calculated to be 7170 cal./gram mole. Kobe and Lynn (193) list the critical temperature as 584° K. and the critical pressure as 102 atmospheres.

CADMIUM

Craig and coworkers (75) have recently measured the heat capacity from 12° to 320° K. and have reported the entropy at 298° K. as 12.37 ± 0.01 e.u. These heat capacity data lead to an enthalpy at 298° K. of 1491 cal./gram atom. Kelley's values (185) for the heat capacity of the solid and liquid, the melting point of 594° K., and heat of melting of 1450 cal./gram atom have been used. The last value is in good agreement with the recent measurements of Oelsen and coworkers (253, 254).

The thermodynamic properties of the monatomic gas have been calculated from spectroscopic data given by Landolt-Bornstein (208). Kelley (180) has selected 1038° K. for the boiling point and 23,870 cal./gram atom for the heat of vaporization. This leads to a heat of sublimation at 298° K. of 26,750 cal./gram atom. Recent measurements of O'Donnell (251) and Kotov (202) are in good agreement with Kelley's selected value, giving heats of sublimation of 26,620 and 26,910 cal./gram atom, respectively.

CALCIUM

Kelley (186) selects the entropy at 298° K. as 9.95 ± 0.10 e. u., relying almost entirely on the low temperature data of Clusius and Vaughen (70). From these data we calculate an enthalpy at 298° K. of 1380 cal./gram atom. The solid and liquid heat capacity data given by Kubaschewski (205) have been used. He lists 713° K. for the solid state transition with a heat of transition of 270 ± 40 cal./gram atom and 1123° K. as the melting point, with a heat of melting of 2070 ± 80 cal./gram atom.

Thermodynamic properties of the ideal monatomic gas have been calculated from the spectroscopic information reported by Moore (241). Vapor pressure measurements have been made by Douglas (88), Hartmann and Schneider (146), Pilling (261), Priselkov and

Nesmeianov (263), Rudberg (277), Tomlin (329), and Ruff and Hartmann (280). The pressures reported by Rudberg appear to be too low, while those of Ruff and Hartmann seem to increase too rapidly with increase of temperature. The remaining measurements are in good agreement. We calculate a heat of sublimation at 298° K. of 42,200 cal./gram atom, a boiling point of 1765° K., and a heat of vaporization at 1765° K. of 35,840 cal./gram atom.

CARBON

DeSorbo and Tyler (85) have recently measured the heat capacity from 13° to 300° K., and calculate the entropy at 298° K. to be 1.372 ± 0.005 e. u., and an enthalpy at 298° K. of 251 cal./gram atom. Thermodynamic properties of the solid and the ideal monatomic gas have been taken from the compilation of Rossini and coworkers (273). Thermodynamic properties of the ideal diatomic gas have been calculated from the spectroscopic data of Herzberg (152). Our calculated entropy at 298° K. agrees with that calculated by Kelley (186), but is $R \ln 3$ less than the value calculated by Gordon (129). According to Brewer (34) additional low lying electronic states are to be expected, so that the present treatment must be considered approximate. Thermodynamic properties for the ideal triatomic gas have been calculated from the estimated molecular constants listed by Glockler (128).

The heat of sublimation of graphite to ideal monatomic gas has been the subject of numerous investigations. Recent work (55, 56, 87, 158) has given increasing support to a value in the vicinity of 170 kcal./gram atom. According to Brewer and Kane (37) and Thorn and Winslow (326) the experimental conditions sometimes prevent reaching a true equilibrium between graphite and all the gaseous species. This may be responsible for the divergence of the values found by the various experimental methods. Thus, experiments to date probably yield a reliable value for the heat of sublimation of the ideal monatomic species only. The only values for the heats of sublimation of higher species at the present time have come from the mass spectrometer measurements of Chupka and Inghram. We have used rounded values of 200,000 cal./gram mole for the diatomic and triatomic species.

The best value for the heat of sublimation of graphite to ideal monatomic gas can be obtained from a consideration of the following reactions:

$$C(gr) + 1/2 O_2(g) \rightarrow CO(g) \tag{1}$$

$$CO(g) \rightarrow C(g) + O(g) \tag{2}$$

$$O(g) \rightarrow 1/2 O_2(g) \tag{3}$$

$$C(gr) \rightarrow C(g) \tag{4}$$

The heat of Reaction 1 is from Rossini and coworkers (274), that of Reaction 2 is from Douglas (87), and that of Reaction 3 is from Brix and Herzberg (41). This gives for Reaction 4 at 298° K. a value of 170,890 ± 500 cal./gram atom. In view of the uncertainties that have been mentioned, we estimate the total vapor pressure reaches 1 atmosphere at a temperature of about 4000° K.

CERIUM

Parkinson, Simon, and Spedding (257) have measured the heat capacity from 2° to 180° K. and report the entropy at 298° K. to be 16.64 e. u. and an enthalpy at 298° K. of 1742 cal./gram atom. We have adjusted Kelley's (185) solid heat capacity equation so that it joins smoothly with the low temperature data. Spedding and Daane (314) report a transition at 1027° K. and a melting point of 1077° K. We have estimated the heats accompanying these phase changes, as well as the liquid heat capacity. Ahmann (4) and Brewer (35) have measured the vapor pressure, and differ by nearly one order of magnitude. Taking an average of their data and estimating the gaseous spectroscopic contribution, we find a normal boiling point of 3200° K., with an associated heat of vaporization of 75,000 cal./gram atom.

CESIUM

Based on the low temperature measurements of Dauphinee, Martin, and Preston-Thomas (82), we calculate an entropy at 298° K. of 20.16 e. u. and an enthalpy at 298° K. of 1859 cal./gram atom. Clusius and Stern (69) have measured the melting point as 301.8° K. We have averaged the values for heat of melting reported by Kelley (185), Dauphinee, Martin, and Preston-Thomas (82), and Clusius and Stern (69) to obtain 510 cal./gram atom. The liquid heat capacity measurements of Dauphinee, Martin, and Preston-Thomas (82) have been extrapolated to the boiling point.

Evans, Jacobson, Munson, and Wagman (105) have calculated the thermodynamic properties of the ideal monatomic and diatomic gases and list 10,380 cal./mole for the heat of dissociation of the diatomic gas. Vapor pressure measurements have been made by Scott (290), Ruff and Johannsen (281), Taylor and Langmuir (324), Fuchtbauer and Bartels (116), Kroner (204), and by Hackspill (141). The data of the last four sets of workers are in excellent agreement, and lead to a heat of sublimation at 298° K. of 18,670 cal./gram atcm for the ideal monatomic species, 26,630 cal./mole for the ideal diatomic species, a norn al boiling point of 958° K., and a heat of vaporization to equilibrium vapor at 958° K. of 15,750 cal./gram atom.

CHLORINE

Based on the measurements of Giauque and Powell (124), Rossini and coworkers (274) give the melting point, 172.16° K.; heat of melting, 1531 cal /mole; normal boiling point, 239.10° K.; and heat of vaporization, 4878 cal./mole. The critical temperature, 417° K., and critical pressure, 76.1 atmospheres, listed by Kobe and Lynn (193) have been adopted. Evans, Munson, and Wagman (106) have calculated the thermodynamic properties of ideal monatomic and diatomic gases. They select 57,880 cal./gram mole for the dissociation energy at 298° K. Note that the reference state table is based on 2 gram atomic weights.

CHROMIUM

Low temperature measurements by Weertman, Burk, and Goldman (342) and by Anderson (12) are in fair agreement and give an entropy at 298° K. of 5.70 e. u. and an enthalpy of 973 cal./gram atom. Recent heat capacity work by Armstrong and Grayson-Smith (17) on a very pure sample has been adopted and has been extended to join Kelley's (185) equation smoothly at about 1300° K. Kelley's equation has been extrapolated to the transition. In working with multicomponent systems as well as very pure chromium, Bloom, Putnam, and Grant (28) have found evidence of a transition at 2113° ± 15° K. and a melting point of 2176° ± 10° K. The data of Grube and Knabe (137) on the palladium-chromium system lead to a calculated heat of melting of 3300 ± 200 cal./gram atom. However, the directly measured value of Umino (332) is 3650 cal./gram atom and probably includes the heat of the transition, so we have selected the difference, 350 cal./gram atom for the heat of the transition, leaving 3300 ± 200 cal./gram atom for the heat of melting. The heat capacity of the solid above the transition has been assumed to have the same value as the liquid, which Kelley (179) reports as 9.70 cal./degree/gram atom based on Umino's data.

Thermodynamic properties for the ideal monatomic gas state have been calculated using the spectroscopic energy levels listed by Moore (241). The vapor pressure has been reported by Speiser, Johnston, and Blackburn (317), Gulbransen and Andrew (139), and Baur and Brunner (23). The data of Baur and Brunner appear to be too high, since there is good agreement in the data of the first two sources. We calculate the heat of sublimation at 298° K. to be 95,000 cal./gram atom, the normal boiling point of 2915° K., and a heat of vaporization at the normal boiling point of 83,360 cal./gram atom.

COBALT

From the data of Duyckaerts (*92*) as well as their own low temperature measurements from 15° to 270° K., Clusius and Schachinger (*68*) calculate an entropy at 298° K. of 7.18 e. u. We calculate from these same data an enthalpy at 298° K. of 1146 cal./gram atom. Since the latest compilation of Kelley (*185*), Armstrong and Grayson-Smith (*17*) have measured the heat capacity of a very pure sample up to 1073° K. They report a peak in the heat capacity curve from 720° to 755° K. which we interpret as a sluggish phase change. We select 720° K. as the ideal transformation temperature which would be obtained with infinitely slow heating. Armstrong and Grayson-Smith obtain 60 cal./gram atom as the heat of this transition by direct integration of the peak. The heat capacity data of Armstrong and Grayson-Smith have been used and extrapolated to the Curie point. The Curie point is given by the "Metals Handbook" (*216*) as 1388° K. and by Meyer and Taglang (*235*) as 1404° K. We interpret these data as a lambda point at 1395° K. of undefined shape and add 130 cal./gram atom at this temperature, the value selected by Kelley (*185*) for this discontinuity. The melting point was determined as 1768° ± 1° K. by Van Dusen and Dahl (*336*). Kelley (*185*) lists 3640 cal./gram atom for the heat of melting and also gives the liquid heat capacity.

We have calculated the thermodynamic properties of the ideal monatomic gas from the spectroscopic data of Moore (*241*). Vapor pressure has been measured by Dancy (*80*), Ruff and Keilig (*282*), Kornev and Golubkin (*199*), and by Edwards, Johnston, and Ditmars (*98*). We have the most confidence in the measurements of Edwards, Johnston, and Ditmars, although the data from the other three sources form a different consistent pattern. Combining the measurements of Edwards, Johnston, and Ditmars with the other thermodynamic data, we find the heat of sublimation at 298° K. to be 101,600 cal./gram atom, a normal boiling point of 3150° K., and a heat of vaporization at the normal boiling point of 91,400 cal./gram atom.

COPPER

Kelley (*186*) has calculated the entropy as 7.97 ± 0.02 e. u. based on heat capacities from several sources including measurements to 1° K. Giauque and Meads (*123*) give the enthalpy at 298° K. as 1201 cal./gram atom. Solid and liquid heat capacity and heat of melting of 3120 cal./gram atom have been taken from Kelley's (*185*) compilation. Rossini and coworkers (*274*) give 1356° K. as the melting point.

Thermodynamic functions for the ideal monatomic gas have been calculated from spectroscopic data given by Moore (*241*). Older vapor pressure measurements of Harteck (*145*) and of Marshall, Dornte, and Norton (*223*) agree with the more recent measurements of Hersh (*151*) and of Edwards, Johnston, and Ditmars (*99*). From these data we find the heat of sublimation at 298° K. to be 81,100 cal./gram atom, a normal boiling point of 2855° K., and a heat of vaporization at 2855° K. of 72,800 cal./gram atom.

DYSPROSIUM

Griffel, Skochdopole, and Spedding (*136*) have measured the heat capacity from 15° to 300° K., and report an entropy at 298° K. of 17.78 e. u. and an enthalpy at 298° K. of 2116 cal./gram atom. Spedding and Daane (*314*) indicate 1773° K. as the approximate melting point and have measured the vapor pressure at 1390° K. as 0.01 mm. of mercury (*313*). They calculate a heat of vaporization at 1390° K. of 66,700 cal./gram atom. From these data we estimate the boiling point as 2600° K. and the heat of vaporization at the boiling point as 60,000 cal./gram atom. The solid, liquid, and gas heat capacity and heat of melting are all estimated and are intended for use only until measured values become available.

ERBIUM

Skochdopole, Griffel, and Spedding (*310*) have measured the heat capacity from 15° to 320° K., and calculate an entropy at 298° K. of 17.48 e. u. and an enthalpy at 298° K. of 1763 cal./gram atom. Spedding and Daane (*314*) indicate a melting point of about 1800° K. and from their suggested volatility we estimate a normal boiling point of 2900° K. The remaining data have been estimated by comparison with related metals and should be used only until measured values are available.

EUROPIUM

Skochdopole, Griffel, and Spedding (*310*) have compared measured entropies for the rare earths with theoretically predicted values. Although they do not predict a value for europium, they believe it is somewhat higher than its immediate periodic table neighbors. On this basis, we adopt a value of 17 e. u. for the entropy of europium at 298° K. Spedding and Daane (*314*) remark that europium is the most volatile of the rare earths. Landolt-Bornstein (*208*) report available spectroscopic terms from which we have calculated the thermodynamic properties of the ideal monatomic gas. The remaining data listed for this element are estimated and are consistent with the above known facts. These data are intended for use only until measured values become available.

FLUORINE

Hu, White, and Johnston (*162*) have determined the low temperature thermal data for fluorine and report a solid state transition at 45.55° K. with a heat of 173.90 cal./mole, the melting point at 53.54° K. with a heat of melting of 121.98 ± 0.5 cal./mole, the boiling point at 85.02° K., and a heat of vaporization at 84.71° K. of 1563.98 ± 3 cal./mole. The resulting calorimetric entropy of the gas at 85.02° K. is in excellent agreement with that calculated by statistical methods. Correcting the heat of vaporization to 760 mm. and 85.02° K. gives 1562 ± 4 cal./mole. The critical temperature, 144.2° K., and pressure, 55 atmospheres, have been taken from Cady and Hildebrand (*50*). Thermodynamic functions for the ideal monatomic and diatomic gases as well as the dissociation energy are from the work of Evans, Munson, and Wagman (*106*). Note that the reference state represents 2 gram atomic weights for this element.

FRANCIUM

These data are completely estimated by comparison with the other alkali metals and are intended to serve only until measured data become available.

GADOLINIUM

Griffel, Skochdopole, and Spedding (*135*) have measured the heat capacity from 15° to 355° K., and report an entropy at 298° K. of 15.77 e. u. and an enthalpy of 2172 cal./gram atom. We have estimated the solid and liquid heat capacities as well as the heat of melting. Spedding and Daane (*314*) report approximately 1600° K. for the melting point and a volatility which places the normal boiling point in the vicinity of 3000° K. Gaseous spectroscopic data from Russell (*284*) permit calculation of the thermodynamic properties of the ideal monatomic gas. Based on these data, we calculate a heat of sublimation at 298° K. of 82,500 cal./gram atom and a heat of vaporization at the normal boiling point of 74,500 cal./gram atom.

GALLIUM

An entropy of 9.82 ± 0.05 e. u. and an enthalpy of 1331 cal./gram atom at 298° K., based on measurements from 15° to 323° K., have been calculated by Adams, Johnston, and Kerr (*1*). Their values for the heat of melting, 1335 cal./gram atom, melting point 303° K., and liquid heat capacity are also employed. Speiser and Johnston (*316*) have estimated the liquid heat capacity to be 6.65 cal./degree/gram atom in the high temperature region.

Thermodynamic properties for the ideal monatomic gas have been computed from the spectroscopic energy levels reported by Moore (*241*). Vapor pressures reported by Harteck (*145*) are somewhat lower than the more recent measurements of Speiser and Johnston. Giving the latter workers the most weight we calculate a heat of sublimation at 298° K. of 65,000 cal./gram atom, a normal boiling point of 2510° K., and a heat of vaporization at the normal boiling point of 61,200 cal./gram atom.

GERMANIUM

Estermann and Weertman (*104*) and Hill and Parkinson (*155*) have recently measured low temperature heat capacities of very pure samples, covering the temperature range from 0° to 200° K. with good agreement. The extension of the heat capacity curve to the melting point was accomplished by direct analogy with similar measured data for silicon and gray tin. This results in a smooth curve reaching a value of 7.0 cal./degree/gram atom at the melting point. Integration leads to an entropy at 298° K. of 7.43 ± 0.10 e. u., in good agreement with the value of 7.40 e. u. derived by Coughlin (*73*), and an enthalpy at 298° K. of 1105 cal./gram atom. Hassion, Thurmond, and Trumbore (*147*) have measured the melting point under a variety of conditions and report 1210.4° K. Wittig (*348*) and Greiner (*133*) have measured the heat of melting as 7100 and 8100 cal./gram atom, respectively. An average of 7600 cal./gram atom has been adopted. We have estimated the heat capacity of the liquid to be 7.0 cal./degree/gram atom.

We have assumed that the gas is ideal and monatomic and have calculated the thermodynamic properties based on the energy levels of Moore (*241*). Searcy (*291*) and Searcy and Freeman (*293*) have measured the vapor pressure, while Honig (*161*) has studied the vapor species in a mass spectrometer. These data are consistent with a heat of sublimation at 298° K. of 90,000 cal./gram atom, a normal boiling point of 3100° K , and a heat of vaporization at 3100° K. of 79,900 cal./gram atom.

GOLD

Geballe and Giauque (*119*) have recently measured the heat capacity from 15° to 300° K., and report the entropy and enthalpy at 298° K. to be 11.32 ± 0.02 e. u. and 1434 cal./gram atom, respectively. The solid and liquid heat capacity and the heat of melting have been taken from the compilation of Kelley (*185*). Stimson (*319*) has listed 1336.15° K., the defined melting point, as a primary calibration point on the International Temperature Scale. From spectroscopic data listed by Landolt-Bornstein (*208*), we have calculated the thermodynamic properties of the ideal monatomic gas. Vapor pressure data given by Hall (*142*) lead to a heat of sublimation at 298° K. of 84,700 cal./gram atom, a normal boiling point of 2980° K., and a heat of vaporization at the normal boiling point of 77,540 cal./gram atom.

HAFNIUM

Low temperature heat capacities have been measured by Cristescu and Simon (*76*) from 13° to 210° K., and by Weertman, Burk, and Goldman (*342*) from 50° to 200° K. Since the latter workers have not substantiated the anomaly reported by the former workers, we have adopted the values of the latter group and have extrapolated them to absolute zero with a Debye function. From this information, we calculate the entropy at 298° K. to be 10.91 e. u. and the enthalpy at 298° K. to be 1448 cal./gram atom. We have estimated the heat capacity of the solid above 298° K. and of the liquid. A transition point has been reported by Duwez (*91*) and by Fast (*110*). The melting point has been reported by Adenstedt (*2*), Litton (*213*), and Zwikker (*352*). Considerable disagreement is evidenced by these values. There is probably a transition in the vicinity of the melting point, but in view of the uncertainty existing, we have elected to minimize the necessary estimations by considering a single phase change at the melting point and combining any transitional heat with the heat of melting. It seems to us that the most reasonable melting point is that given by Adenstedt, 2250° K. Estimating the entropy of melting at 2.3 e. u., we calculate the heat of melting to be 5200 cal./gram atom.

From data given in the Landolt-Bornstein Tabellen (208), we have calculated the thermodynamic functions for the ideal monatomic gas. Richardson (268) has roughly measured the normal boiling point to be 5400° K., which is in good agreement with the estimate by Brewer (35) of 5500° K., which we have used. From these data we calculate a heat of sublimation at 298°.K. of 168,000 cal./gram atom, and at the normal boiling point a heat of vaporization of 158,000 cal./gram atom.

HELIUM

The solid is not stable at one atmosphere, and can only be obtained at elevated pressures. In the range from 0° to 1° K., the required pressure is reported by Simon and Swenson (304) as 25 atmospheres. At a pressure of 103 atmospheres, Keesom (174) reports the melting point to be 3.5° K., with an associated heat of 5 cal./gram atom. Keesom also reports the second order transition (lambda point) at 2.186° K., and the normal boiling point at 4.216° K. with the associated heat of vaporization of 20 cal./gram atom. Thermodynamic properties for the ideal monatomic gas have been calculated at the National Bureau of Standards (295). Kobe and Lynn (193) report the critical temperature as 5.3° K. and the critical pressure as 2.26 atmospheres.

HOLMIUM

Skochdopole, Griffel, and Spedding (310) have estimated the entropy at 300° K. to be 17.81 e. u., very close to the entropy of dysprosium. Spedding and Daane (314) give the approximate melting point of 1773° K., and place its volatility similar to that of dysprosium. This element appears to be very similar to dysprosium. The values listed are, therefore, based on dysprosium and are to be used only until measured data are available.

HYDROGEN

Woolley, Scott, and Brickwedde (350) have compiled the thermodynamic properties for normal hydrogen. They list the melting point as 13.95_7° K. and give the heat of melting measured by Simon and Lange (302) as 28.0 ± 0.15 cal./mole. Unpublished vapor pressure measurements by Brickwedde and Scott cited in the compilation lead to a normal boiling point of 20.390° K., the presently accepted value. A new determination of the temperature scale using gas thermometry by Moessen, Aston, and Ascah (238) will, if adopted in 1960 by the International Committee of Weights and Measures, lead to a value of 20.365° K. The value measured by Simon and Lange for the heat of vaporization is 215.8 ± 1.1 cal./mole.

White, Friedman, and Johnston (343) have measured the critical constants for normal hydrogen and have found 33.24_4° K. and 12.797 atmospheres. Woolley, Scott, and Brickwedde have presented data on the dissociation energy and the thermodynamic properties for the ideal diatomic gas, including contributions from nuclear spin. We have omitted the spin entropy in compiling our tables. Thermodynamic properties for the ideal monatomic gas have been computed at the National Bureau of Standards (295). Note that the reference state represents 2 gram atomic weights for this element.

INDIUM

Clusius and Schachinger (66) have measured the heat capacity from 12° to 273° K., and Clement and Quinnell (58) from 1.7° to 21.3° K., from which can be derived the entropy at 298° K. of 13.82 e. u. and an enthalpy at 298° K. of 1578 cal./gram atom. Roth, Meyer, and Zeumer (275) have reported data for the solid heat capacity, melting point, heat of melting, and liquid heat capacity. Oelsen (253) has measured the heat of melting and liquid heat capacity and Oelsen, Oelsen, and Thiel (*254*) give a value for the heat of melting. From these sources we have selected our heat capacity data and the heat of melting of 780 cal./gram atom. Valentiner (*333*) has accurately measured the melting point as 429.32° K.

Thermodynamic properties for the ideal monatomic gas have been calculated from energy levels listed in the Landolt-Bornstein Tabellen (*208*). Kohlmeyer and Spandau (*194*) have measured the normal boiling point directly and report 2273° ± 10° K. Anderson (*14*) has measured the vapor pressure from 1000° to 1348° K. His results extrapolate to a normal boiling point of 2364° K. We have selected the heat of sublimation at 298° K. to be 57,000 cal./gram atom, which leads to an average normal boiling point of 2320° K. and an associated heat of vaporization of 54,100 cal./gram atom.

IODINE

On the basis of literature values of low temperature heat capacities, Kelley (*186*) calculates an entropy at 298° K. of 27.9 e. u., in good agreement with calculations of Giauque (*120*). Giauque also reports an enthalpy of 3178 cal./gram mole and a heat of sublimation of 14,877 cal./gram mole, both at 298° K. Kelley (*185*) gives an equation for the heat capacity of the solid to the melting point of 386.8° K., the heat of melting as 3770 cal./mole, the heat capacity of the liquid to the normal boiling point at 456° K., and the heat of vaporization at the boiling point of 9970 cal./mole. Thermodynamic properties of the ideal monatomic and diatomic species as well as the dissociation energy are given by Evans, Munson, and Wagman (*106*). Note that the reference state represents 2 gram atomic weights for this element.

IRIDIUM

The entropy at 298° K. has been estimated by Lewis and Gibson (*212*) to be 8.7 ± 0.5 e. u. Kelley (*185*) has given an equation for the solid heat capacity from 298° to 1800° K., which we have extrapolated to the melting point and have assumed that the heat capacity of the liquid is the same as that of the solid at the melting point. Based on the work of Henning and Wensel (*150*) and of Morris and Scholes (*242*), Vines (*338*) selects a melting point of 2727° K. For a face-centered cubic lattice, we adopt an entropy of melting of 2.3 e. u., which leads to a heat of melting of 6300 cal./gram atom.

Thermodynamic functions for the ideal monatomic gas have been calculated from spectroscopic data listed in the Landolt-Bornstein Tabellen (*208*). Brewer (*33*) believes a former estimate of 4800° K. for the normal boiling point is somewhat high, and we have selected 4400° K. This results in a heat of sublimation at 298° K. of 150,000 cal./gram atom and a heat of vaporization at 4400° K. of 134,700 cal./gram atom.

IRON

Kelley (*186*) gives the entropy at 298° K. as 6.49 ± 0.03 e. u., based on measurements from 1° K. upward. From these data an enthalpy at 298° K. of 1070 cal./gram atom can be derived. An excellent review on the high temperature thermal properties of iron is given by Darken and Smith (*81*) and we have used their data exclusively. We prefer their treatment of the heat capacity in the vicinity of the Curie point. According to this view, there is no change in phase in this temperature region and hence no heat of transition. There is a very sharp peak or lambda point in the heat capacity curve at 1033° K. and the measured data have been integrated directly to obtain the derived values. Bona fide transitions occur at 1183° and 1673° K., with associated heats of 215 and 165 cal./gram atom, respectively. The melting point is listed as 1812° K. with a heat of melting of 3670 cal./gram atom. Liquid heat capacity data of Darken and Smith have been extrapolated to the boiling point.

Thermodynamic functions for the ideal monatomic gas have been calculated from spectroscopic data reported by Moore (*241*). The vapor pressure of iron has been measured by Jones, Langmuir, and Mackay (*170*), Marshall, Dornte, and Norton (*223*) and

Edwards, Johnston, and Ditmars (98). While agreement between the first two sets of observations is good, truly pure iron has only been produced within the last few years. Thus, we believe the slightly lower pressures reported by Edwards, Johnston, and Ditmars are more nearly correct. Their data yield a heat of sublimation at 298° K. of 99,830 cal./gram atom, a normal boiling point of 3160° K., and a heat of vaporization of 83,900 cal./gram atom.

KRYPTON

Clusius (60) reports 115.9° K. for the melting point and 391 cal./gram atom for the heat of melting. Michels, Wassenaar, and Zwietering (237) have measured the vapor pressure and find 119.75° K. for the normal boiling point. Clusius, Kruis, and Konnertz (64) have measured the heat of vaporization at the normal boiling point as 2158 cal./gram atom. Kobe and Lynn (193) give 209.4° K. as the critical temperature and 54.3 atmospheres for the critical pressure. Thermodynamic functions for the ideal monatomic gas have been calculated at the National Bureau of Standards (295).

LANTHANUM

Parkinson, Simon, and Spedding (257) have measured the heat capacity from 2° to 180° K., and calculate the entropy at 298° K. as 13.60 e.u. and the enthalpy at 298° K. as 1569 cal./gram atom. Kelley (185) reports the heat capacity of the solid above room temperature. Spedding and Daane (314) have reported the melting point at 1193° K. We estimate the heat of melting to be 2700 cal./gram atom. Kelley (187) has estimated the liquid heat capacity. Thermodynamic functions for the ideal monatomic gas have been calculated from the spectroscopic data reported in the Landolt-Bornstein Tabellen (208). Daane (78) has measured the vapor pressure from 1600° to 1900° K. By selecting a smoothed value in the middle of this range, we derive a heat of sublimation at 298° K. of 99,600 cal./gram atom. These data extrapolate to a normal boiling point of 3640° K. and a heat of vaporization at the normal boiling point of 95,500 cal./gram atom.

LEAD

Based on seven sets of measurements covering the range from 2° to 303° K., Kelley (186) computes the entropy at 298° K. to be 15.49 ± 0.05 e.u., while Meads, Forsythe, and Giauque (230) report the enthalpy at 298° K. to be 1644 cal./gram atom. Data for the solid and liquid heat capacity, melting point, and heat of melting have been adopted from the work of Douglas and Dever (90). Thermodynamic functions of the ideal monatomic gas have been computed from the energy levels listed in the Landolt-Bornstein Tabellen (208).

Vapor pressures have been measured by Baur and Brunner (23), Harteck (145), Rodebush and Dixon (269, 270), Fischer (111), von Leitgebel (211), Egerton (100), von Wartenberg (341), Ingold (164), Greenwood (130–132), and Ruff and Bergdahl (279). Measurements reported in the last six references are high in comparison with the remaining measurements and, in agreement with Kelley (180), we believe these high results are unreliable. Of the first seven references, we have given the most weight to the results of Rodebush and Dixon in calculating a heat of sublimation at 298° K. of 46,800 cal./gram atom. Extrapolation gives a normal boiling point of 2024° K. and a heat of vaporization at the normal boiling point of 42,880 cal./gram atom.

LITHIUM

Evans, Jacobson, Munson, and Wagman (105) have critically reviewed the literature and have selected a consistent set of values. They find the entropy at 298° K. to be 6.753 e. u., the enthalpy at 298° K. as 1092.2 cal./gram atom, the melting point as 453.70° K., and the heat of melting to be 722.8 cal./gram atom. They present the solid and liquid heat capacities as well as the thermodynamic properties of the ideal monatomic and diatomic gases. They have also summarized the vapor pressure data and derive heats of sublimation at 298° K. of 38,440 cal./gram atom and 50,470 cal./gram mole for the monatomic and diatomic gases, respectively. From these data we calculate that this system reaches a total pressure of one atmosphere at 1604° K., at which temperature the heat of vaporization to equilibrium gas is 32,190 cal./gram atom.

LUTETIUM

Skochdopole, Griffel, and Spedding (*310*) have estimated the entropy at 300° K. as 11.79 e. u. Spedding and Daane (*314*) report the melting point in the range from 1923° to 2023° K. and place the volatility between samarium and thulium. Klinkenberg (*192*) gives the available spectroscopic data from which we calculate the thermodynamic properties of the ideal monatomic gas. The remaining data are estimated, are consistent with the above facts and are intended for use only until measured information is available.

MAGNESIUM

The third law entropy based on measurements from 12° to 320° K. by Craig and coworkers (*75*) is 7.81 e. u. at 298° K. Using their heat capacities we calculate an enthalpy at 298° K. of 1195 cal./gram atom. In addition to the heat capacity data reviewed by Kelley (*185*), we have considered the values given by Kubaschewski (*205, 206*) and the measurements of McDonald and Stull (*228*). The heat of melting is 2140 cal./gram atom from McDonald and Stull. Rossini and coworkers (*274*) have selected a melting point of 923° K.

Thermodynamic functions for the ideal monatomic gas have been calculated from the energy levels reported by Moore (*241*). Vapor pressure data measured by Baur and Brunner (*23*), Hartmann and Schneider (*146*), Greenwood (*130*), von Leitgebel (*211*), Schneider and Esch (*285*), Vetter and Kubaschewski (*337*), Ruff and Hartmann (*280*), and Coleman and Egerton (*72*) are in reasonably good agreement, except for Coleman and Egerton who are somewhat high. Giving the most weight to the results of Hartmann and Schneider, we calculate a heat of sublimation at 298° K. of 35,600 cal./gram atom, a normal boiling point of 1390° K., and a heat of vaporization at the normal boiling point of 30,750 cal./gram atom.

MANGANESE

Shomate (*300*), Kelley (*177*), Armstrong and Grayson-Smith (*16*), Elson, Smith, and Wilhelm (*102*), and Booth, Hoare, and Murphy (*29*) have reported low temperature heat capacity data. From these data we calculate an entropy and enthalpy at 298° K. of 7.65 e. u. and 1194 cal./gram atom, respectively. Above 298° K. Armstrong and Grayson-Smith (*17*) and Naylor (*244*) have reported heat capacity measurements which we regard as equally reliable. We adopt an average of these data to the first transition. Naylor finds transitions at 1000°, 1374°, and 1410° K., with accompanying transitional heats of 535, 545, and 430 cal./gram atom, respectively. The heat capacity between 1000° K. and the melting point has been adjusted to give the enthalpy found by Naylor. The melting point, heat of melting, and the liquid heat capacity have been given by Kelley (*185*) as 1517° K., 3500 cal./gram atom, and 11.00 cal./degree/gram atom, respectively.

Thermodynamic functions for the ideal monatomic gas have been calculated from the energy levels listed by Moore (*241*). Brewer (*33*) has reported the heat of sublimation at 298° K. as 66,730 cal./gram atom, which leads to a normal boiling point of 2314° K. and a heat of vaporization at 2314° K. of 52,520 cal./gram atom.

MERCURY

Busey and Giauque (*48*) have measured the heat capacities and transitional heats from about 15° to 300° K. Their melting point of 234.29° K. is in good agreement with that of Wilhelm (*346*) who found 234.287° K. and proposed this transition as a secondary thermometric calibration point. Busey and Giauque (48) find 548.6 cal./gram atom for the heat of melting, 18.19 e.u. for the entropy of the liquid state at 298° K., and an enthalpy at 298° K. of 2232 cal./gram atom. Liquid heat capacities of Busey and Giauque have been adopted. They have extended their measurements by adjusting the values of Douglas, Ball, and Ginnings (89) to join smoothly with their own.

Thermodynamic functions for the ideal monatomic gas were calculated. Energy levels listed by Landolt-Bornstein Tabellen (208) indicate that below 3000° K. there is no electronic contribution. Busey and Giauque have reviewed the vapor pressure data and find the normal boiling point at 629.88° K., the heat of vaporization to the ideal monatomic gas at the normal boiling point of 14,137 cal./gram atom, while the heat of vaporization at 298° K. is 14,652 cal./gram atom. Beale (25) has recently measured the heat of vaporization as 13,595 ± 23 cal./gram atom. Beale (24) points out that this heat of vaporization can only be made consistent with the other thermodynamic properties by assuming a much larger gas imperfection than that derived by Busey and Giauque from vapor pressure data. Experimental data on mercury vapor are needed to resolve the question.

MOLYBDENUM

Simon and Zeidler (305) have measured the low temperature heat capacity, which leads to an entropy and enthalpy at 298° K. of 6.83 ± 0.05 e.u. and 1092 cal./gram atom, respectively. Using the Shomate method (301), enthalpy measurements of Kothen (200) and Redfield and Hill (265) have been combined with the values selected by Kelley (185) to give the heat capacity of the solid to the melting point. We have adopted a melting point of 2890° ± 10° K., which is an average of the values selected by Rossini and coworkers (274) and by Kelley (182). Brewer (35) estimates the heat of melting to be 6600 cal./gram atom. We estimate the liquid heat capacity to be 10.00 cal./degree/mole.

Thermodynamic properties of the ideal monatomic gas have been calculated from energy levels given in the Landolt-Bornstein Tabellen (208) and by Trees and Harvey (330). The vapor pressure has been measured by Jones, Langmuir, and Mackay (170) and by Edwards, Johnston, and Blackburn (97). These data have been averaged to obtain a heat of sublimation at 298° K. of 157,500 cal./gram atom, a normal boiling point of 5100° K., and a heat of vaporization at the normal boiling point of 142,000 cal./gram atom.

NEODYMIUM

Parkinson, Simon, and Spedding (257) have measured the heat capacity from 2° to 180° K., and report an entropy at 298° K. of 17.50 e.u. and an enthalpy of 1804 cal./gram atom. Spedding and Miller (315) have measured the heat capacity from 273° to 673° K. and support the equation given by Kelley (185). Spedding and Daane (314) report a transition point at 1141° K. and the melting point at 1297° K., and also give vapor pressure data. We have estimated the heats of transition and melting and the heat capacities of the solid above the transition and of the liquid. Using spectroscopic data from Klinkenberg (190) and Schuurmans (289) we have calculated the thermodynamic functions of the ideal monatomic gas. From these data we calculate a heat of sublimation at 298° K. of 76,800 cal./gram atom, a normal boiling point of 3360° K., and a heat of vaporization at the normal boiling point of 67,800 cal./gram atom.

NEON

Clusius (60) reports 24.55° K. as the melting point, with 80.1 cal./gram atom as the heat of melting. Henning and Otto (149) have measured the vapor pressure and find the normal boiling point at 27.07° K. From the heat of sublimation calculated by Clusius (59), we calculate the heat of vaporization at the normal boiling point to be 422 cal./gram atom. Thermodynamic functions for the ideal monatomic gas have been calculated at the National Bureau of Standards (295). Kobe and Lynn (193) report 45.5° K. for the critical temperature and 26.9 atmospheres for the critical pressure.

NICKEL

Busey and Giauque (47) have measured the third law entropy and enthalpy at 298° K. to be 7.137 e. u. and 1144 cal./gram atom, respectively. The heat capacity data selected by Sykes and Wilkinson (323) are in good agreement with the results of Neel (245) and Krauss and Warncke (203) up to the lambda point at 630° K. Above the lambda point, both Neel and Persoz (259) are from 6 to 7% above the coincident data of Sykes and Wilkinson, Kelley (185), and Krauss and Warncke. We have adopted the heat capacity data of Sykes and Wilkinson up to about 850° K., where it joins smoothly with Kelley's equation. Van Dusen and Dahl (336) have determined the melting point at 1728° ± 1° K., while Kelley lists the heat of melting as 4210 cal./gram atom. Kelley's value of 9.20 cal./degree/gram atom has been used for the heat capacity throughout the liquid range.

Thermodynamic functions for the ideal monatomic gas have been calculated from the energy levels listed by Moore (241). Our calculations based on the vapor pressure data of Johnston and Marshall (169) give a heat of sublimation at 298° K. of 101,260 cal./gram atom, a normal boiling point of 3110° K., and a heat of vaporization of 88,870 cal./gram atom.

NIOBIUM

Brown, Zemansky, and Boorse (45) have measured the heat capacity up to 12° K. and also in the range from 65° to 75° K. We have used these meager data with a Debye function to calculate the entropy and enthalpy at 298° K. as 8.73 e.u. and 1264 cal./gram atom, respectively. Kelley (185) lists the heat capacity for the solid above 298° K. Reimann and Grant (266) have determined the melting point as 2770° K. We estimate the heat of melting of 6400 cal./gram atom and the liquid heat capacity.

Thermodynamic properties of the ideal monatomic gas have been calculated using energy levels listed by Moore (241). From the rate of evaporation measurements of Reimann and Grant (266), we calculate a heat of sublimation at 298° K. of 177,500 cal./gram atom. Estimating the gaseous heat capacity to be 1 cal./degree/gram atom less than the liquid heat capacity in the range from 3000° to 5000° K., we calculate a normal boiling point of 5200° K. and a heat of vaporization at the normal boiling point of 166,500 cal./gram atom.

NITROGEN

Giauque and Clayton (121) have measured the low temperature heat capacity and give 55 cal./mole for the heat of transition and 172 cal./mole for the heat of melting. Corrected for changes in temperature scale, the transition temperature is 35.62° K. Furukawa and McCoskey (117) give the triple point as 63.18° K. Armstrong (15) has measured the vapor pressure and finds a normal boiling point of 77.36° K. Giauque and Clayton and Furukawa and McCoskey have measured the heat of vaporization. We have adopted an average value of 1335 cal./mole. The critical temperature of 126.26° K., and the critical pressure of 33.54 atmospheres have been measured by White, Friedman, and Johnston (344).

Wagman and coworkers (339) report the thermodynamic properties for the ideal diatomic gas, while the National Bureau of Standards (295) has published calculations of the thermodynamic functions for the ideal monatomic gas. The value of 225,100 cal./gram mole selected by Gaydon (118) for the dissociation energy at 0° K. is supported by the recent work of Douglas (87), Hendrie (148), Burns (46), Toennies and Greene (328), and Altshuller (6). This appears to conclude a voluminous literature on this subject. Note that the reference state represents 2 gram atomic weights for this element.

OSMIUM

The entropy at 298° K. has been estimated by Lewis and Gibson (212) to be 7.8 ± 0.5 e. u. Kelley (185) has given an equation for the solid heat capacity from 298° to 1800° K. which we have extrapolated to the melting point. We have assumed that the heat capacity of the liquid is the same as that of the solid at the melting point. We have adopted the value of 3000° K. for the melting point based on the estimate of Vines (338). For a hexagonal close-packed lattice we adopt 2.3 e. u. for the entropy of melting, which leads to a heat of melting of 7000 cal./gram atom. Thermodynamic functions for the ideal monatomic gas have been calculated from energy levels listed in the Landolt-Bornstein Tabellen (208). At the suggestion of Brewer (33) we have lowered previous estimates of the boiling point to 4500° K., and have calculated a heat of sublimation at 298° K. of 160,000 cal./gram atom and a heat of vaporization at 4500° K. of 150,000 cal./gram atom.

OXYGEN

Hoge (159) has reviewed the literature and has assigned the transition points, 23.886° and 43.800° K., and the melting point, 54.363° K., as well as the critical temperature, 154.78° K., and the critical pressure, 50.14 atmospheres. Giauque and Johnston (122) have measured the heats of these transitions: 22.42 cal./mole at 23.886° K., 177.6 cal./mole at 43.800° K., and 106.3 cal./mole for the heat of melting. The presently accepted International Temperature Scale defines 90.190° K. (−182.97° C.) as the normal boiling point (319). A new absolute determination of 90.154° K. (using 0° C. = 273.16° K.) by Aston and Moessen (18) will be subject to review by the International Committee of Weights and Measures in 1960. Furukawa and McCoskey (117) have measured the heat of vaporization and have reviewed previous data. We have adopted an average value of 1630 cal./mole. Thermodynamic properties for the ideal diatomic gas have been calculated from spectroscopic data by Woolley (349). The National Bureau of Standards (295) has published calculations of the thermodynamic properties of the ideal monatomic gas. The dissociation energy of the ideal diatomic gas at 0° K. is given by Brix and Herzberg (41) as 117,960 ± 40 cal./gram mole. Note that the reference state represents 2 gram atomic weights for this element.

PALLADIUM

Based on the measurements of Clusius and Schachinger (68) as well as their own measurements, Pickard and Simon (260) calculate the entropy at 298° K. to be 9.05 e. u. and the enthalpy as 1308 cal./gram atom. We have adopted the solid heat capacity values above 298° K. of Kelley (185). Rossini and coworkers (274) have selected 1823° K. for the melting point and 4000 cal./gram atom for the heat of melting. We have estimated the heat capacity of the liquid to be the same as the solid at the melting point, 8.30 cal./degree/gram atom. Thermodynamic properties of the ideal monatomic gas have been computed from the spectroscopic data of Shenstone (298). From Brewer's (35) estimate of the vapor pressure, we calculate a heat of sublimation at 298° K. of 94,000 cal./gram atom, a normal boiling point of 3400° K., and a heat of vaporization at the normal boiling point of 90,000 cal./gram atom.

PHOSPHORUS

Farr (109), of the Tennessee Valley Authority, has compiled a resume of the physical and thermodynamic properties of the allotropic forms of phosphorus. Based on entropy calculations from low temperature heat capacity measurements, Stephenson (318) believes that red crystalline triclinic phosphorus (T.V.A. designation V) is the most stable form at room temperature. This point of view is buttressed by the x-ray work of Roth, DeWitt, and Smith (276). Consequently we have selected red phosphorus V as the reference state up to its sublimation point at 704° K.

Stephenson reports the entropy of the red triclinic crystals at 298° K. as 5.46 e. u. Farr has reported the heat capacity of this form to the sublimation point as well as a melting point of about 870° K. Spectroscopic data by Moore (241) on the monatomic species and by Herzberg (152, 153) for the diatomic and tetratomic species have been used to compute the thermodynamic functions of the ideal monatomic, diatomic, and tetratomic gases. From vapor pressure measurements reported by Farr, we calculate the heat of sublimation of red phosphorus V to the tetratomic ideal gas at 298° K. to be 30,820 cal./mole of P_4. The heat of sublimation of red phosphorus V to the ideal diatomic gas at 298° K. is calculated to be 42,725 cal./mole P_2, based on the heat of the dissociation of P_4 to P_2 of 54,630 cal./mole of P_4, derived from the measurements of Stock, Gibson, and Stamm (*320*). From Gaydon's (*118*) dissociation energy of 116,000 cal./mole of P_2 at 0° K. to ideal monatomic gas, we calculate a heat of sublimation of red phosphorus V at 298° K. to ideal monatomic gas of 79,800 cal./gram atom. At the normal sublimation point, 704° K., the vapor is completely composed of the tetratomic species. We calculate a heat of sublimation at 704° K. of 7200 cal./gram atom. Since in most of the temperature range from 704° to 3000° K. the diatomic form is predominant, we have selected the ideal diatomic gas as the reference state in this region. Note that the table for the reference state is for 1 gram atomic weight.

Stephenson reports the entropy of the white α (cubic)-form at 298° K. as 9.80 e. u. Kelley (*185*) lists the heat capacity of the solid and liquid forms and the heat of melting of 150 cal./gram atom at the melting point of 317.4° K. The heat of sublimation of the white α-form at 298° K. to ideal tetratomic gas is 14,100 cal./mole of P_4, based on the measurements of MacRae and Van Voorhis (*218*), Centnerszwer (*52*), and Fishbeck and Eich (*113*). A slightly higher value is obtained from the vapor pressure data listed by Farr (*109*) and may be due to nonideal behavior at high pressures. Farr lists the normal boiling point of liquid white phosphorus as 554° K., and we calculate a heat of vaporization to P_4 vapor at this temperature of 2960 cal./gram atom. In the temperature range from 600° to 800° K. liquid white phosphorus is rapidly converted to red phosphorus. The heat of formation at 298° K. of white α from red V derived from the data presented here is 4180 cal./gram atom, in good agreement with the value of 4200 cal./gram atom selected by Yost and Russell (*351*) from calorimetric measurements.

PLATINUM

From low temperature measurements by Kok and Keesom (*195*) and by Simon and Zeidler (*305*), Kelley (*186*) calculates the entropy at 298° K. as 10.00 ± 0.05 e. u. and we calculate an enthalpy at 298° K. of 1384 cal./gram atom. Kelley (*185*) also gives an equation for the solid heat capacity from 298° to 1800° K., which we have extrapolated to the melting point. We assume the heat capacity of the liquid to be the same as that of the solid at the melting point. Kelley (*182*) and Rossini and coworkers (*274*) are in substantial agreement on a melting point of 2043° K. and a heat of melting of 4700 cal./gram atom.

Thermodynamic properties of the ideal monatomic gas have been calculated from energy levels listed in the Landolt-Bornstein Tabellen (*208*). Jones, Langmuir, and Mackay (*170*) have measured the vapor pressure. We calculate a heat of sublimation at 298° K. of 134,800 cal./gram atom, a normal boiling point of 4100° K., and a heat of vaporization at the normal boiling point of 122,000 cal./gram atom.

POLONIUM

Maxwell (*224*) and Beamer and Maxwell (*26*) have measured the melting point as 527° K. and find a transition at about 370° K. The sluggish nature of the transition suggests a small heat of transition which can be neglected. Brooks (*44*) has measured the vapor pressure from 711° to 1018° K., which can best be fit by assuming both diatomic and monatomic species to be present in the vapor. This view finds support in that the diatomic form is important in bismuth and tellurium, neighboring elements in the periodic table. The thermodynamic functions of the ideal gases as well as the entropy, heat capacity, and heat of melting of the solid and the heat capacity of the liquid are all estimated. These estimates were used in calculating the heats of sublimation at 298° K. of the monatomic and diatomic species as 34,450 and 32,900 cal./mole, respectively. The boiling point is 1235° K. and the heat of vaporization to equilibrium gas at 1235° K. is 14,400 cal./gram atom.

POTASSIUM

The low temperature measurements of Dauphinee, Martin, and Preston-Thomas (82) and of Wallace, Craig, and Krier (340) are in excellent agreement and lead to an entropy and enthalpy at 298° K. of 15.39 e.u. and 1695 cal./gram atom, respectively. Evans, Jacobson, Munson, and Wagman (105) have critically reviewed the literature and have selected a consistent set of values. We have used their values for the heat capacities of the condensed states. They report 336.4° K. for the melting point and 554 cal./gram atom for the heat of fusion. They present complete thermodynamic functions for the ideal monatomic and diatomic gases as well as the dissociation energy. Employing their evaluation of the vapor pressure data and adjusting for the above new entropy value, we calculate the heats of sublimation at 298° K. to be 21,420 cal./gram atom and 30,580 cal./gram mole to the ideal monatomic and diatomic gases, respectively. We calculate a normal boiling point of 1039° K. and an associated heat of vaporization of 18,530 cal./gram atom of equilibrium gas. This boiling point is higher than the 1027° K. recently reported by Makansi, Madsen, Selke, and Bonilla (221).

PRASEODYMIUM

Parkinson, Simon, and Spedding (257) have measured the heat capacity from 2° to 180° K., and report the entropy at 298° K. to be 17.45 e. u., and the enthalpy as 1697 cal./gram atom. We have extrapolated the solid heat capacity to the transition point at 1071° K. and the melting point at 1208° K., both of which are reported by Spedding and Daane (314). We estimate the heat of this transition to be 320 cal./gram atom and the heat of melting to be 2400 cal./gram atom. We have estimated the heat capacity of the liquid. Daane (78) has measured the vapor pressure from 1425° to 1692° K. and, reports 3290° ± 90° K. for the normal boiling point and 79,500 ± 1100 cal./gram atom for the heat of vaporization by a second law extrapolation. Spectroscopic data are not available to make a third law check of these values.

PROMETHIUM

Skochdopole, Griffel, and Spedding (310) estimate the entropy at 300° K. to be 17.25 e. u. All other values are estimated and are intended to serve only until measured values are available.

PROTACTINIUM

All data are estimated and are intended to serve only until measured values are available.

RADIUM

Rossini and coworkers (274) list 973° K. for the melting point. Landolt-Bornstein Tabellen (208) present spectroscopic data for the ideal monatomic gas. The remainder of these data are estimated and are intended to serve only until measured values are available.

RADON

Rossini and coworkers (274) estimate 202° K. as the melting point, 693 cal./gram atom as the heat of melting, 211° K. as the normal boiling point, and 3920 cal./gram atom as the associated heat of vaporization. Thermodynamic properties of the ideal monatomic gas have been calculated at the National Bureau of Standards (295).

RHENIUM

Smith, Oliver, and Cobble (*312*) have measured the low temperature heat capacity and report 8.887 e. u. and 1307 cal./gram atom for the entropy and enthalpy at 298° K., respectively. Kelley's (*185*) solid heat capacity equation, based on data to 1500° K., has been extrapolated to 3000° K. Sims, Craighead, and Jaffee (*306*) have measured the melting point and report 3453° ± 20° K. Estimating the entropy of melting to be 2.3 e.u., which is reasonable for a hexagonal close-packed structure, we calculate 7900 cal./gram atom for the heat of melting. Thermodynamic functions for the ideal monatomic gas have been calculated using spectroscopic data given by Klinkenberg (*191*). Sherwood, Rosenbaum, Blocher, and Campbell (*299*) have measured the vapor pressure and estimate the liquid heat capacity at 10.8 cal./degree/gram atom. We have calculated a heat of sublimation at 298° K. of 185,650 cal./gram atom, a normal boiling point of 5900° K., and an associated heat of vaporization of 169,000 cal./gram atom.

RHODIUM

Lewis and Gibson (*212*) have estimated the entropy at 298° K. to be 7.6 ± 0.5 e. u. Kelley (*185*) gives an equation for the heat capacity of the solid which we have extrapolated to the melting point. We have assumed the heat capacity of the liquid to have the same constant value as the solid at the melting point. The melting point selected by Vines (*338*) is confirmed by the recent work of Oriani and Jones (*255*) at 2239° ± 3° K. For a face-centered cubic lattice we employ an entropy of melting of 2.3 e. u., which leads to a heat of melting of 5200 cal./gram atom. Thermodynamic functions for the ideal monatomic gas have been calculated from the spectroscopic data of Molnar and Hitchcock (*239*). We estimate the normal boiling point as 4000° K., leading to a heat of sublimation at 298° K. of 133,000 cal./gram atom and a heat of vaporization at the normal boiling point of 118,400 cal./gram atom.

RUBIDIUM

From the low temperature measurements of Dauphinee, Martin, and Preston-Thomas (*82*), we calculate an entropy and enthalpy at 298° K. of 18.22 e. u. and 1790 cal./gram atom, respectively. These workers report 560 cal./gram atom for the heat of melting. Rossini and coworkers (*274*) select 312.0° K. for the melting point. We have estimated an average heat capacity for the liquid range. Evans, Jacobson, Munson, and Wagman (*105*) present the thermodynamic properties of the ideal monatomic and diatomic gases, as well as the dissociation energy of the diatomic gas. Vapor pressure has been measured by Scott (*290*), Hackspill (*141*), Ruff and Johannsen (*281*), and Killian (*188*). At 298° K. we calculate the heat of sublimation to the ideal diatomic gas as 27,550 cal./gram mole and the heat of sublimation to the ideal monatomic gas as 19,600 cal./gram atom. The total pressure in the gas phase reaches one atmosphere at 974° K., at which temperature the heat of vaporization to equilibrium gas is 16,540 cal./gram atom.

RUTHENIUM

Lewis and Gibson (*212*) have estimated the entropy at 298° K. to be 6.9 ± 0.5 e. u. We adopt Kelley's (*185*) values for the solid heat capacity and the heats and temperatures of the transitions: at 1308° K. a heat of 60 cal./gram atom; at 1473° K. a second order transition (no heat change); and at 1773° K. a heat of 320 cal./gram atom. Brewer (*35*) estimates the melting point as 2700° K. and the heat of melting as 6100 cal./gram atom. We estimate a boiling point of 4000° K., which leads to a heat of sublimation at 298° K. of 144,000 cal./gram atom and a heat of vaporization at the normal boiling point of 135,700 cal./gram atom. The heat capacity of the liquid and the gas are assumed to be equal in the range from 3000° to 4000° K.

SAMARIUM

Skochdopole, Griffel, and Spedding (*310*) estimate the entropy at 300° K. to be 16.32 e. u. We have estimated the heat capacities of the solid and liquid states. Spedding and Daane (*314*) report a transition at 1190° K. and the melting point at 1325° K. We have estimated the heats of these phase changes. Spectroscopic data from Brix (*40*) and Albertson (*5*) have been used to calculate the thermodynamic functions of the ideal monatomic gas. Spedding (*313*) indicates that the vapor pressure reaches 0.01 mm. of mercury at a temperature less than 1073° K. Assuming it to be 0.01 mm. at 1000° K., we calculate a heat of sublimation at 298° K. of 50,000 cal./gram atom, a normal boiling point of 1860° K., and a heat of vaporization at the normal boiling point of 45,800 cal./gram atom.

SCANDIUM

The entropy at 298° K. has been estimated to be 9.0 e. u. by Brewer (*35*). Kelley (*187*) has estimated the heat capacity of the solid and the liquid as well as the melting point, 1673° K., and the heat of melting, 3850 cal./gram atom. We estimate the normal boiling point to be 2750° K., which may be in error by several hundred degrees. Assuming the gas to be ideal and monatomic, we have calculated the thermodynamic functions from the energy levels given by Moore (*241*). From these data, we calculate the heat of sublimation at 298° K. to be 82,000 cal./gram atom and the heat of vaporization at the normal boiling point as 72,850 cal./gram atom.

SELENIUM

DeSorbo (*84*) has recently measured the heat capacity from 15° to 300° K., and calculates an entropy and enthalpy at 298° K. of 10.15 ± 0.05 e. u. and 1319.2 cal./gram atom, respectively. The low temperature heat capacity has been extended linearly to the melting point using the measured data of Monval (*240*) and Borelius and Paulson (*30*). We adopt the values of Kelley (*185*) for the melting point, 490° K., the heat of melting, 1300 cal./gram atom, and the heat capacity of the liquid. Thermodynamic functions for the ideal monatomic gas have been calculated from the energy levels listed by Moore (*241*), while those for the ideal diatomic gas are based on the spectroscopic data given by Herzberg (*152*). The heat capacity of the ideal hexatomic gas has been estimated.

Vapor pressures have been measured by Brooks (*43*), deSelincourt (*296*), Niwa and Sibata (*249*), Neumann and Lichtberger (*248*), and Preuner and Brockmöller (*262*). An entropy of 110 e. u. at 298° K. for the hexatomic gas and heats of sublimation at 298° K. of 35,380 and 34,120 cal./mole for the hexatomic and diatomic species, respectively, were selected to give the best fit with the vapor pressure data. Gaydon (*118*) gives the dissociation energy of the diatomic gas as 64,600 cal./mole, from which we calculate the heat of sublimation at 298° K. of the ideal monatomic gas to be 49,400 cal./gram atom. At the normal boiling point, 958° K., we calculate a heat of vaporization of 1 gram atom of selenium to equilibrium gas to be 6290 cal. Note that the values given for the reference state are based on 1 gram atom of selenium and that the diatomic gas is selected as the reference state above the boiling point.

SILICON

Using the low temperature heat capacity data of Pearlman and Keesom (*258*), Nernst and Schwers (*246*), Magnus (*219*), and Anderson (*8*), we calculate a third law entropy at 298° K. of 4.53 ± 0.05 e. u. and an enthalpy of 769 cal./gram atom. Since the measured data for the solid heat capacity of Serebrennikov and Gel'd (*297*) and Magnus (*219*) are in agreement, we have chosen the equation given by the former. Hansen and coworkers (*144*) have measured the melting point to be $1683° \pm 5°$ K., while Korber and Oelsen (*198*) give the value 11,100 cal./gram atom for the heat of melting. We estimate the heat capacity of the liquid state to be equal to that of the solid state at the melting point and obtain the value 7.0 cal./degree/gram atom.

Thermodynamic properties of the ideal monatomic gas were calculated using energy levels listed by Moore (*241*). Although Honig (*161*) has detected polyatomic species in silicon vapor, there is not sufficient information available to calculate the thermodynamic functions of these species. Honig gives the heat of sublimation of the monatomic species as 105,000 cal./gram atom and calculates a boiling point of about 2950° K., considerably higher than earlier determinations of Ruff and Konschak (*283*) and Baur and Brunner (*23*). This high value is supported by the spectroscopic work of Barrow and Rowlinson (*21*). Without more spectroscopic data on the polyatomic species we cannot check the boiling point or calculate the heat of vaporization to equilibrium gas by third law methods.

SILVER

Based on five different sets of measurements from 1° to 303° K., Kelley (*186*) calculates the entropy at 298° K. as 10.20 ± 0.05 e.u., while Meads, Forsythe, and Giauque (*230*) calculate an enthalpy at 298° K. of 1373 cal./gram atom. The measurements of Lyashenko (*215*) have been considered along with sources listed by Kelley (*185*) in selecting the solid heat capacity from 298° K. to the melting point. The defined melting point on the International Temperature Scale as described by Stimson (*319*) is 1233.95° K. The heat of melting, 2,700 cal./gram atom, is a rounded value reached by considering those reported by Kubaschewski and coworkers (*206*), by Wittig (*347*), and by Kelley (*179, 182, 185*). The liquid heat capacity has been estimated as 7.5 cal./degree/gram atom.

Thermodynamic functions of the ideal monatomic gas have been calculated from the spectroscopic data listed in Landolt-Bornstein Tabellen (*208*). Kelley (*180*) selects the vapor pressure data of Harteck (*145*) as being the most reliable of the older data. Harteck is in fair agreement with the measurements of Fischer (*111*) and McCabe and Birchenall (*226b*), while Lyubimov and Granovskaya (*217*) are too low and Baur and Brunner (*23*) are too high. We find a heat of sublimation at 298° K. of 68,400 cal./gram atom, a normal boiling point of 2450° K., and a heat of vaporization at the normal boiling point of 60,960 cal./gram atom. Searcy, Freeman, and Michel (*294*) have recently indicated that polyatomic species may be important in silver vapor.

SODIUM

Low temperature measurements of Dauphinee, Martin, and Preston-Thomas (*82*), Simon and Zeidler (*305*) and Parkinson and Quarrington (*256*) were used to calculate an entropy and enthalpy at 298° K. of 12.21 e. u. and 1532 cal./gram atom, respectively. Published values of other thermodynamic properties have been reviewed by Evans, Jacobson, Munson, and Wagman (*105*). They select the melting point to be 370.97° K. and the heat of melting as 621.8 cal./gram atom. They present data on the solid and liquid heat capacity, thermodynamic functions of the ideal monatomic and diatomic gases, and the dissociation energy. Consistent with this information we find the heats of sublimation at 298° K. to ideal monatomic and diatomic gases as 25,900 and 33,800 cal./mole, respectively. The total pressure reaches one atmosphere at 1163° K. and the heat of vaporization to equilibrium gas at this temperature is 21,280 cal./gram atom. A more comprehensive review of the physical and thermodynamic properties of sodium has been compiled by Thomson and Garelis (*325*).

STRONTIUM

Kelley (*186*) estimates the entropy at 298° K. as 12.5 ± 0.5 e. u. The solid heat capacity above room temperature was estimated by comparison with calcium. Eastman, Cubicciotti, and Thurmond (*93*) have reported a transition point at 862° K. and a melting point of 1043° K., in good agreement with the review of Kubaschewski, Brizgys, Huchler, Jauch, and Reinartz (*206*). Kubaschewski and coworkers (*206*) have estimated the heat of melting to be 2200 cal./gram atom. We estimate the heat of the transition to be 200 cal./gram atom by comparison with calcium. Thermodynamic functions of the ideal monatomic gas have been calculated from the energy levels listed by Moore (*241*). Vapor pressures measured by Hartmann and Schneider (*146*) and by Priselkov and Nesmeianov (*263*) are in fair agreement, and lead to a heat of sublimation at 298° K. of 39,100 cal./gram atom, a normal boiling point of 1640° K., and a heat of vaporization of 33,200 cal./gram atom.

SULFUR

Eastman and McGavock (*94*) have measured the heat capacity of the solid from 12° to 366° K., from which can be derived for the rhombic form at 298° K. an entropy and enthalpy of 7.62 e. u. and 1053 cal./gram atom, respectively. Braune and Moller (*31*) have measured the heat capacity of the liquid and have reviewed previous work. They list a heat of transition from rhombic to monoclinic of 90 cal./gram atom at 368.6° K. and the heat of melting as 337 cal./gram atom at 392° K. The boiling point of sulfur is defined on the International Temperature Scale as 717.75° K. (444.60° C.) as described by Stimson (*319*).

Guthrie, Scott, and Waddington (*140*) have calculated thermodynamic functions of the octatomic gas as well as the heat of sublimation of this form to be 24,350 cal./mole at 298° K., while Evans and Wagman (*107*) present data for the diatomic form, including the heat of sublimation of 30,840 cal./mole at 298° K. As noted by Guthrie, Scott, and Waddington, these data cannot be reconciled with the vapor density data for sulfur recently determined by Braune, Peter, and Neveling (*32*). Luft (*214*) has attempted to correlate the data by postulating several gaseous species between the octatomic and the diatomic forms, but does not appear to give enough weight to the heat of sublimation of the diatomic gas derived by Evans and Wagman. We believe additional data are needed to define completely the sulfur vapor phase and have, therefore, presented only the data for octatomic, diatomic, and monatomic forms. Evans and Wagman give the thermodynamic functions of the monatomic gas, while the dissociation energy of the diatomic form was taken from the work of St. Pierre and Chipman (*321*). An approximate value of the heat of vaporization can be derived using the equilibrium constants derived by Braune, Peter, and Neveling. At the normal boiling point the value calculated is 2300 cal./gram atom.

TANTALUM

Low temperature data have been given by Kelley (*178*) Keesom and Desirant (*175*), Simon and Ruhemann (*303*), and Clusius and Gutierrez Losa (*62*), from which we calculate an entropy and enthalpy at 298° of 9.90 e. u. and 1358 cal./gram atom, respectively. Hoch (*157*), Jaeger and Veenstra (*167*), and Magnus and Holzmann (*220*) have measured the heat capacity of the solid to 2939°, 1828° and 1173° K., respectively. We have smoothed these data by the method of Shomate (*301*) and have extrapolated them to 3000° K. Brewer (*35*) lists the melting point as 3270° K. and the heat of melting as 7500 cal./gram atom.

Thermal properties of the ideal monatomic gas have been calculated from the spectroscopic data of Van Den Berg, Klinkenberg, and Van Den Bosch (*334*). Edwards, Johnston, and Blackburn (*95*), Langmuir and Malter (*209*), and Fiske (*114*) have measured the vapor pressure. Data reported by the first two sources are in good agreement but lower than that of Fiske. Using the data of the first two sources, we calculate a heat of sublimation to ideal monatomic gas at 298° K. of 186,800 cal./gram atom. Assuming the heat capacity of the liquid and the gas above 3000° K. to be equal, we compute a normal boiling point of 5700° K. and an accompanying heat of vaporization of 180,000 cal./gram atom.

TECHNETIUM

Brewer (*35*) has estimated the entropy at 298° K. to be 8.0 e. u., the heat capacity of the solid, the melting point as 2400° K., and the heat of melting as 5500 cal./gram atom. The spectroscopic data of Meggers (*232*) have been employed to calculate the thermodynamic functions of the ideal monatomic gas. Using Brewer's estimate of the vapor pressure at 3000° K. and estimating ΔC_p to be 3.5 cal./degree/gram atom, we calculate a heat of sublimation at 298° K. of 155,000 cal./gram atom, a normal boiling point of 4900° K., and an accompanying heat of vaporization of 138,000 cal./gram atom.

TELLURIUM

Based on the low temperature measurements of Slansky and Coulter (*311*) and of Anderson (*13*), Kelley (*186*) calculates the entropy at 298° K. as 11.88 ± 0.10 e. u., and we compute the enthalpy at 298° K. as 1463 cal./gram atom. Kubaschewski (*205*) has given the heat capacity of the solid and liquid states and lists 723° K. as the melting point and 4180 ± 130 cal./gram atom as the heat of melting. Spectroscopic energy levels listed in Landolt-Bornstein Tabellen (*208*) have been employed to calculate the thermodynamic properties of the ideal monatomic gas. The computations of Kelley (*185, 186*) on the thermodynamic properties of the ideal diatomic gas have been extended to 3000° K. Gaydon (*118*) gives the dissociation energy of the diatomic gas at 0° K. as 53,000 cal./gram mole.

The vapor pressures measured by Brooks (*43*) fall between those of Schneider and Schupp (*286*) and Doolan and Partington (*86*), and extrapolate nicely to those of Niwa and Sibata (*250*), who show the gas to be diatomic in the temperature range from 593° to 683° K. From the foregoing information we calculate the heat of sublimation at 298° K. to ideal monatomic and diatomic gases as 46,500 and 39,600 cal./gram mole, respectively. The total vapor pressure reaches one atmosphere at 1260° K. and the accompanying heat of vaporization to equilibrium gas is 12,100 cal./gram atom.

TERBIUM

Skochdopole, Griffel, and Spedding (*310*) estimate the entropy at 300° K. as 17.5 e. u. Spedding and Daane (*314*) indicate the melting point is between 1673° and 1773° K. The remainder of the values are estimated by analogy with neighboring elements and should be used only until measured data are available.

THALLIUM

The entropy at 298° K. is 15.35 ± 0.06 e. u. as given by Kelley (*186*). From the measurements of Hicks (*154*) we calculate the enthalpy at 298° K. as 1632 cal./gram atom. Recent measurements of the solid and liquid heat capacities and heats of transitions have been made by Kubaschewski and coworkers (*205, 206*) and by Oelsen and coworkers (*253, 254*). Considering also the review of Kelley (*185*), we have selected the heat capacities, the transition point of 507° K. with associated heat of transition of 90 cal./gram atom, as well as a melting point of 577° K. and a heat of melting of 1020 cal./gram atom. Meggers and Murphy (*234*) have reported spectroscopic data which we have used to calculate the thermodynamic functions of the ideal monatomic gas. Vapor pressures have been reported by Gibson (*125*), von Leitgebel (*211*), and Coleman and Egerton (*72*). We calculate a heat of sublimation at 298° K. of 43,000 cal./gram atom, a normal boiling point of 1740° K., and an accompanying heat of vaporization of 38,740 cal./gram atom.

THORIUM

Griffel and Skochdopole (*134*) have measured the heat capacity from 20° to 300° K., and report an entropy and enthalpy at 298° K. of 12.760 e. u. and 1556 cal./gram atom, respectively. Kelley (*185*) has reported an equation for the solid heat capacity which we have adjusted to fit the low temperature data and extrapolated to the melting point. Chiotti (*53, 54*) has recently shown the presence of a solid state transition from a face-centered to a body-centered lattice at 1673° K. and has determined the melting point as 1968° K. We have estimated the heat of transition and heat of melting as 670 and 3740 cal./gram atom, respectively. The liquid heat capacity is also estimated. Based on the work of Zwikker (*352*), Brewer (*35*) has calculated the normal boiling point at 4500° K. and a heat of vaporization of 130,000 cal./gram atom. The term values of thorium gas have not yet been determined and no thermodynamic functions for the gas can be calculated.

THULIUM

Skochdopole, Griffel, and Spedding (*310*) estimate the entropy at 300° K. as 17.10 e. u. Spedding and Daane (*314*) place the melting point between 1823° and 1923° K. and believe the volatility is between that of dysprosium and lutetium. Meggers (*231*) lists the available spectroscopic data. The information listed has been estimated and is consistent with the above known facts. These values are intended for use only until measured data are available.

TIN

On the basis of five sets of low temperature data covering the range from 1° to 287° K., Kelley (*186*) calculates an entropy value of 12.29 e. u. at 298° K. for white tin, and we calculate the enthalpy as 1507 cal./gram atom. Jovanovic (*172*) has measured the heat of transition of gray to white tin at 292° K. and gives 535 ± 8 cal./gram atom. The heat capacity of the solid and liquid have been adopted from the compilation of Kelley (*185*), who also gives the heat of melting as 1720 cal./gram atom at 505° K. Thermodynamic functions of the ideal monatomic gas have been calculated from the energy levels listed in Landolt-Bornstein Tabellen (*208*).

Of the vapor pressure measurements reviewed by Baughan (*22*), only those of Harteck (*145*) are in agreement with the recent measurements of Brewer and Porter (*38*) and Searcy and Freeman (*292*). Searcy and Freeman have demonstrated that tin vapor is monatomic. We adopt a rounded value of 72,000 cal./gram atom for the heat of sublimation at 298° K., and further calculate a normal boiling point of 2960° K. with an accompanying heat of vaporization of 69,400 cal./gram atom.

TITANIUM

Kothen and Johnston (*201*) have measured the heat capacity of a high purity sample from 15° to 305° K., and report an entropy and enthalpy at 298° K. of 7.33 ± 0.02 e. u. and 1150 cal./gram atom, respectively. Kothen (*200*) and Jaeger, Rosenbohm, and Fonteyne (*166*) have measured the heat content, from which we derive the heat capacity of the solid and the heat of transition of 950 cal./gram atom. Edwards, Johnston, and Ditmars (*99*) have confirmed McQuillan's (*229*) value of 1155° K. for the transition temperature. Of the recent values listed for the melting point (*3, 143, 226a, 255, 287*), we have taken an average value of 1950° ± 20° K. No direct measurement of the heat of melting has been made, but the average entropy of melting for body-centered cubic elements is about 1.9 e. u. On this basis we have used a heat of melting of 3700 cal./gram atom. The heat capacity of the liquid has been estimated.

Thermodynamic functions of the ideal monatomic gas have been calculated from the energy levels of Moore (*241*), and are in good agreement with those of Kolsky and Gilles (*196*). Vapor pressure measurements have been made by Edwards, Johnston, and Ditmars (*99*), who have corrected the results of Blocher and Campbell (*27*) and Carpenter and Mair (*51*). From these data we calculate a heat of sublimation at 298° K. of 112,600 cal./gram atom, a normal boiling point of 3550° K., and an associated heat of vaporization of 102,500 cal./gram atom.

TUNGSTEN

Kelley (*186*) reports the entropy at 298° K. as 8.04 ± 0.10 e. u. From the same data we calculate the enthalpy at 298° K. as 1216 cal./gram atom. Solid heat content measurements of Magnus and Holzmann (*220*), Jaeger and Rosenbohm (*165*), and Hoch (*157*) are in good agreement and have been smoothed by the method of Shomate (*301*) and extrapolated to 3000° K. Brewer (*35*) lists the melting point as 3650° K. and the heat of melting as 8420 cal./gram atom. Thermodynamic functions for the ideal monatomic gas have been calculated from energy levels listed in Landolt-Bornstein Tabellen (*208*). From the vapor pressure measurements of Jones, Langmuir, and Mackay (*170*) we calculate a heat of sublimation at 298° K. of 200,000 cal./gram atom. By estimating the heat capacity of liquid and gas from 3000° to 6000° K. at 8.5 and 7.5 cal./degree/gram atom, respectively, we find a normal boiling point of 5800° K. and an accompanying heat of vaporization of 191,000 cal./gram atom.

URANIUM

Jones, Gordon, and Long (*171*) have measured the heat capacity from 15° to 300° K., and report an entropy of 12.03 ± 0.03 e. u. at 298° K. We calculate the enthalpy at 298° K. to be 1559 cal./gram atom. We have adopted the heat capacity of the solid and the heats and temperatures of transition as measured by Ginnings and Corruccini (*126*). The melting point of 1406° ± 2° K. seems well established from the measurements of Dahl and Cleaves (*79*), Udy and Boulger (*331*), and Buzzard, Liss, and Fickle (*49*). We have estimated the heat capacity of the liquid to be the same as the solid at the melting point.

Van Den Bosch and Van Den Berg (*335*) have reported spectroscopic energy levels from which we have calculated the thermodynamic functions of the ideal monatomic gas. From the vapor pressure measurements of Rauh and Thorn (*264*), we calculate the heat of sublimation at 298° K. to be 117,160 cal./gram atom, the heat of fusion at the melting point to be 3700 cal./gram atom, the normal boiling point as 4200° K., and the heat of vaporization at the normal boiling point as 101,000 cal./gram atom.

VANADIUM

Anderson (*11*) has measured the heat capacity from about 50° to 300° K. and computes an entropy at 298° K. of 7.01 ± 0.10 e. u. We calculate a heat content of 1122 cal./gram atom from his data. The heat capacity of the solid has been derived from the data of Jaeger and Veenstra (*167*). Recent measurements of the melting point by Oriani and Jones (*255*) and by Adenstedt, Pequignot, and Raymer (*3*) are in reasonable agreement and we adopt the value 2190° ± 10° K. In the absence of a measured value for the heat of melting, we have used 1.9 e. u., the average entropy of melting for body-centered cubic elements, and calculate a heat of melting of 4200 cal./gram atom. By comparison with chromium and titanium, the heat capacity of the liquid has been estimated to be 9.50 cal./degree/gram atom.

Thermodynamic functions for the ideal monatomic gas state have been calculated from the spectroscopic energy levels of Moore (*241*). From the vapor pressure measurements of Edwards, Johnston, and Blackburn (*96*), we calculate a heat of sublimation at 298° K. of 122,750 cal./gram atom, a normal boiling point of 3650° K., with an accompanying heat of vaporization of 109,600 cal./gram atom.

XENON

A melting point of 161.3° K. and heat of melting of 549 cal./gram atom have been reported by Clusius and Riccoboni (*65*). Michels and Wassenaar (*236*) have measured the vapor pressure and find the normal boiling point at 165.04° K. in good agreement with the measurements of Clusius and Wiegand (*71*), who also report 3021 cal./gram atom for the heat of vaporization at this temperature. Kobe and Lynn (*193*) adopt 256.57° K. as the critical temperature and 58.0 atmospheres as the critical pressure. Thermodynamic functions of the ideal gas have been calculated at the National Bureau of Standards (*295*).

YTTERBIUM

Spedding and Daane (*314*) report a transition at 1071° K. and the melting point at 1097° K. They indicate the volatility to be between europium and samarium. Thermodynamic functions of the ideal monatomic gas have been calculated from energy levels given in Landolt-Bornstein Tabellen (*208*). The information listed for the reference state is consistent with the above facts and has been estimated. It is intended that this information will serve only until measured data are available.

YTTRIUM

Brewer (*35*) estimates the entropy at 298° K. as 11.0 e. u. Kelley (*187*) has estimated the heat capacities of the solid and liquid, as well as the melting point of 1773° K. and the heat of melting of 4100 cal./gram atom. Brewer (*35*) has estimated the normal boiling point of 3500° K. The energy levels listed by Moore (*241*) have been used to calculate the thermodynamic functions of the ideal monatomic gas. Consistent with these estimates, we have computed the heat of sublimation at 298° K. to be 102,000 cal./gram atom, and a heat of vaporization at the normal boiling point of 94,000 cal./gram atom.

ZINC

Kelley (*186*) lists the entropy at 298° K. as 9.95 ± 0.05 e. u. Barrow and coworkers (*20*) have calculated the enthalpy at 298° K. to be 1349 cal./gram atom. Kelley (*186*) also reports the heat capacity of the solid and the liquid states, based on numerous sources. His value of the melting point, 692.7° K., and his heat of melting, 1765 cal./gram atom, have been recently confirmed by Kubaschewski and coworkers (*206*). Thermodynamic functions for the ideal monatomic gas have been calculated from the energy levels of Moore (*241*). Barrow and coworkers (*20*) have reviewed the vapor pressure information and find the heat of sublimation at 298° K. to be 31,180 cal./gram atom. We calculate a normal boiling point of 1181° K. with an accompanying heat of vaporization of 27,560 cal./gram atom.

ZIRCONIUM

Skinner and Johnston (*309*) have measured the heat capacity from 14° to 300° K. and Todd (*327*) has measured the range from 51° to 298° K. The entropies at 298° K. are in good agreement and we adopt the value 9.29 ± 0.04 e. u. Skinner and Johnston (*309*) report an enthalpy at 298° K. of 1313 cal./gram atom. Skinner (*307*) has used his own measurements and those of Coughlin and King (*74*) to calculate the thermodynamic functions for the solid state. Skinner finds a transition at 1143° K. with an associated heat change of 1040 cal./gram atom. Adenstedt (*2*) and Oriani and Jones (*255*) are in good agreement that the melting point is 2125° K. No direct measurement of the heat of melting has been made, but the average entropy of melting for body-centered cubic elements is about 1.9 e. u. On this basis, we have used a heat of melting of 4000 cal./gram atom. The heat capacity of the liquid has been estimated.

Thermodynamic functions of the ideal monatomic gas state have been calculated from the spectroscopic data of Moore (*241*). Values recently reported by Kolsky and Gilles (*197*) are in agreement. Skinner, Edwards, and Johnston (*308*) have reported the only vapor pressure data measured from 1949° to 2054° K. Their data lead to a heat of sublimation at 298° K. of 146,000 cal./gram atom, a normal boiling point of 4650° K., and an accompanying heat of vaporization of 139,000 cal./gram atom.

Tabulated Values of Thermodynamic Properties

ACTINIUM

ACTINIUM   Ac   Solid from 298° to 1470°, Liquid from 1470° to 3000°.

REFERENCE STATE

Gfw 227. GRAMS CAL./GFW.

$(H°_{298.15} - H°_0)$ =

M.P. (1,470) °K
ΔH$_m$ (3,400) CAL./GFW.

B.P. (3,600) °K
ΔH$_v$ (95,000) CAL./GFW.

S.P.   °K
ΔH$_s$ CAL./GFW.

T.P.   °K
ΔH$_t$ CAL./GFW.

T.P.   °K
ΔH$_t$ CAL./GFW.

T$_c$ =   °K
P$_c$ =   ATM.

T TEMPERATURE °K	C°$_P$ HEAT CAPACITY CAL./DEG./GFW.	H°$_T$ - H°$_{298.15}$ HEAT CONTENT CAL./GFW.	S°$_T$ ENTROPY CAL./DEG./GFW.	$-(F°_T - H°_{298.15})$ FREE ENERGY FUNCTION CAL./DEG./GFW.	FORMATION FROM REFERENCE STATE		
					HEAT ΔH°$_f$ CAL./GFW.	FREE ENERGY ΔF°$_f$ CAL./GFW.	LOG$_{10}$ K$_P$
298	6.50	0	15.00	15.00			
300	6.50	12	15.04	15.00			
400	6.70	670	16.94	15.27			
500	6.90	1350	18.45	15.75			
600	7.10	2050	19.73	16.32			
700	7.30	2770	20.84	16.89			
800	7.50	3510	21.83	17.45			
900	7.70	4270	22.72	17.98			
1000	7.90	5050	23.54	18.49			
1100	8.10	5850	24.30	18.99			
1200	8.30	6670	25.02	19.47			
1300	8.50	7510	25.69	19.92			
1400	8.70	8370	26.33	20.36			
1500	8.00	12650	29.24	20.81			
1600	8.00	13450	29.76	21.36			
1700	8.00	14250	30.25	21.87			
1800	8.00	15050	30.70	22.34			
1900	8.00	15850	31.14	22.80			
2000	8.00	16650	31.55	23.23			
2100	8.00	17450	31.94	23.64			
2200	8.00	18250	32.31	24.02			
2300	8.00	19050	32.66	24.38			
2400	8.00	19850	33.00	24.73			
2500	8.00	20650	33.33	25.07			
2600	8.00	21450	33.64	25.39			
2700	8.00	22250	33.95	25.71			
2800	8.00	23050	34.24	26.01			
2900	8.00	23850	34.52	26.30			
3000	8.00	24650	34.79	26.58			

THERMODYNAMIC PROPERTIES OF THE ELEMENTS

ALUMINUM

ALUMINUM	Al		Solid from 298° to 932°, Liquid from 932° to 2720°, Ideal Monatomic Gas from 2720° to 3000°.
Gfw	26.98	GRAMS	
REFERENCE STATE			
$(H°_{298.15} - H°_0) = 1,094$		CAL./GFW.	
M.P.	932	°K	
ΔH_m	2,550	CAL./GFW.	
B.P.	2,720	°K	
ΔH_v	70,200	CAL./GFW.	
S.P.		°K	
ΔH_s		CAL./GFW.	
T.P.		°K	
ΔH_t		CAL./GFW.	
T.P.		°K	
ΔH_t		CAL./GFW.	
$T_c =$		°K	
$P_c =$		ATM.	

T TEMPERATURE °K	$C°_P$ HEAT CAPACITY CAL./DEG./GFW.	$H°_T - H°_{298.15}$ HEAT CONTENT CAL./GFW.	$S°_T$ ENTROPY CAL./DEG./GFW.	$-(F°_T - H°_{298.15})/T$ FREE ENERGY FUNCTION CAL./DEG./GFW.	FORMATION FROM REFERENCE STATE		
					HEAT $\Delta H°_f$ CAL./GFW.	FREE ENERGY $\Delta F°_f$ CAL./GFW.	LOG$_{10}$ K$_P$
298	5.82	0	6.77	6.77			
300	5.83	11	6.80	6.77			
400	6.12	600	8.49	6.99			
500	6.43	1230	9.91	7.45			
600	6.72	1890	11.11	7.96			
700	7.02	2580	12.17	8.49			
800	7.31	3310	13.15	9.02			
900	7.61	4060	14.03	9.52			
1000	7.00	7330	17.53	10.20			
1100	7.00	8030	18.19	10.89			
1200	7.00	8730	18.80	11.53			
1300	7.00	9430	19.36	12.11			
1400	7.00	10130	19.88	12.65			
1500	7.00	10830	20.36	13.14			
1600	7.00	11530	20.81	13.61			
1700	7.00	12230	21.24	14.05			
1800	7.00	12930	21.64	14.46			
1900	7.00	13630	22.02	14.85			
2000	7.00	14330	22.38	15.22			
2100	7.00	15030	22.72	15.57			
2200	7.00	15730	23.04	15.89			
2300	7.00	16430	23.35	16.21			
2400	7.00	17130	23.65	16.52			
2500	7.00	17830	23.94	16.81			
2600	7.00	18530	24.21	17.09			
2700	7.00	19230	24.48	17.36			
2800	4.97	89970	50.50	18.37			
2900	4.97	90460	50.67	19.48			
3000	4.97	90960	50.84	20.52			

ALUMINUM

ALUMINUM Al
IDEAL MONATOMIC GAS

Gfw 26.98 GRAMS
$(H°_{298.15} - H°_0) = 1,654$ CAL./GFW.

Reference State for Calculating $\Delta H°_f$, $\Delta F°_f$, and $\text{Log}_{10} K_p$: Solid from 298° to 932°, Liquid from 932° to 2720°, Ideal Monatomic Gas from 2720° to 3000°.

TEMPERATURE °K	$C°_P$ HEAT CAPACITY CAL./DEG./GFW.	$H°_T - H°_{298.15}$ HEAT CONTENT CAL./GFW.	$S°_T$ ENTROPY CAL./DEG./GFW.	$-(F°_T - H°_{298.15})/T$ FREE ENERGY FUNCTION CAL./DEG./GFW.	FORMATION FROM REFERENCE STATE		
					HEAT $\Delta H°_f$ CAL./GFW.	FREE ENERGY $\Delta F°_f$ CAL./GFW.	$\text{LOG}_{10} K_p$
298	5.11	0	39.30	39.30	77500	67801	-49.701
300	5.11	9	39.33	39.30	77498	67739	-49.351
400	5.05	517	40.79	39.50	77417	64497	-35.242
500	5.02	1020	41.92	39.88	77290	61285	-26.789
600	5.00	1521	42.83	40.30	77131	58099	-21.164
700	4.99	2021	43.60	40.72	76941	54940	-17.154
800	4.99	2520	44.27	41.12	76710	51814	-14.156
900	4.98	3018	44.85	41.50	76458	48720	-11.831
1000	4.98	3516	45.38	41.87	73686	45836	-10.018
1100	4.98	4014	45.85	42.21	73484	43058	-8.555
1200	4.98	4512	46.29	42.53	73282	40294	-7.339
1300	4.98	5009	46.68	42.83	73079	37563	-6.315
1400	4.97	5507	47.05	43.12	72877	34839	-5.439
1500	4.97	6004	47.40	43.40	72674	32114	-4.679
1600	4.97	6501	47.72	43.66	72471	29415	-4.017
1700	4.97	6999	48.02	43.91	7,2269	26743	-3.438
1800	4.97	7496	48.30	44.14	72066	24078	-2.923
1900	4.97	7993	48.57	44.37	71863	21418	-2.463
2000	4.97	8490	48.83	44.59	71660	18760	-2.049
2100	4.97	8987	49.07	44.80	71457	16122	-1.677
2200	4.97	9484	49.30	44.99	71254	13482	-1.339
2300	4.97	9981	49.52	45.19	71051	10860	-1.031
2400	4.97	10478	49.73	45.37	70848	8256	-.751
2500	4.97	10975	49.94	45.55	70645	5645	-.493
2600	4.97	11472	50.13	45.72	70442	3050	-.256
2700	4.97	11969	50.32	45.89	70239	471	-.038
2800	4.97	12466	50.50	46.05	0	0	0
2900	4.97	12963	50.67	46.20	0	0	0
3000	4.97	13460	50.84	46.36	0	0	0

M.P. °K
ΔH_m CAL./GFW.

B.P. °K
ΔH_v CAL./GFW.

S.P. °K
ΔH_s CAL./GFW.

T.P. °K
ΔH_t CAL./GFW.

T.P. °K
ΔH_t CAL./GFW.

$T_c =$ °K
$P_c =$ ATM.

THERMODYNAMIC PROPERTIES OF THE ELEMENTS

ANTIMONY

ANTIMONY   Sb

REFERENCE STATE: Solid from 298° to 903°, Liquid from 903° to 1910°, Ideal Diatomic Gas from 1910° to 3000°.

Gfw	121.76	GRAMS
$(H°_{298.15} - H°_0)$ =	1,410	CAL./GFW.
M.P.	903	°K
ΔH_m	4,740	CAL./GFW.
B.P.	1,910	°K
ΔH_v	16,230	CAL./GFW.
S.P.		°K
ΔH_s		CAL./GFW.
T.P.		°K
ΔH_t		CAL./GFW.
T.P.		°K
ΔH_t		CAL./GFW.
T_c =		°K
P_c =		ATM.

T TEMPERATURE °K	$C°_P$ HEAT CAPACITY CAL./DEG./GFW.	$H°_T - H°_{298.15}$ HEAT CONTENT CAL./GFW.	$S°_T$ ENTROPY CAL./DEG./GFW.	$-(F°_T - H°_{298.15})/T$ FREE ENERGY FUNCTION CAL./DEG./GFW.	FORMATION FROM REFERENCE STATE		
					HEAT $\Delta H°_f$ CAL./GFW.	FREE ENERGY $\Delta F°_f$ CAL./GFW.	$LOG_{10} K_p$
298	6.03	0	10.92	10.92			
300	6.03	11	10.96	10.93			
400	6.21	625	12.72	11.16			
500	6.38	1250	14.11	11.61			
600	6.55	1890	15.28	12.13			
700	6.73	2550	16.30	12.66			
800	6.90	3240	17.22	13.17			
900	7.08	3950	18.06	13.68			
1000	7.50	9440	24.10	14.66			
1100	7.50	10190	24.81	15.55			
1200	7.50	10940	25.46	16.35			
1300	7.50	11690	26.06	17.07			
1400	7.50	12440	26.62	17.74			
1500	7.50	13190	27.13	18.34			
1600	7.50	13940	27.62	18.91			
1700	7.50	14690	28.07	19.43			
1800	7.50	15440	28.50	19.93			
1900	7.50	16190	28.90	20.38			
2000	4.47	35775	38.89	21.01			
2100	4.47	36220	39.11	21.87			
2200	4.47	36670	39.32	22.66			
2300	4.47	37115	39.52	23.39			
2400	4.47	37560	39.71	24.06			
2500	4.47	38010	39.89	24.69			
2600	4.47	38455	40.07	25.28			
2700	4.47	38900	40.23	25.83			
2800	4.47	39350	40.40	26.35			
2900	4.47	39800	40.55	26.83			
3000	4.47	40240	40.71	27.30			

ANTIMONY

ANTIMONY	Sb_2		Reference State for Calculating ΔH_f°, ΔF_f°, and
IDEAL DIATOMIC GAS			$Log_{10}Kp$: Solid from 298° to 903°, Liquid from 903° to 1910°, Ideal Diatomic Gas from 1910° to 3000°.

								FORMATION FROM REFERENCE STATE		
Gfw 243.52	GRAMS	T	C_p°	$H_T^\circ - H_{298.15}^\circ$	S_T°	$-(F^\circ - H_0^\circ)/T_{298.15}$		HEAT ΔH_f°	FREE ENERGY ΔF_f°	$LOG_{10} K_p$
$(H_{298.15}^\circ - H_0^\circ) = 2,363$	CAL./GFW.	TEMPERATURE °K	HEAT CAPACITY CAL./DEG./GFW.	HEAT CONTENT CAL./GFW.	ENTROPY CAL./DEG./GFW.	FREE ENERGY FUNCTION CAL./DEG./GFW.		CAL./GFW.	CAL./GFW.	
		298	8.69	0	60.90	60.90		56400	44753	-32.806
		300	8.69	16	60.96	60.91		56394	44682	-32.553
		400	8.79	890	63.47	61.25		56040	40828	-22.309
		500	8.85	1775	65.44	61.89		55675	37065	-16.202
		600	8.87	2660	67.05	62.62		55280	33386	-12.161
M.P. °K		700	8.89	3545	68.42	63.36		54845	29771	-9.295
ΔH_m CAL./GFW.		800	8.90	4435	69.60	64.06		54355	26227	-7.165
		900	8.91	5325	70.65	64.74		53825	22748	-5.524
		1000	8.92	6220	71.60	65.38		43740	20340	-4.445
		1100	8.92	7112	72.45	65.99		43132	18019	-3.580
B.P. °K		1200	8.93	8005	73.23	66.56		42525	15753	-2.869
ΔH_v CAL./GFW.		1300	8.93	8897	73.94	67.10		41917	13551	-2.278
		1400	8.93	9790	74.60	67.61		41310	11406	-1.780
		1500	8.93	10683	75.22	68.10		40703	9263	-1.349
		1600	8.93	11575	75.79	68.56		40095	7215	-.985
S.P. °K		1700	8.93	12468	76.33	69.00		39488	5165	-.664
ΔH_s CAL./GFW.		1800	8.93	13361	76.84	69.42		38881	3169	-.384
		1900	8.94	14255	77.32	69.82		38275	1187	-.136
		2000	8.94	15149	77.78	70.21		0	0	0
T.P. °K		2100	8.94	16042	78.22	70.59		0	0	0
ΔH_t CAL./GFW.		2200	8.94	16936	78.63	70.94		0	0	0
		2300	8.94	17830	79.03	71.28		0	0	0
		2400	8.94	18723	79.41	71.61		0	0	0
T.P. °K		2500	8.94	19617	79.78	71.94		0	0	0
ΔH_t CAL./GFW.		2600	8.94	20510	80.13	72.25		0	0	0
		2700	8.94	21404	80.46	72.54		0	0	0
$T_c =$ °K		2800	8.94	22298	80.79	72.83		0	0	0
$P_c =$ ATM.		2900	8.94	23192	81.10	73.11		0	0	0
		3000	8.94	24086	81.41	73.39		0	0	0

THERMODYNAMIC PROPERTIES OF THE ELEMENTS

ANTIMONY

ANTIMONY Sb_4
IDEAL TETRATOMIC GAS

Gfw 487.04
$(H°_{298.15} - H°_0) =$

Reference State for Calculating $\Delta H_f°$, $\Delta F_f°$, and $\text{Log}_{10} K_p$: Solid From 298° to 903°, Liquid from 903° to 1910°, Ideal Diatomic Gas from 1910° to 3000°.

T TEMPERATURE °K	$C_p°$ HEAT CAPACITY CAL./DEG./GFW.	$H°_T - H°_{298.15}$ HEAT CONTENT CAL./GFW.	$S°_T$ ENTROPY CAL./DEG./GFW.	$-(F°-H°_{298.15})$ FREE ENERGY FUNCTION CAL./DEG./GFW.	FORMATION FROM REFERENCE STATE		
					HEAT $\Delta H_f°$ CAL./GFW.	FREE ENERGY $\Delta F_f°$ CAL./GFW.	$\text{LOG}_{10} K_p$
298	19.35	0	83.65	83.65	49000	37079	-27.180
300	19.35	36	83.79	83.67	48992	37007	-26.961
400	19.55	1980	89.38	84.43	48480	33080	-18.075
500	19.65	3940	93.75	85.87	47940	29285	-12.801
600	19.72	5910	97.34	87.49	47350	25618	-9.332
700	19.76	7890	100.39	89.12	46690	22057	-6.887
800	19.79	9860	103.05	90.73	45900	18564	-5.071
900	19.81	11840	105.35	92.20	45040	15241	-3.701
1000	19.82	13830	107.45	93.62	25070	14020	-3.064
1100	19.83	15810	109.35	94.98	24050	12929	-2.568
1200	19.83	17790	111.05	96.23	23030	11978	-2.181
1300	19.84	19770	112.65	97.45	22010	11077	-1.862
1400	19.84	21760	114.15	98.61	21000	10262	-1.602
1500	19.84	23745	115.45	99.62	19985	9590-	-1.397
1600	19.84	25730	116.75	100.67	18970	8938	-1.220
1700	19.85	27715	117.95	101.65	17955	8316	-1.069
1800	19.85	29700	119.15	102.65	16940	7670	-.931
1900	19.85	31685	120.15	103.48	15925	7280	-.837
2000	19.85	33670	121.25	104.42	-60430	8190	.894

M.P. °K	
ΔH_m CAL./GFW.	
B.P. °K	
ΔH_v CAL./GFW.	
S.P. °K	
ΔH_s CAL./GFW.	
T.P. °K	
ΔH_t CAL./GFW.	
T.P. °K	
ΔH_t CAL./GFW.	
T_c °K	
P_c ATM.	

ANTIMONY

ANTIMONY   Sb   Reference State for Calculating ΔH_f°, ΔF_f°, and
IDEAL MONATOMIC GAS $Log_{10} K_p$: Solid from 298° to 903°, Liquid from 903°
     to 1910°, Ideal Diatomic Gas from 1910° to 3000°.

Gfw 121.76   GRAMS
$(H^\circ_{298.15} - H^\circ_0) = 1,481$ CAL./GFW.

	T TEMPERATURE °K	C_P° HEAT CAPACITY CAL./DEG./GFW.	$H^\circ_T - H^\circ_{298.15}$ HEAT CONTENT CAL./GFW.	S°_T ENTROPY CAL./DEG./GFW.	$-(F^\circ - H^\circ_{298.15})/T$ FREE ENERGY FUNCTION CAL./DEG./GFW.	FORMATION FROM REFERENCE STATE		
						HEAT ΔH_f° CAL./GFW.	FREE ENERGY ΔF_f° CAL./GFW.	LOG$_{10}$ K$_P$
	298	4.97	0	43.06	43.06	62700	53117	−38.937
	300	4.97	9	43.09	43.06	62698	53059	−38.656
M.P.	400	4.97	506	44.52	43.26	62581	49861	−27.245
ΔH_m	500	4.97	1003	45.63	43.63	62453	46693	−20.410
	600	4.97	1500	46.53	44.03	62310	43560	−15.868
	700	4.97	1996	47.30	44.45	62146	40446	−12.628
B.P.	800	4.97	2493	47.96	44.85	61953	37361	−10.207
ΔH_v	900	4.97	2990	48.55	45.23	61740	34299	−8.329
	1000	4.97	3487	49.07	45.59	56747	31777	−6.945
	1100	4.97	3984	49.55	45.93	56494	29280	−5.817
	1200	4.98	4482	49.98	46.25	56242	26818	−4.884
S.P.	1300	4.99	4980	50.38	46.55	55990	24374	−4.098
ΔH_s	1400	5.00	5480	50.75	46.84	55740	21958	−3.428
	1500	5.03	5981	51.09	47.11	55491	19551	−2.848
	1600	5.06	6485	51.42	47.37	55245	17165	−2.344
T.P.	1700	5.09	6993	51.73	47.62	55003	14781	−1.900
ΔH_t	1800	5.14	7504	52.02	47.86	54764	12428	−1.508
	1900	5.19	8021	52.30	48.08	54531	10071	−1.158
	2000	5.26	8543	52.57	48.30	35468	8108	−.885
	2100	5.32	9072	52.82	48.50	35552	6761	−.703
	2200	5.40	9609	53.07	48.71	35639	5389	−.535
T.P.	2300	5.48	10152	53.32	48.91	35737	3997	−.379
ΔH_t	2400	5.56	10705	53.55	49.09	35845	2629	−.239
	2500	5.65	11265	53.78	49.28	35955	1230	−.107
	2600	5.74	11834	54.00	49.45	36079	139	−.011
	2700	5.83	12413	54.22	49.63	36213	−1560	.126
$T_c =$ °K	2800	5.92	13000	54.43	49.79	36350	−2934	.228
$P_c =$ ATM.	2900	6.01	13596	54.64	49.96	36496	−4365	.328
	3000	6.09	14201	54.85	50.12	36661	−5759	.419

THERMODYNAMIC PROPERTIES OF THE ELEMENTS

ARGON

ARGON A

REFERENCE STATE

Ideal Monatomic Gas from 298° to 3000°.

Gfw	39.944	GRAMS	
$(H°_T - H°_0)_{298.15}$ =	1,481	CAL./GFW.	
M.P.	83.78	°K	
ΔH_m	281.	CAL./GFW.	
B.P.	87.29	°K	
ΔH_v	1,558.	CAL./GFW.	
S.P.		°K	
ΔH_s		CAL./GFW.	
T.P.		°K	
ΔH_t		CAL./GFW.	
T.P.		°K	
ΔH_t		CAL./GFW.	
T_c =	151.	°K	
P_c =	48.0	ATM.	

T TEMPERATURE °K	$C°_P$ HEAT CAPACITY CAL./DEG./GFW.	$H°_T - H°_{298.15}$ HEAT CONTENT CAL./GFW.	$S°_T$ ENTROPY CAL./DEG./GFW.	$-(F°_T - H°_{298.15})/T$ FREE ENERGY FUNCTION CAL./DEG./GFW.	FORMATION FROM REFERENCE STATE		
					HEAT $\Delta H°_f$ CAL./GFW.	FREE ENERGY $\Delta F°_f$ CAL./GFW.	LOG$_{10}$ K$_P$
298	4.97	0	36.98	36.98			
300	4.97	9	37.01	36.98			
400	4.97	506	38.44	37.18			
500	4.97	1003	39.55	37.55			
600	4.97	1500	40.46	37.96			
700	4.97	1996	41.22	38.37			
800	4.97	2493	41.89	38.78			
900	4.97	2990	42.47	39.15			
1000	4.97	3487	43.00	39.52			
1100	4.97	3984	43.47	39.85			
1200	4.97	4480	43.90	40.17			
1300	4.97	4977	44.30	40.48			
1400	4.97	5474	44.67	40.76			
1500	4.97	5971	45.01	41.03			
1600	4.97	6468	45.33	41.29			
1700	4.97	6964	45.63	41.54			
1800	4.97	7461	45.92	41.78			
1900	4.97	7958	46.18	42.00			
2000	4.97	8455	46.44	42.22			
2100	4.97	8952	46.68	42.42			
2200	4.97	9448	46.91	42.62			
2300	4.97	9945	47.13	42.81			
2400	4.97	10442	47.34	42.99			
2500	4.97	10939	47.55	43.18			
2600	4.97	11436	47.74	43.35			
2700	4.97	11932	47.93	43.52			
2800	4.97	12429	48.11	43.68			
2900	4.97	12926	48.28	43.83			
3000	4.97	13423	48.45	43.98			

ARSENIC

ARSENIC As
Solid from 298° to 886°, Ideal Diatomic Gas from 886° to 3000°.

REFERENCE STATE

Gfw	74.91 GRAMS	
$(H°_{298.15} - H°_0) = 1,226$	CAL./GFW.	
M.P. 1,090	°K	
ΔH_m (6,620)	CAL./GFW.	
B.P.	°K	
ΔH_v	CAL./GFW.	
S.P. 886	°K	
ΔH_s 7,630	CAL./GFW.	
T.P.	°K	
ΔH_t	CAL./GFW.	
T.P.	°K	
ΔH_t	CAL./GFW.	
$T_c =$	°K	
$P_c =$	ATM.	

T TEMPERATURE °K	$C°_P$ HEAT CAPACITY CAL./DEG./GFW.	$H°_T - H°_{298.15}$ HEAT CONTENT CAL./GFW.	$S°_T$ ENTROPY CAL./DEG./GFW.	$-(F°_T - H°_{298.15})/T$ FREE ENERGY FUNCTION CAL./DEG./GFW.	FORMATION FROM REFERENCE STATE HEAT $\Delta H°_f$ CAL./GFW.	FORMATION FROM REFERENCE STATE FREE ENERGY $\Delta F°_f$ CAL./GFW.	$LOG_{10} K_P$
298	5.90	0	8.40	8.40			
300	5.90	10	8.43	8.40			
400	6.12	610	10.16	8.64			
500	6.34	1240	11.56	9.08			
600	6.56	1880	12.73	9.60			
700	6.78	2540	13.75	10.13			
800	7.01	3230	14.68	10.65			
900	4.44	26630	33.39	3.81			
1000	4.44	27070	33.86	6.79			
1100	4.45	27520	34.28	9.27			
1200	4.45	27960	34.67	11.37			
1300	4.46	28410	35.03	13.18			
1400	4.46	28850	35.36	14.76			
1500	4.46	29300	35.67	16.14			
1600	4.46	29740	35.95	17.37			
1700	4.46	30190	36.22	18.47			
1800	4.46	30640	36.48	19.46			
1900	4.46	31080	36.72	20.37			
2000	4.46	31530	36.95	21.19			
2100	4.47	31970	37.17	21.95			
2200	4.47	32420	37.37	22.64			
2300	4.47	32870	37.57	23.28			
2400	4.47	33310	37.76	23.89			
2500	4.47	33760	37.94	24.44			
2600	4.47	34210	38.12	24.97			
2700	4.47	34650	38.29	25.46			
2800	4.47	35100	38.45	25.92			
2900	4.47	35550	38.61	26.36			
3000	4.47	35990	38.76	26.77			

ARSENIC

ARSENIC As₂

Reference State for Calculating ΔH_f°, ΔF_f°, and $\log_{10} K_p$:
Solid from 298° to 886°, Ideal Diatomic Gas from 298° to 3000°.

IDEAL DIATOMIC GAS

Gfw 149.82 GRAMS CAL./GFW.
$(H^\circ_{298.15} - H^\circ_0) = 2,251$

					FORMATION FROM REFERENCE STATE		
T TEMPERATURE °K	C_p° HEAT CAPACITY CAL./DEG./GFW.	$H_T^\circ - H_{298.15}^\circ$ HEAT CONTENT CAL./GFW.	S_T° ENTROPY CAL./DEG./GFW.	$-(F_T^\circ - H_{298.15}^\circ)/T$ FREE ENERGY FUNCTION CAL./DEG./GFW.	HEAT ΔH_f° CAL./GFW.	FREE ENERGY ΔF_f° CAL./GFW.	$\log_{10} K_p$
298	8.36	0	57.19	57.19	48000	35957	-26.358
300	8.37	15	57.24	57.24	47980	35866	-26.130
400	8.59	865	59.68	57.52	47645	31901	-17.431
500	8.71	1730	61.62	58.16	47250	28000	-12.239
600	8.78	2605	63.21	58.87	46845	24195	-8.813
700	8.82	3485	64.57	59.60	46405	20456	-6.387
800	8.85	4365	65.74	60.29	45905	16801	-4.590
900	8.87	5255	66.78	60.95	0	0	0
1000	8.88	6140	67.72	61.58	0	0	0
1100	8.89	7030	68.56	62.17	0	0	0
1200	8.90	7920	69.34	62.74	0	0	0
1300	8.91	8810	70.05	63.28	0	0	0
1400	8.91	9700	70.71	63.79	0	0	0
1500	8.91	10590	71.33	64.27	0	0	0
1600	8.92	11480	71.90	64.73	0	0	0
1700	8.92	12370	72.44	65.17	0	0	0
1800	8.92	13270	72.95	65.58	0	0	0
1900	8.92	14160	73.43	65.98	0	0	0
2000	8.93	15050	73.89	66.37	0	0	0
2100	8.93	15940	74.33	66.74	0	0	0
2200	8.93	16840	74.74	67.09	0	0	0
2300	8.93	17730	75.14	67.44	0	0	0
2400	8.93	18620	75.52	67.77	0	0	0
2500	8.93	19520	75.88	68.08	0	0	0
2600	8.93	20410	76.24	68.39	0	0	0
2700	8.93	21300	76.57	68.69	0	0	0
2800	8.93	22200	76.90	68.98	0	0	0
2900	8.93	23090	77.21	69.25	0	0	0
3000	8.93	23980	77.51	69.52			

M.P.	°K	
ΔH_m	CAL./GFW.	
B.P.	°K	
ΔH_v	CAL./GFW.	
S.P.	°K	
ΔH_s	CAL./GFW.	
T.P.	°K	
ΔH_t	CAL./GFW.	
T.P.	°K	
ΔH_t	CAL./GFW.	
$T_c =$	°K	
$P_c =$	ATM.	

ARSENIC

ARSENIC As_4

IDEAL TETRATOMIC GAS

Gfw 299.64
$(H°_{298.15} - H°_0) =$

Reference State for Calculating $\Delta H_f°$, $\Delta F_f°$, and $Log_{10} K_p$:
Solid from 298° to 886°, Ideal Diatomic Gas from 886° to 3000°.

T TEMPERATURE °K	$C_p°$ HEAT CAPACITY CAL./DEG./GFW.	$H_T° - H_{298.15}°$ HEAT CONTENT CAL./GFW.	$S_T°$ ENTROPY CAL./DEG./GFW.	$-(F_T° - H_{298.15}°)/T$ FREE ENERGY FUNCTION CAL./DEG./GFW.	FORMATION FROM REFERENCE STATE		
					HEAT $\Delta H_f°$ CAL./GFW.	FREE ENERGY $\Delta F_f°$ CAL./GFW.	$LOG_{10} K_p$
298	18.48	0	75.00	75.00	34500	22154	-16.239
300	18.50	34	75.12	75.01	34494	22074	-16.082
400	19.10	1920	80.54	75.74	33980	18020	-9.846
500	19.35	3840	84.83	77.15	33380	14085	-6.156
600	19.49	5790	88.37	78.72	32770	10300	-3.752
700	19.58	7740	91.39	80.34	32080	6607	-2.062
800	19.64	9700	94.00	81.08	31280	3056	-.834
900	19.69	11670	96.30	83.34	-60350	-26816	6.512
1000	19.72	13640	98.40	84.76	-60140	-23100	5.048
1100	19.74	15610	100.30	86.11	-59970	-19468	3.868
1200	19.76	17590	102.00	87.35	-59750	-15734	2.865
1300	19.78	19560	103.60	88.56	-59580	-12104	2.035
1400	19.80	21540	105.00	89.62	-59360	-8344	1.302
1500	19.81	23520	106.40	90.72	-59180	-4760	.693
1600	19.82	25500	107.70	91.77	-58960	-1200	.163
1700	19.82	27490	108.90	92.73	-58770	2396	-.308
1800	19.83	29470	110.00	93.63	-58590	6066	-.736
1900	19.83	31450	111.10	94.55	-58370	9612	-1.105
2000	19.84	33430	112.10	95.39	-58190	13210	-1.443

M.P. °K
ΔH_m CAL./GFW.

B.P. °K
ΔH_v CAL./GFW.

S.P. °K
ΔH_s CAL./GFW.

T.P. °K
ΔH_t CAL./GFW.

T.P. °K
ΔH_t CAL./GFW.

$T_c =$ °K
$P_c =$ ATM.

ARSENIC

ARSENIC As
Gfw 74.91
$(H^°_{298.15} - H^°_0) = 1,481$
IDEAL MONATOMIC GAS

Reference State for Calculating $\Delta H^°_f$, $\Delta F^°_f$, and $Log_{10} Kp$:
Solid from 298° to 886°, Ideal Diatomic Gas from 298° to 3000°.

	GRAMS CAL./GFW.
M.P. °K	
ΔH_m CAL./GFW.	
B.P. °K	
ΔH_v CAL./GFW.	
S.P. °K	
ΔH_s CAL./GFW.	
T.P. °K	
ΔH_t CAL./GFW.	
T.P. °K	
ΔH_t CAL./GFW.	
T_c = °K	
P_c = ATM.	

T TEMPERATURE °K	$C^°_P$ HEAT CAPACITY CAL./DEG./GFW.	$H^°_T - H^°_{298.15}$ HEAT CONTENT CAL./GFW.	$S^°_T$ ENTROPY CAL./DEG./GFW.	$-(F^°_T - H^°_{298.15})/T$ FREE ENERGY FUNCTION CAL./DEG./GFW.	FORMATION FROM REFERENCE STATE		
					HEAT $\Delta H^°_f$ CAL./GFW.	FREE ENERGY $\Delta F^°_f$ CAL./GFW.	$LOG_{10} K_P$
298	4.97	0	41.61	41.61	69000	59098	-43.321
300	4.97	9	41.64	41.61	68999	59036	-43.011
400	4.97	506	43.07	41.81	68896	55732	-30.453
500	4.97	1003	44.18	42.18	68763	52453	-22.928
600	4.97	1500	45.09	42.59	68620	49204	-17.924
700	4.97	1996	45.85	43.00	68456	45986	-14.358
800	4.97	2493	46.52	43.41	68263	42791	-11.690
900	4.97	2990	47.10	43.78	45360	33021	-8.019
1000	4.97	3487	47.62	44.14	45417	31657	-6.919
1100	4.97	3984	48.10	44.48	45464	30262	-6.013
1200	4.97	4481	48.53	44.80	45521	28889	-5.261
1300	4.97	4978	48.93	45.11	45568	27498	-4.623
1400	4.98	5475	49.30	45.39	45625	26109	-4.076
1500	4.99	5973	49.64	45.66	45673	24718	-3.601
1600	5.00	6472	49.96	45.92	45732	23316	-3.184
1700	5.01	6973	50.27	46.17	45783	21898	-2.815
1800	5.04	7475	50.55	46.40	45835	20509	-2.489
1900	5.06	7980	50.83	46.63	45900	19091	-2.195
2000	5.10	8488	51.09	46.85	45958	17678	-1.931
2100	5.14	9000	51.34	47.06	46030	16273	-1.693
2200	5.18	9516	51.58	47.26	46096	14834	-1.473
2300	5.24	10037	51.81	47.45	46167	13415	-1.274
2400	5.30	10563	52.03	47.63	46253	12005	-1.093
2500	5.36	11096	52.25	47.82	46336	10561	-.923
2600	5.43	11635	52.46	47.99	46425	9141	-.768
2700	5.50	12181	52.67	48.16	45531	7705	-.623
2800	5.57	12735	52.87	48.33	46635	6259	-.488
2900	5.65	13296	53.06	48.48	45746	4841	-.364
3000	5.73	13866	53.26	48.64	46876	3376	-.245

ASTATINE

ASTATINE At_2 **GRAMS** 420.*

$(H°_{298.15} - H°_0) =$ _____ CAL./GFW.

REFERENCE STATE: Solid from 298° to 575°, Liquid from 575° to 650°, Ideal Diatomic Gas from 650° to 3000°.

		T °K TEMPERATURE	$C°_p$ HEAT CAPACITY CAL./DEG./GFW.	$H°_T - H°_{298.15}$ HEAT CONTENT CAL./GFW.	$S°_T$ ENTROPY CAL./DEG./GFW.	$-(F°_T - H°_{298.15})/T$ FREE ENERGY FUNCTION CAL./DEG./GFW.	FORMATION FROM REFERENCE STATE		
							HEAT $\Delta H°_f$ CAL./GFW.	FREE ENERGY $\Delta F°_f$ CAL./GFW.	$LOG_{10} K_p$
M.P. (575) °K		298	14.00	0	29.09	29.09			
ΔH_m (5,700) CAL./GFW.		300	14.00	26	29.09	29.09			
		400	14.00	1426	33.12	29.56			
		500	14.00	2826	36.24	30.59			
B.P. (650) °K		600	20.00	10080	48.96	32.16			
ΔH_v (21,600) CAL./GFW.		700	9.03	25210	73.66	37.65			
		800	9.05	26120	74.87	42.22			
		900	9.06	27020	75.94	45.92			
		1000	9.08	27930	76.89	48.96			
		1100	9.09	28840	77.76	51.55			
		1200	9.10	29745	78.55	53.77			
S.P. °K		1300	9.11	30660	79.28	55.70			
ΔH_s CAL./GFW.		1400	9.13	31570	79.95	57.40			
		1500	9.14	32480	80.59	58.94			
		1600	9.16	33400	81.18	60.31			
		1700	9.17	34310	81.73	61.55			
T.P. °K		1800	9.19	35230	82.26	62.69			
ΔH_t CAL./GFW.		1900	9.20	36150	82.75	63.73			
		2000	9.22	37070	83.23	64.70			
		2100	9.23	38000	83.68	65.59			
		2200	9.24	38920	84.11	66.42			
T.P. °K		2300	9.26	39840	84.52	67.20			
ΔH_t CAL./GFW.		2400	9.27	40770	84.91	67.93			
		2500	9.28	41700	85.29	68.61			
		2600	9.30	42630	85.65	69.26			
		2700	9.31	43560	86.01	69.88			
$T_c =$ °K		2800	9.32	44490	86.34	70.46			
$P_c =$ ATM.		2900	9.33	45420	86.67	71.01			
		3000	9.35	46350	86.99	71.54			

*Isotope of longest known half life.

THERMODYNAMIC PROPERTIES OF THE ELEMENTS

ASTATINE

ASTATINE At_2
IDEAL DIATOMIC GAS

Gfw 420.*
$(H°_{298.15} - H°_0)$ =   GRAMS CAL./GFW.

Reference State for Calculating $\Delta H°_f$, $\Delta F°_f$, and $Log_{10} K_p$: Solid from 298° to 575°, Liquid from 575° to 650°, Ideal Diatomic Gas from 650° to 3000°.

T TEMPERATURE °K	$C°_p$ HEAT CAPACITY CAL./DEG./GFW.	$H°_T - H°_{298.15}$ HEAT CONTENT CAL./GFW.	$S°_T$ ENTROPY CAL./DEG./GFW.	$-(F°_T - H°_{298.15})/T$ FREE ENERGY FUNCTION CAL./DEG./GFW.	FORMATION FROM REFERENCE STATE		
					HEAT ΔH_f CAL./GFW.	FREE ENERGY ΔF_f CAL./GFW.	$LOG_{10} K_p$
298	8.90	0	66.00	66.00	21600	10568	-7.746
300	8.90	16	66.06	66.01	21590	10502	-7.651
400	8.96	909	68.63	66.36	21083	6879	-3.758
500	9.00	1807	70.63	67.02	20581	3386	-1.480
600	9.02	2708	72.27	67.76	14228	242	-.088
700	9.03	3611	73.66	68.51	0	0	0
800	9.05	4515	74.87	69.23	0	0	0
900	9.06	5420	75.94	69.92	0	0	0
1000	9.08	6327	76.89	70.57	0	0	0
1100	9.09	7236	77.76	71.19	0	0	0
1200	9.10	8145	78.55	71.77	0	0	0
1300	9.11	9056	79.28	72.32	0	0	0
1400	9.13	9968	79.95	72.83	0	0	0
1500	9.14	10881	80.59	73.34	0	0	0
1600	9.16	11796	81.18	73.81	0	0	0
1700	9.17	12713	81.73	74.26	0	0	0
1800	9.19	13631	82.26	74.69	0	0	0
1900	9.20	14551	82.75	75.10	0	0	0
2000	9.22	15472	83.23	75.50	0	0	0
2100	9.23	16395	83.68	75.88	0	0	0
2200	9.24	17318	84.11	76.24	0	0	0
2300	9.26	18243	84.52	76.59	0	0	0
2400	9.27	19170	84.91	76.93	0	0	0
2500	9.28	20097	85.29	77.26	0	0	0
2600	9.30	21026	85.65	77.57	0	0	0
2700	9.31	21957	86.01	77.88	0	0	0
2800	9.32	22888	86.34	78.17	0	0	0
2900	9.33	23821	86.67	78.46	0	0	0
3000	9.35	24754	86.99	78.74	0	0	0

M.P.   °K
ΔH_m CAL./GFW.

B.P.   °K
ΔH_v CAL./GFW.

S.P.   °K
ΔH_s CAL./GFW.

T.P.   °K
ΔH_t CAL./GFW.

T.P.   °K
ΔH_t CAL./GFW.

T_c =   °K
P_c =   ATM.

*Isotope of Longest Known Half Life.

ASTATINE

Reference State for Calculating ΔH_f°, ΔF_f°, and $\log_{10} K_p$: Solid from 298° to 575°, Liquid from 575° to 650°, Ideal Diatomic Gas from 650° to 3000°.

ASTATINE At **IDEAL MONATOMIC GAS**

G_{fw} 210.* GRAMS
$(H^\circ_{298.15} - H^\circ_0) = 1,481$ CAL./GFW.

T TEMPERATURE °K	C°_p HEAT CAPACITY CAL./DEG./GFW.	$H^\circ_T - H^\circ_{298.15}$ HEAT CONTENT CAL./GFW.	S°_T ENTROPY CAL./DEG./GFW.	$-(F^\circ_T - H^\circ_{298.15})$ FREE ENERGY FUNCTION CAL./DEG./GFW.	HEAT ΔH°_f CAL./GFW.	FREE ENERGY ΔF°_f CAL./GFW.	$\log_{10} K_p$
29H	4.97	0	44.68	44.68	22000	13002	−9.531
300	4.97	9	44.71	44.68	21996	12948	−9.433
400	4.97	506	46.14	44.88	21793	9961	−5.442
500	4.97	1003	47.25	45.25	21590	7025	−3.070
600	4.97	1500	48.15	45.65	18460	4258	−1.551
700	4.97	1996	48.92	46.07	11390	2927	−.913
800	4.97	2493	49.58	46.47	11433	1721	−.470
900	4.97	2990	50.17	46.85	11480	500	−.121
1000	4.97	3487	50.69	47.21	11523	717	.156
1100	4.97	3984	51.16	47.54	11566	1942	.385
1200	4.97	4480	51.60	47.87	11607	3177	.578
1300	4.97	4977	51.99	48.17	11649	4406	.740
1400	4.97	5474	52.36	48.45	11690	5642	.880
1500	4.97	5971	52.70	48.72	11730	6870	1.001
1600	4.97	6468	53.03	48.99	11770	8134	1.111
1700	4.97	6964	53.33	49.24	11807	9375	1.205
1800	4.97	7461	53.61	49.47	11845	10619	1.289
1900	4.98	7958	53.88	49.70	11882	11868	1.365
2000	4.98	8456	54.14	49.92	11920	13120	1.433
2100	4.99	8955	54.38	50.12	11957	14377	1.496
2200	5.00	9454	54.61	50.32	11995	15615	1.551
2300	5.02	9955	54.83	50.51	12033	16878	1.603
2400	5.03	10458	55.05	50.70	12073	18143	1.652
2500	5.05	10962	55.25	50.87	12113	19387	1.694
2600	5.07	11468	55.45	51.04	12155	20657	1.736
2700	5.09	11976	55.64	51.21	12197	21904	1.772
2800	5.11	12486	55.83	51.38	12242	23206	1.811
2900	5.13	12998	56.01	51.53	12287	24456	1.843
3000	5.16	13512	56.18	51.68	12335	25705	1.872

M.P. °K
ΔH_m CAL./GFW.

B.P. °K
ΔH_v CAL./GFW.

S.P. °K
ΔH_s CAL./GFW.

T.P. °K
ΔH_t CAL./GFW.

T.P. °K
ΔH_t CAL./GFW.

$T_c =$ °K
$P_c =$ ATM.

THERMODYNAMIC PROPERTIES OF THE ELEMENTS

BARIUM

BARIUM	Ba		Solid I from 298° to 643°, Solid II from 643° to 983°, Liquid from 983° to 1910°, Ideal Monatomic Gas from 1910° to 3000°.
REFERENCE STATE			

G$_w$	137.36	GRAMS		
(H°$_{298.15}$ − H°$_0$) =		CAL./GFW.		
M.P.	983	°K		
ΔH$_m$	1,830	CAL./GFW.		
B.P.	1,910	°K		
ΔH$_v$	36,070	CAL./GFW.		
S.P.		°K		
ΔH$_s$		CAL./GFW.		
T.P.	643	°K		
ΔH$_t$	140	CAL./GFW.		
T.P.		°K		
ΔH$_t$		CAL./GFW.		
T$_c$ =		°K		
P$_c$ =		ATM.		

					FORMATION FROM REFERENCE STATE		
T TEMPERATURE °K	C°$_p$ HEAT CAPACITY CAL./DEG./GFW.	H°$_T$ − H°$_{298.15}$ HEAT CONTENT CAL./GFW.	S°$_T$ ENTROPY CAL./DEG./GFW.	−(F°$_T$ − H°$_{298.15}$)/T FREE ENERGY FUNCTION CAL./DEG./GFW.	HEAT Δ H°$_f$ CAL./GFW.	FREE ENERGY Δ F°$_f$ CAL./GFW.	LOG$_{10}$ K$_p$
298	6.30	0	15.50	15.50			
300	6.31	11	15.54	15.51			
400	6.64	656	17.40	15.76			
500	7.00	1334	18.91	16.25			
600	7.30	2046	20.21	16.80			
700	7.40	2921	21.57	17.40			
800	8.10	3694	22.60	17.99			
900	8.80	4537	23.59	18.55			
1000	7.50	7247	26.38	19.14			
1100	7.50	7997	27.10	19.83			
1200	7.50	8747	27.75	20.47			
1300	7.50	9497	28.35	21.05			
1400	7.50	10247	28.91	21.60			
1500	7.50	10997	29.42	22.09			
1600	7.50	11747	29.91	22.57			
1700	7.50	12497	30.36	23.01			
1800	7.50	13247	30.79	23.44			
1900	7.50	13997	31.20	23.84			
2000	6.81	50721	50.43	25.07			
2100	7.30	51438	50.78	26.29			
2200	7.81	52191	51.13	27.41			
2300	8.34	53001	51.49	28.45			
2400	8.87	53871	51.86	29.42			
2500	9.36	54753	52.22	30.32			
2600	9.92	55747	52.61	31.17			
2700	10.40	56754	52.99	31.97			
2800	10.84	57826	53.38	32.73			
2900	11.22	58909	53.76	33.45			
3000	11.56	60060	54.15	34.13			

BARIUM

BARIUM	Ba	Reference State for Calculating ΔH_f°, ΔF_f°, and Log $_{10}K_p$:
Gfw 137.36	GRAMS	Solid I from 298° to 643°, Solid II from 643° to 983°, Liquid
IDEAL MONATOMIC GAS		from 983° to 1910°, Ideal Monatomic Gas from 1910° to 3000°.
$(H^\circ_{298.15} - H^\circ_0) = 1,481$	CAL./GFW.	

							FORMATION FROM REFERENCE STATE		
T TEMPERATURE °K	C_p° HEAT CAPACITY CAL./DEG./GFW.	$H^\circ_T - H^\circ_{298.15}$ HEAT CONTENT CAL./GFW.	S°_T ENTROPY CAL./DEG./GFW.	$-(F^\circ - H^\circ_{298.15})$ FREE ENERGY FUNCTION CAL./DEG./GFW.			HEAT ΔH_f° CAL./GFW.	FREE ENERGY ΔF_f° CAL./GFW.	LOG$_{10}$ K_p
298	4.97	0	40.67	40.67			41736	34231	-25.093
300	4.97	9	40.70	40.67			41734	34186	-24.906
400	4.97	506	42.12	40.86			41586	31698	-17.320
500	4.97	1003	43.23	41.23			41405	29245	-12.783
600	4.97	1500	44.14	41.64			41190	26832	-9.774
700	4.97	1996	44.91	42.06			40811	24473	-7.641
800	4.97	2493	45.57	42.46			40535	22159	-6.054
900	4.97	2990	46.15	42.83			40189	19885	-4.829
1000	4.98	3488	46.68	43.20			37977	17677	-3.863
1100	5.00	3987	47.15	43.53			37726	15671	-3.113
1200	5.03	4488	47.59	43.85			37477	13669	-2.489
1300	5.08	4991	47.99	44.16			37230	11698	-1.966
1400	5.17	5503	48.37	44.44			36992	9748	-1.521
1500	5.31	6026	48.73	44.72			36765	7800	-1.136
1600	5.50	6567	49.08	44.98			36556	5884	-.803
1700	5.74	7128	49.42	45.23			36367	3965	-.509
1800	6.08	7724	49.76	45.47			36213	2067	-.250
1900	6.41	8340	50.09	45.71			36079	188	-.021
2000	6.81	8999	50.43	45.94			0	0	0
2100	7.30	9710	50.78	46.16			0	0	0
2200	7.81	10465	51.13	46.38			0	0	0
2300	8.34	11272	51.49	46.59			0	0	0
2400	8.87	12132	51.86	46.81			0	0	0
2500	9.36	13034	52.22	47.01			0	0	0
2600	9.92	14012	52.61	47.23			0	0	0
2700	10.40	15029	52.99	47.43			0	0	0
2800	10.84	16089	53.38	47.64			0	0	0
2900	11.22	17195	53.76	47.84			0	0	0
3000	11.56	18334	54.15	48.04			0	0	0

M.P.	°K	
ΔH_m	CAL./GFW.	
B.P.	°K	
ΔH_v	CAL./GFW.	
S.P.	°K	
ΔH_s	CAL./GFW.	
T.P.	°K	
ΔH_t	CAL./GFW.	
T.P.	°K	
ΔH_t	CAL./GFW.	
$T_c =$	°K	
$P_c =$	ATM.	

BERYLLIUM

BERYLLIUM Be Solid to 1556°, Liquid from 1556° to 2750°,
Ideal Monatomic Gas from 2750° to 3000°K.

REFERENCE STATE

		GRAMS	
Gfw	9.013		
$(H°_{298.15} - H°_0)$	468		CAL./GFW.

		°K		CAL./GFW.
M.P.	1,556			
ΔH_m	2,800			
B.P.	2,750			
ΔH_v	70,400			
S.P.				
ΔH_s				
T.P.				
ΔH_t				
T.P.				
ΔH_t				
$T_c =$				
$P_c =$		ATM.		

T TEMPERATURE °K	$C°_P$ HEAT CAPACITY CAL./DEG./GFW.	$H°_T - H°_{298.15}$ HEAT CONTENT CAL./GFW.	$S°_T$ ENTROPY CAL./DEG./GFW.	$-(F°_T - H°_{298.15})$ FREE ENERGY FUNCTION CAL./DEG./GFW.	HEAT $\Delta H°_f$ CAL./GFW.	FREE ENERGY $\Delta F°_f$ CAL./GFW.	$LOG_{10} K_p$
298	3.93	0	2.28	2.28			
300	3.95	7	2.30	2.28			
400	4.74	442	3.54	2.44			
500	5.26	942	4.66	2.78			
600	5.60	1486	5.65	3.18			
700	5.84	2059	6.53	3.59			
800	6.06	2654	7.32	4.01			
900	6.29	3271	8.05	4.42			
1000	6.53	3912	8.73	4.82			
1100	6.78	4578	9.36	5.20			
1200	7.02	5268	9.96	5.57			
1300	7.26	5982	10.53	5.93			
1400	7.50	6720	11.08	6.28			
1500	7.74	7482	11.60	6.62			
1600	7.50	11050	13.90	7.00			
1700	7.50	11800	14.35	7.41			
1800	7.50	12550	14.78	7.81			
1900	7.50	13300	15.19	8.19			
2000	7.50	14050	15.57	8.55			
2100	7.50	14800	15.94	8.90			
2200	7.50	15550	16.29	9.23			
2300	7.50	16300	16.62	9.54			
2400	7.50	17050	16.94	9.84			
2500	7.50	17800	17.25	10.13			
2600	7.50	18550	17.54	10.41			
2700	7.50	19300	17.82	10.68			
2800	5.00	90336	43.67	11.41			
2900	5.01	90837	43.85	12.53			
3000	5.02	91338	44.02	13.58			

BERYLLIUM

BERYLLIUM   Be   Reference State for Calculating ΔH_f°, ΔF_f°, and $Log_{10}K_p$: Solid to 1556°, Liquid from 1556° to 2750°, Ideal Monatomic Gas from 2750° to 3000°.

IDEAL MONATOMIC GAS

G_{fw} 9.013   GRAMS   CAL./GFW.
$(H^\circ_{298.15} - H^\circ_0) = 1,481$ CAL./GFW.

T TEMPERATURE °K	C_p° HEAT CAPACITY CAL./DEG./GFW.	$H^\circ_T - H^\circ_{298.15}$ HEAT CONTENT CAL./GFW.	S°_T ENTROPY CAL./DEG./GFW.	$-(F^\circ_T - H^\circ_{298.15})$ FREE ENERGY FUNCTION CAL./DEG./GFW.	FORMATION FROM REFERENCE STATE		
					HEAT ΔH°_f CAL./GFW.	FREE ENERGY ΔF°_f CAL./GFW.	$LOG_{10} K_p$
298	4.97	0	32.55	32.55	77900	68874	−50.488
300	4.97	9	32.58	32.55	77902	68818	−50.138
400	4.97	506	34.01	32.75	77964	65776	−35.941
500	4.97	1003	35.11	33.11	77961	62736	−27.423
600	4.97	1500	36.02	33.52	77914	59692	−21.744
700	4.97	1996	36.79	33.94	77837	56655	−17.689
800	4.97	2493	37.45	34.34	77739	53635	−14.653
900	4.97	2990	38.03	34.71	77619	50637	−12.297
1000	4.97	3487	38.56	35.08	77475	47645	−10.413
1100	4.97	3984	39.03	35.41	77306	44669	−8.875
1200	4.97	4480	39.46	35.73	77112	41712	−7.597
1300	4.97	4977	39.86	36.04	76895	38766	−6.517
1400	4.97	5474	40.23	36.32	76654	35844	−5.595
1500	4.97	5971	40.57	36.59	76389	32934	−4.798
1600	4.97	6468	40.89	36.85	73318	30134	−4.116
1700	4.97	6964	41.19	37.10	73064	27436	−3.527
1800	4.97	7461	41.48	37.34	72811	24751	−3.005
1900	4.97	7958	41.75	37.57	72558	22094	−2.541
2000	4.97	8455	42.00	37.78	72305	19445	−2.124
2100	4.97	8952	42.24	37.98	72052	16822	−1.750
2200	4.97	9448	42.48	38.19	71798	14180	−1.408
2300	4.97	9945	42.70	38.38	71545	11561	−1.098
2400	4.97	10443	42.91	38.56	71293	8965	−.816
2500	4.98	10940	43.11	38.74	71040	6390	−.558
2600	4.98	11438	43.31	38.92	70788	3786	−.318
2700	4.99	11937	43.49	39.07	70537	1228	−.099
2800	5.00	12436	43.67	39.23	0	0	0
2900	5.01	12937	43.85	39.39	0	0	0
3000	5.02	13438	44.02	39.55	0	0	0

M.P. °K
ΔH_m CAL./GFW.

B.P. °K
ΔH_v CAL./GFW.

S.P. °K
ΔH_s CAL./GFW.

T.P. °K
ΔH_t CAL./GFW.

T.P. °K
ΔH_t CAL./GFW.

$T_c =$ °K
$P_c =$ ATM.

THERMODYNAMIC PROPERTIES OF THE ELEMENTS

BISMUTH

BISMUTH   B1   Solid from 298° to 544.5°, Liquid from 544.5° to 1832°, Ideal Monatomic Gas from 1832° to 3000°.

REFERENCE STATE

Gfw	209.00	GRAMS
$(H°_{298.15} - H°_0) =$	1,536	CAL./GFW.
M.P.	544.5	°K
ΔH_m	2,600.	CAL./GFW.
B.P.	1,832.	°K
ΔH_v	36,200.	CAL./GFW.
S.P.		°K
ΔH_s		CAL./GFW.
T.P.		°K
ΔH_t		CAL./GFW.
T.P.		°K
ΔH_t		CAL./GFW.
$T_c =$		°K
$P_c =$		ATM.

T TEMPERATURE °K	$C°_P$ HEAT CAPACITY CAL./DEG./GFW.	$H°_T - H°_{298.15}$ HEAT CONTENT CAL./GFW.	$S°_T$ ENTROPY CAL./DEG./GFW.	$-(F°_T - H°_{298.15})$ FREE ENERGY FUNCTION CAL./DEG./GFW.	HEAT $\Delta H°_f$ CAL./GFW.	FREE ENERGY $\Delta F°_f$ CAL./GFW.	$LOG_{10} K_P$
298	6.11	0	13.58	13.58			
300	6.11	11	13.62	13.59			
400	6.65	650	15.45	13.83			
500	7.19	1340	16.99	14.31			
600	7.50	4680	23.12	15.32			
700	7.50	5430	24.28	16.53			
800	7.50	6180	25.28	17.56			
900	7.50	6930	26.16	18.46			
1000	7.50	7680	26.95	19.27			
1100	7.50	8430	27.67	20.01			
1200	7.50	9180	28.32	20.67			
1300	7.50	9930	28.92	21.29			
1400	7.50	10680	29.47	21.85			
1500	7.50	11430	29.99	22.37			
1600	7.50	12180	30.48	22.87			
1700	7.50	12930	30.93	23.33			
1800	7.50	13680	31.36	23.76			
1900	5.00	55460	53.87	24.69			
2000	5.01	55970	54.13	26.15			
2100	5.03	56470	54.38	27.49			
2200	5.04	56970	54.61	28.72			
2300	5.07	57480	54.84	29.85			
2400	5.09	57980	55.05	30.90			
2500	5.12	58490	55.26	31.87			
2600	5.15	59010	55.46	32.77			
2700	5.19	59530	55.66	33.62			
2800	5.23	60050	55.85	34.41			
2900	5.27	60570	56.03	35.15			
3000	5.32	61100	56.21	35.85			

BISMUTH

BISMUTH   **Bi**
Gfw 209.00   GRAMS
$(H°_{298.15} - H°_0) = 1,481$ CAL./GFW.
IDEAL MONATOMIC GAS

Reference State for Calculating $\Delta H°_f$, $\Delta F°_f$, and $\text{Log}_{10} K_p$: Solid from 298° to 544.5°, Liquid from 544.5° to 1832°, Ideal Monatomic Gas from 1832° to 3000°.

T TEMPERATURE °K	$C°_p$ HEAT CAPACITY CAL./DEG./GFW.	$H°_T - H°_{298.15}$ HEAT CONTENT CAL./GFW.	$S°_T$ ENTROPY CAL./DEG./GFW.	$-(F°_T - H°_{298.15})/T$ FREE ENERGY FUNCTION CAL./DEG./GFW.	FORMATION FROM REFERENCE STATE		
					HEAT $\Delta H°_f$ CAL./GFW.	FREE ENERGY $\Delta F°_f$ CAL./GFW.	$\text{LOG}_{10} K_p$
298	4.97	0	44.67	44.67	47500	38231	−28.025
300	4.97	9	44.70	44.67	47498	38174	−27.812
400	4.97	506	46.13	44.87	47356	35084	−19.170
500	4.97	1003	47.24	45.24	47163	32038	−14.004
600	4.97	1500	48.14	45.64	44320	29308	−10.676
700	4.97	1996	48.91	46.06	44066	26825	−8.375
800	4.97	2493	49.57	46.46	43813	24381	−6.661
900	4.97	2990	50.16	46.84	43560	21960	−5.332
1000	4.97	3487	50.68	47.20	43307	19577	−4.278
1100	4.97	3984	51.16	47.54	43054	17215	−3.420
1200	4.97	4480	51.59	47.86	42800	14876	−2.709
1300	4.97	4977	51.99	48.17	42547	12556	−2.111
1400	4.97	5474	52.35	48.44	42294	10262	−1.602
1500	4.97	5971	52.70	48.72	42041	7976	−1.162
1600	4.98	6469	53.02	48.98	41789	5725	−.781
1700	4.98	6967	53.32	49.23	41537	3474	−.446
1800	4.99	7465	53.60	49.46	41285	1253	−.152
1900	5.00	7964	53.87	49.68	0	0	0
2000	5.01	8465	54.13	49.90	0	0	0
2100	5.03	8966	54.38	50.12	0	0	0
2200	5.04	9470	54.61	50.31	0	0	0
2300	5.07	9975	54.84	50.51	0	0	0
2400	5.09	10483	55.05	50.69	0	0	0
2500	5.12	10994	55.26	50.87	0	0	0
2600	5.15	11508	55.46	51.04	0	0	0
2700	5.19	12025	55.66	51.21	0	0	0
2800	5.23	12546	55.85	51.37	0	0	0
2900	5.27	13071	56.03	51.53	0	0	0
3000	5.32	13600	56.21	51.68	0	0	0

M.P. °K
ΔH_m CAL./GFW.

B.P. °K
ΔH_v CAL./GFW.

S.P. °K
ΔH_s CAL./GFW.

T.P. °K
ΔH_t CAL./GFW.

T.P. °K
ΔH_t CAL./GFW.

$T_c =$ °K
$P_c =$ ATM.

THERMODYNAMIC PROPERTIES OF THE ELEMENTS

BISMUTH

BISMUTH Bi_2
IDEAL DIATOMIC GAS

Gfw 418.00 GRAMS
$(H°_{298.15} - H°_0) = 2,453$ CAL./GFW.

Reference State for Calculating $\Delta H_f°$, $\Delta F_f°$, and $Log_{10} K_P$: Solid from 298° to 544.5°, Liquid from 544.5° to 1832°, Ideal Monatomic Gas from 1832° to 2000°.

						FORMATION FROM REFERENCE STATE		
T TEMPERATURE °K	$C°_P$ HEAT CAPACITY CAL./DEG./GFW.	$H°_T - H°_{298.15}$ HEAT CONTENT CAL./GFW.	$S°_T$ ENTROPY CAL./DEG./GFW.	$-(F°_T - H°_{298.15})/T$ FREE ENERGY FUNCTION CAL./DEG./GFW.	HEAT $\Delta H°_f$ CAL./GFW.	FREE ENERGY $\Delta F°_f$ CAL./GFW.	$LOG_{10} K_P$
298	8.83	0	65.40	65.40	55300	43899	-32.180
300	8.83	16	65.46	65.41	55294	43828	-31.931
400	8.88	900	68.00	65.75	54900	40060	-21.889
500	8.90	1790	69.98	66.40	54410	36410	-15.915
600	8.91	2685	71.62	67.15	48625	33397	-12.165
700	8.92	3575	72.99	67.89	48015	30914	-9.652
800	8.93	4465	74.18	68.60	47405	28509	-7.788
900	8.93	5360	75.23	69.28	46800	26181	-6.358
1000	8.93	6250	76.17	69.92	46190	23920	-5.228
1100	8.93	7140	77.02	70.53	45580	21732	-4.318
1200	8.94	8040	77.80	71.10	44980	19588	-3.567
1300	8.94	8930	78.51	71.65	44370	17499	-2.942
1400	8.94	9830	79.18	72.16	43770	15434	-2.409
1500	8.94	10720	79.79	72.65	43160	13445	-1.959
1600	8.94	11615	80.37	73.12	42555	11499	-1.570
1700	8.94	12510	80.91	73.56	41950	9565	-1.229
1800	8.94	13400	81.42	73.98	41340	7680	-.932
1900	8.94	14295	81.91	74.39	-41325	7752	-.891
2000	8.94	15190	82.36	74.77	-41450	10350	-1.130

M.P. °K
ΔH_m CAL./GFW.

B.P. °K
ΔH_v CAL./GFW.

S.P. °K
ΔH_s CAL./GFW.

T.P. °K
ΔH_t CAL./GFW.

T.P. °K
ΔH_t CAL./GFW.

T_c = °K
P_c = ATM.

BORON

BORON B Solid to 2300°, Liquid from 2300° to 3000°.

REFERENCE STATE

Gfw	10.82	GRAMS
$(H°_{298.15} - H°_0)$ =	292	CAL./GFW.
M.P.	2,300	°K
ΔH_m	(5,300)	CAL./GFW.
B.P.	4,200	°K
ΔH_v	128,800	CAL./GFW.
S.P.		°K
ΔH_s		CAL./GFW.
T.P.		°K
ΔH_f		CAL./GFW.
T.P.		°K
ΔH_f		CAL./GFW.
T_c =		°K
P_c =		ATM.

T TEMPERATURE °K	$C_p°$ HEAT CAPACITY CAL./DEG./GFW.	$H°_T - H°_{298.15}$ HEAT CONTENT CAL./GFW.	$S°_T$ ENTROPY CAL./DEG./GFW.	$-(F°_T - H°_{298.15})$ FREE ENERGY FUNCTION CAL./DEG./GFW.	FORMATION FROM REFERENCE STATE HEAT $\Delta H°_f$ CAL./GFW.	FORMATION FROM REFERENCE STATE FREE ENERGY $\Delta F°_f$ CAL./GFW.	$LOG_{10} K_p$
298	2.63	0	1.40	1.40			
300	2.65	5	1.41	1.40			
400	3.45	311	2.29	1.52			
500	4.05	688	3.13	1.76			
600	4.57	1120	3.91	2.05			
700	5.00	1599	4.65	2.37			
800	5.35	2117	5.34	2.70			
900	5.65	2668	5.99	3.03			
1000	5.90	3246	6.60	3.36			
1100	6.10	3846	7.17	3.68			
1200	6.27	4465	7.71	3.99			
1300	6.42	5099	8.22	4.30			
1400	6.55	5748	8.70	4.60			
1500	6.67	6409	9.15	4.88			
1600	6.78	7082	9.59	5.17			
1700	6.90	7765	10.00	5.44			
1800	7.00	8460	10.40	5.70			
1900	7.10	9165	10.78	5.96			
2000	7.20	9880	11.15	6.21			
2100	7.30	10605	11.50	6.45			
2200	7.40	11340	11.84	6.69			
2300	7.50	17380	14.47	6.92			
2400	7.50	18130	14.79	7.24			
2500	7.50	18880	15.10	7.55			
2600	7.50	19630	15.39	7.84			
2700	7.50	20380	15.68	8.14			
2800	7.50	21130	15.95	8.41			
2900	7.50	21880	16.22	8.68			
3000	7.50	22630	16.47	8.93			

THERMODYNAMIC PROPERTIES OF THE ELEMENTS

BORON

BORON B **Gfw** 10.82 **GRAMS**
IDEAL MONATOMIC GAS
$(H°_{298.15} - H°_0)$ 1,510 CAL./GFW.

Reference State for Calculating $\Delta H°_f$, $\Delta F°_f$, and $\log_{10} K_p$: Solid to 2300°, Liquid from 2300° to 3000°.

T TEMPERATURE °K	$C°_p$ HEAT CAPACITY CAL./DEG./GFW.	$H°_T - H°_{298.15}$ HEAT CONTENT CAL./GFW.	$S°_T$ ENTROPY CAL./DEG./GFW.	$-(F°_T - H°_{298.15})/T$ FREE ENERGY FUNCTION CAL./DEG./GFW.	FORMATION FROM REFERENCE STATE		
					HEAT $\Delta H°_f$ CAL./GFW.	FREE ENERGY $\Delta F°_f$ CAL./GFW.	$\log_{10} K_p$
298	4.97	0	36.65	36.65	141000	130490	-95.655
300	4.97	9	36.68	36.65	141004	130423	-95.020
400	4.97	506	38.11	36.85	141195	126867	-69.322
500	4.97	1003	39.22	37.22	141315	123270	-53.885
600	4.97	1500	40.12	37.62	141380	119654	-43.587
700	4.97	1997	40.89	38.04	141398	116030	-36.229
800	4.97	2494	41.55	38.44	141377	112409	-30.711
900	4.97	2990	42.14	38.82	141322	108787	-26.418
1000	4.97	3487	42.66	39.18	141241	105181	-22.989
1100	4.97	3984	43.14	39.52	141138	101571	-20.182
1200	4.97	4481	43.57	39.84	141016	97984	-17.846
1300	4.97	4978	43.97	40.15	140879	94404	-15.872
1400	4.97	5475	44.33	40.42	140727	90845	-14.182
1500	4.97	5971	44.68	40.70	140562	87267	-12.715
1600	4.97	6468	45.00	40.96	140386	83730	-11.436
1700	4.97	6965	45.30	41.21	140200	80190	-10.309
1800	4.97	7462	45.58	41.44	140002	76678	-9.309
1900	4.97	7959	45.85	41.67	139794	73161	-8.414
2000	4.97	8455	46.11	41.89	139575	69655	-7.611
2100	4.97	8952	46.35	42.09	139347	66162	-6.885
2200	4.97	9449	46.58	42.29	139109	62681	-6.226
2300	4.97	9946	46.80	42.48	133566	59207	-5.625
2400	4.97	10443	47.01	42.66	133313	55985	-5.097
2500	4.97	10939	47.21	42.84	133059	52784	-4.614
2600	4.97	11436	47.41	43.02	132806	49554	-4.165
2700	4.97	11933	47.60	43.19	132553	46369	-3.753
2800	4.97	12430	47.78	43.35	132300	43176	-3.369
2900	4.97	12927	47.95	43.50	132047	40030	-3.016
3000	4.97	13423	48.12	43.65	131793	36843	-2.684

M.P. °K
ΔH_m CAL./GFW.

B.P. °K
ΔH_v CAL./GFW.

S.P. °K
ΔH_s CAL./GFW.

T.P. °K
ΔH_t CAL./GFW.

T.P. °K
ΔH_t CAL./GFW.

T_c = °K
P_c = ATM.

BROMINE

BROMINE Br_2
Liquid from 298° to 331.4°, Ideal Diatomic Gas from 331.4° to 3000°.

REFERENCE STATE

	GRAMS							FORMATION FROM REFERENCE STATE		
	CAL./GFW.		T TEMPERATURE °K	C_p° HEAT CAPACITY CAL./DEG./GFW.	$H_T^\circ - H_{298.15}^\circ$ HEAT CONTENT CAL./GFW.	S_T° ENTROPY CAL./DEG./GFW.	$-(F^\circ - H^\circ_{298.15})$ FREE ENERGY FUNCTION CAL./DEG./GFW.	HEAT ΔH_f° CAL./GFW.	FREE ENERGY ΔF_f° CAL./GFW.	$LOG_{10} K_p$
Gfw 159.832			298	17.00	0	36.25	36.25			
$(H_{298.15}^\circ - H_0^\circ) =$			300	17.00	31	36.36	36.26			
M.P. 265.95	°K		400	8.78	8337	61.20	40.36			
ΔH_m 2,520.	CAL./GFW.		500	8.86	9219	63.17	44.74			
			600	8.91	10107	64.79	47.95			
			700	8.94	11000	66.17	50.46			
B.P. 331.4	°K		800	8.97	11895	67.36	52.50			
ΔH_v 7,170.	CAL./GFW.		900	8.99	12794	68.42	54.21			
			1000	9.01	13694	69.37	55.68			
			1100	9.03	14595	70.23	56.97			
S.P.	°K		1200	9.04	15498	71.02	58.11			
ΔH_s	CAL./GFW.		1300	9.06	16404	71.74	59.13			
			1400	9.07	17310	72.41	60.05			
			1500	9.08	18218	73.04	60.90			
			1600	9.09	19126	73.62	61.67			
T.P.	°K		1700	9.11	20035	74.17	62.39			
ΔH_t	CAL./GFW.		1800	9.12	20948	74.70	63.07			
			1900	9.13	21859	75.19	63.69			
			2000	9.14	22774	75.66	64.28			
			2100	9.15	23687	76.10	64.83			
			2200	9.16	24602	76.53	65.35			
T.P.	°K		2300	9.17	25520	76.94	65.85			
ΔH_t	CAL./GFW.		2400	9.18	26438	77.33	66.32			
			2500	9.20	27356	77.70	66.76			
			2600	9.21	28276	78.06	67.19			
			2700	9.22	29197	78.41	67.60			
$T_c =$ 584.	°K		2800	9.23	30119	78.74	67.99			
$P_c =$ 102.	ATM.		2900	9.24	31043	79.07	68.37			
			3000	9.25	31967	79.38	68.73			

THERMODYNAMIC PROPERTIES OF THE ELEMENTS

BROMINE

BROMINE Br$_2$
Gfw 159.832 GRAMS
(H°$_{298.15}$ − H°$_0$) = 2,325 CAL./GFW.
IDEAL DIATOMIC GAS

Reference State for Calculating ΔH_f°, ΔF_f°,
and $\log_{10} K_p$: Liquid from 298° to 331.4°,
Ideal Diatomic Gas from 331.4° to 3000°.

T TEMPERATURE °K	C°$_p$ HEAT CAPACITY CAL./DEG./GFW.	H°$_T$ − H°$_{298.15}$ HEAT CONTENT CAL./GFW.	S°$_T$ ENTROPY CAL./DEG./GFW.	−(F°$_T$ − H°$_{298.15}$) FREE ENERGY FUNCTION CAL./DEG./GFW.	FORMATION FROM REFERENCE STATE		
					HEAT Δ H°$_f$ CAL./GFW.	FREE ENERGY Δ F°$_f$ CAL./GFW.	LOG$_{10}$ K$_p$
298	8.62	0	58.65	58.65	7450	773	−.566
300	8.62	16	58.70	58.65	7434	738	−.537
400	8.78	887	61.20	58.99	0	0	0
500	8.86	1769	63.17	59.64	0	0	0
600	8.91	2657	64.79	60.37	0	0	0
700	8.94	3550	66.17	61.10	0	0	0
800	8.97	4445	67.36	61.81	0	0	0
900	8.99	5344	68.42	62.49	0	0	0
1000	9.01	6244	69.37	63.13	0	0	0
1100	9.03	7145	70.23	63.74	0	0	0
1200	9.04	8048	71.02	64.32	0	0	0
1300	9.06	8954	71.74	64.86	0	0	0
1400	9.07	9860	72.41	65.37	0	0	0
1500	9.08	10768	73.04	65.87	0	0	0
1600	9.09	11676	73.62	66.33	0	0	0
1700	9.11	12585	74.17	66.77	0	0	0
1800	9.12	13498	74.70	67.21	0	0	0
1900	9.13	14409	75.19	67.61	0	0	0
2000	9.14	15324	75.66	68.00	0	0	0
2100	9.15	16237	76.10	68.37	0	0	0
2200	9.16	17152	76.53	68.74	0	0	0
2300	9.17	18070	76.94	69.09	0	0	0
2400	9.18	18988	77.33	69.42	0	0	0
2500	9.20	19906	77.70	69.74	0	0	0
2600	9.21	20826	78.06	70.05	0	0	0
2700	9.22	21747	78.41	70.36	0	0	0
2800	9.23	22669	78.74	70.65	0	0	0
2900	9.24	23593	79.07	70.94	0	0	0
3000	9.25	24517	79.38	71.21			

M.P. °K
Δ H$_m$ CAL./GFW.
B.P. °K
Δ H$_v$ CAL./GFW.
S.P. °K
Δ H$_s$ CAL./GFW.
T.P. °K
Δ H$_t$ CAL./GFW.
T.P. °K
Δ H$_t$ CAL./GFW.
T$_c$ = °K
P$_c$ = ATM.

BROMINE

BROMINE   **Br**

IDEAL MONATOMIC GAS

Gfw 79.916 GRAMS

$(H°_{298.15} - H°_0) = 1,481$ CAL./GFW.

Reference State for Calculating $\Delta H°_f$, $\Delta F°_f$ and $\log_{10} K_p$: Liquid from 298° to 331.4°, Ideal Diatomic Gas from 331.4° to 3000°.

						FORMATION FROM REFERENCE STATE		
T TEMPERATURE °K	$C°_p$ HEAT CAPACITY CAL./DEG./GFW.	$H°_T - H°_{298.15}$ HEAT CONTENT CAL./GFW.	$S°_T$ ENTROPY CAL./DEG./GFW.	$-(F°_T - H°_{298.15})/T$ FREE ENERGY FUNCTION CAL./DEG./GFW.		HEAT $\Delta H°_f$ CAL./GFW.	FREE ENERGY $\Delta F°_f$ CAL./GFW.	$\log_{10} K_p$
298	4.97	0	41.81	41.81		26760	19699	-14.440
300	4.97	9	41.84	41.81		26753	19658	-14.322
400	4.97	506	43.27	42.01		23097	18029	-9.851
500	4.97	1003	44.37	42.37		23153	16763	-7.327
600	4.98	1500	45.28	42.78		23206	15478	-5.638
700	5.00	1999	46.05	43.20		23259	14180	-4.427
800	5.03	2500	46.72	43.60		23312	12880	-3.518
900	5.06	3004	47.31	43.98		23334	11544	-2.803
1000	5.11	3513	47.85	44.34		23393	10223	-2.234
1100	5.15	4026	48.34	44.68		23455	8902	-1.768
1200	5.20	4543	48.79	45.01		23521	7585	-1.381
1300	5.24	5066	49.21	45.32		23591	6249	-1.050
1400	5.28	5592	49.60	45.61		23664	4918	-.767
1500	5.32	6122	49.96	45.88		23740	3580	-.521
1600	5.35	6656	50.31	46.15		23820	2220	-.303
1700	5.38	7192	50.63	46.40		23901	883	-.113
1800	5.40	7731	50.94	46.65		23984	478	.058
1900	5.42	8272	51.23	46.88		24069	1847	.212
2000	5.43	8814	51.51	47.11		24154	3206	.350
2100	5.44	9357	51.77	47.32		24240	4572	.475
2200	5.44	9901	52.03	47.53		24327	5945	.590
2300	5.45	10446	52.27	47.73		24413	7327	.696
2400	5.45	10990	52.50	47.93		24498	8718	.793
2500	5.45	11535	52.72	48.11		24584	10091	.882
2600	5.44	12079	52.93	48.29		24668	11472	.964
2700	5.44	12623	53.14	48.47		24751	12887	1.043
2800	5.43	13166	53.34	48.64		24833	14283	1.114
2900	5.42	13709	53.53	48.81		24914	15686	1.182
3000	5.42	14252	53.71	48.96		24995	17065	1.243

M.P. °K
ΔH_m CAL./GFW.
B.P. °K
ΔH_v CAL./GFW.
S.P. °K
ΔH_s CAL./GFW.
T.P. °K
ΔH_t CAL./GFW.
T.P. °K
ΔH_t CAL./GFW.
$T_c =$ °K
$P_c =$ ATM.

THERMODYNAMIC PROPERTIES OF THE ELEMENTS

CADMIUM

Solid from 298° to 594°, Liquid from 594° to 1038°, Ideal Monatomic Gas from 1038° to 3000°.

CADMIUM	Cd	
Gfw	112.41	GRAMS
REFERENCE STATE		
$(H°_T - H°_{298.15})$ =	1,491	CAL./GFW.
M.P.	594	°K
ΔH_m	1,450	CAL./GFW.
B.P.	1,038	°K
ΔH_v	23,870	CAL./GFW.
S.P.		°K
ΔH_s		CAL./GFW.
T.P.		°K
ΔH_t		CAL./GFW.
T.P.		°K
ΔH_t		CAL./GFW.
$T_c =$		°K
$P_c =$		ATM.

T TEMPERATURE °K	$C°_P$ HEAT CAPACITY CAL./DEG./GFW.	$H°_T - H°_{298.15}$ HEAT CONTENT CAL./GFW.	$S°_T$ ENTROPY CAL./DEG./GFW.	$-(F°_T - H°_{298.15})/T$ FREE ENERGY FUNCTION CAL./DEG./GFW.	FORMATION FROM REFERENCE STATE HEAT $\Delta H°_f$ CAL./GFW.	FORMATION FROM REFERENCE STATE FREE ENERGY $\Delta F°_f$ CAL./GFW.	$LOG_{10} K_P$
298	6.22	0	12.37	12.37			
300	6.23	12	12.41	12.37			
400	6.49	645	14.23	12.62			
500	6.78	1310	15.71	13.04			
600	7.10	3450	19.41	13.66			
700	7.10	4160	20.50	14.56			
800	7.10	4870	21.45	15.37			
900	7.10	5580	22.29	16.09			
1000	7.10	6290	23.04	16.75			
1100	4.97	30730	46.55	18.62			
1200	4.97	31230	46.98	20.96			
1300	4.97	31730	47.38	22.98			
1400	4.97	32220	47.75	24.74			
1500	4.97	32720	48.09	26.28			
1600	4.97	33220	48.41	27.65			
1700	4.97	33710	48.72	28.90			
1800	4.97	34210	49.00	30.00			
1900	4.97	34710	49.27	31.01			
2000	4.97	35200	49.52	31.92			
2100	4.97	35700	49.76	32.76			
2200	4.97	36200	50.00	33.55			
2300	4.97	36690	50.22	34.27			
2400	4.97	37190	50.43	34.94			
2500	4.97	37690	50.63	35.56			
2600	4.97	38190	50.83	36.15			
2700	4.97	38680	51.01	36.69			
2800	4.97	39180	51.19	37.20			
2900	4.97	39680	51.37	37.69			
3000	4.97	40170	51.54	38.15			

CADMIUM

CADMIUM   Cd

IDEAL MONATOMIC GAS

Gfw 112.41 GRAMS	
$(H°_{298.15} - H°_0) = 1,481$ CAL./GFW.	

Reference State for Calculating $\Delta H_f°$, $\Delta F_f°$, and $Log_{10} K_p$: Solid from 298° to 594°, Liquid from 594° to 1038°, Ideal Monatomic Gas from 1038° to 3000°.

T TEMPERATURE °K	$C_p°$ HEAT CAPACITY CAL./DEG./GFW.	$H°_T - H°_{298.15}$ HEAT CONTENT CAL./GFW.	$S°_T$ ENTROPY CAL./DEG./GFW.	$-(F°_T - H°_{298.15})/T$ FREE ENERGY FUNCTION CAL./DEG./GFW.	FORMATION FROM REFERENCE STATE		
					HEAT $\Delta H_f°$ CAL./GFW.	FREE ENERGY $\Delta F_f°$ CAL./GFW.	LOG$_{10}$ K$_p$
298	4.97	0	40.07	40.07	26750	18491	-13.554
300	4.97	9	40.10	40.07	26747	18440	-13.434
400	4.97	506	41.53	40.27	26611	15691	-8.573
500	4.97	1003	42.64	40.64	26443	12978	-5.673
600	4.97	1500	43.54	41.04	24800	10322	-3.760
700	4.97	1996	44.31	41.46	24586	7919	-2.472
800	4.97	2493	44.97	41.86	24373	5557	-1.518
900	4.97	2990	45.56	42.24	24160	3217	-.781
1000	4.97	3487	46.08	42.60	23947	907	-.198
1100	4.97	3984	46.55	42.93	0	0	0
1200	4.97	4480	46.98	43.25	0	0	0
1300	4.97	4977	47.38	43.56	0	0	0
1400	4.97	5474	47.75	43.84	0	0	0
1500	4.97	5971	48.09	44.11	0	0	0
1600	4.97	6468	48.41	44.37	0	0	0
1700	4.97	6964	48.72	44.63	0	0	0
1800	4.97	7461	49.00	44.86	0	0	0
1900	4.97	7958	49.27	45.09	0	0	0
2000	4.97	8455	49.52	45.30	0	0	0
2100	4.97	8952	49.76	45.50	0	0	0
2200	4.97	9448	50.00	45.71	0	0	0
2300	4.97	9945	50.22	45.90	0	0	0
2400	4.97	10442	50.43	46.08	0	0	0
2500	4.97	10939	50.63	46.26	0	0	0
2600	4.97	11436	50.83	46.44	0	0	0
2700	4.97	11932	51.01	46.60	0	0	0
2800	4.97	12429	51.19	46.76	0	0	0
2900	4.97	12926	51.37	46.92	0	0	0
3000	4.97	13423	51.54	47.07	0	0	0

M.P. °K
ΔH_m CAL./GFW.

B.P. °K
ΔH_v CAL./GFW.

S.P. °K
ΔH_s CAL./GFW.

T.P. °K
ΔH_t CAL./GFW.

T.P. °K
ΔH_t CAL./GFW.

T_c =   °K
P_c =   ATM.

CALCIUM

CALCIUM Ca

Solid I from 298° to 713°, Solid II from 713° to 1123°, Liquid from 1123° to 1765°, Ideal Monatomic Gas from 1765° to 3000°.

Gfw	40.08	GRAMS
$(H^°_{298.15} - H^°_0) =$	1,380	CAL./GFW.
M.P.	1,123.	°K
ΔH_m	2,070.	CAL./GFW.
B.P.	1,765.	°K
ΔH_v	35,840.	CAL./GFW.
S.P.		°K
ΔH_s		CAL./GFW.
T.P.	713.	°K
ΔH_t	270.	CAL./GFW.
T.P.		°K
ΔH_t		CAL./GFW.
$T_c =$		°K
$P_c =$		ATM.

TEMPERATURE °K	$C_p^°$ HEAT CAPACITY CAL./DEG./GFW.	$H^°_T - H^°_{298.15}$ HEAT CONTENT CAL./GFW.	$S^°_T$ ENTROPY CAL./DEG./GFW.	$-(F^°_T - H^°_{298.15})/T$ FREE ENERGY FUNCTION CAL./DEG./GFW.	HEAT $\Delta H^°_f$ CAL./GFW.	FREE ENERGY $\Delta F^°_f$ CAL./GFW.	$LOG_{10} K_p$
298	6.30	0	9.95	9.95			
300	6.31	11	9.99	9.96			
400	6.64	659	11.85	10.21			
500	6.98	1340	13.37	10.69			
600	7.31	2054	14.67	11.25			
700	7.64	2802	15.82	11.82			
800	8.08	3848	17.24	12.43			
900	8.78	4691	18.23	13.02			
1000	9.49	5605	19.19	13.59			
1100	10.18	6588	20.13	14.15			
1200	7.40	9465	22.67	14.79			
1300	7.40	10205	23.26	15.41			
1400	7.40	10945	23.81	16.00			
1500	7.40	11685	24.32	16.53			
1600	7.40	12425	24.80	17.04			
1700	7.40	13165	25.25	17.51			
1800	4.99	49660	45.93	18.35			
1900	5.00	50160	46.20	19.80			
2000	5.01	50660	46.45	21.12			
2100	5.03	51160	46.70	22.34			
2200	5.06	51670	46.93	23.45			
2300	5.09	52180	47.16	24.48			
2400	5.16	52690	47.38	25.43			
2500	5.22	53210	47.59	26.31			
2600	5.30	53740	47.80	27.14			
2700	5.40	54270	48.00	27.90			
2800	5.52	54820	48.20	28.63			
2900	5.65	55380	48.39	29.30			
3000	5.80	55950	48.59	29.94			

CALCIUM

CALCIUM   Ca   Reference State for Calculating ΔH_f°, ΔF_f°, and $\text{Log}_{10} K_p$:
Solid I from 298° to 713°, Solid II from 713° to 1123°, Liquid
from 1123° to 1765°, Ideal Monatomic Gas from 1765° to 3000°.

IDEAL MONATOMIC GAS

Gfw 40.08   GRAMS
$(H_{298.15}^\circ - H_0^\circ) = 1,481$ CAL./GFW.

T °K	C_p° HEAT CAPACITY CAL./DEG./GFW.	$H_T^\circ - H_{298.15}^\circ$ HEAT CONTENT CAL./GFW.	S_T° ENTROPY CAL./DEG./GFW.	$-(F^\circ - H_{298.15}^\circ)$ FREE ENERGY FUNCTION CAL./DEG./GFW.	FORMATION FROM REFERENCE STATE			
					HEAT ΔH_f° CAL./GFW.	FREE ENERGY ΔF_f° CAL./GFW.	LOG$_{10}$ K$_p$
298	4.97	0	36.99	36.99	42200	34138	−25.024
300	4.97	9	37.02	36.99	42198	34089	−24.835
400	4.97	506	38.45	37.19	42047	31407	−17.161
500	4.97	1003	39.56	37.56	41863	28768	−12.575
600	4.97	1500	40.47	37.97	41646	26166	−9.531
700	4.97	1996	41.23	38.38	41394	23607	−7.371
800	4.97	2493	41.90	38.79	40845	21117	−5.769
900	4.97	2990	42.48	39.16	40499	18674	−4.534
1000	4.97	3487	43.01	39.53	40082	16262	−3.554
1100	4.97	3984	43.48	39.86	39596	13911	−2.764
1200	4.97	4480	43.91	40.18	37215	11727	−2.135
1300	4.97	4977	44.31	40.49	36972	9607	−1.615
1400	4.97	5474	44.68	40.77	36729	7511	−1.172
1500	4.97	5971	45.02	41.04	36486	5436	−.792
1600	4.97	6468	45.34	41.30	36243	3379	−.461
1700	4.98	6965	45.64	41.55	36000	1337	−.171
1800	4.99	7464	45.93	41.79	0	0	0
1900	5.00	7963	46.20	42.01	0	0	0
2000	5.01	8462	46.45	42.22	0	0	0
2100	5.03	8964	46.70	42.44	0	0	0
2200	5.06	9469	46.93	42.63	0	0	0
2300	5.09	9977	47.16	42.83	0	0	0
2400	5.16	10490	47.38	43.01	0	0	0
2500	5.22	11009	47.59	43.19	0	0	0
2600	5.30	11535	47.80	43.37	0	0	0
2700	5.40	12070	48.00	43.53	0	0	0
2800	5.52	12616	48.20	43.70	0	0	0
2900	5.65	13175	48.39	43.85	0	0	0
3000	5.80	13749	48.59	44.01	0	0	0

M.P. °K
ΔH_m CAL./GFW.

B.P. °K
ΔH_v CAL./GFW.

S.P. °K
ΔH_s CAL./GFW.

T.P. °K
ΔH_t CAL./GFW.

T.P. °K
ΔH_t CAL./GFW.

$T_c =$ °K
$P_c =$ ATM.

CARBON

Solid Graphite from 298° to 3000°.

CARBON	C						FORMATION FROM REFERENCE STATE	
Gfw	12.011	GRAMS						
REFERENCE STATE								
$(H°_{298.15} - H°_0) = 251$		CAL./GFW.						

T TEMPERATURE °K	$C°_P$ HEAT CAPACITY CAL./DEG./GFW.	$H°_T - H°_{298.15}$ HEAT CONTENT CAL./GFW.	$S°_T$ ENTROPY CAL./DEG./GFW.	$-(F°_T - H°_{298.15})$ FREE ENERGY FUNCTION CAL./DEG./GFW.	HEAT $\Delta H°_f$ CAL./GFW.	FREE ENERGY $\Delta F°_f$ CAL./GFW.	LOG$_{10}$ K_P
298	2.07	0	1.37	1.37			
300	2.08	4	1.38	1.37			
400	2.85	251	2.09	1.47			
500	3.50	569	2.80	1.67			
600	4.03	947	3.49	1.92			
700	4.43	1370	4.14	2.19			
800	4.75	1830	4.75	2.47			
900	4.98	2318	5.33	2.76			
1000	5.14	2823	5.86	3.04			
1100	5.27	3344	6.35	3.31			
1200	5.42	3874	6.82	3.60			
1300	5.57	4428	7.26	3.86			
1400	5.67	4990	7.67	4.11			
1500	5.76	5562	8.07	4.37			
1600	5.83	6142	8.44	4.61			
1700	5.90	6728	8.80	4.85			
1800	5.95	7320	9.14	5.08			
1900	6.00	7918	9.46	5.30			
2000	6.05	8520	9.77	5.51			
2100	6.10	9133	10.07	5.73			
2200	6.14	9745	10.35	5.93			
2300	6.18	10360	10.63	6.13			
2400	6.22	10980	10.89	6.32			
2500	6.26	11600	11.14	6.50			
2600	6.30	12230	11.39	6.69			
2700	6.33	12860	11.63	6.87			
2800	6.36	13500	11.86	7.04			
2900	6.39	14140	12.08	7.21			
3000	6.42	14780	12.30	7.38			

M.P.		°K
ΔH_m		CAL./GFW.
B.P.		°K
ΔH_v		CAL./GFW.
S.P.	(4,000)	°K
ΔH_s		CAL./GFW.
T.P.		°K
ΔH_t		CAL./GFW.
T.P.		°K
ΔH_t		CAL./GFW.
T_c =		°K
P_c =		ATM.

CARBON

CARBON   C

IDEAL MONATOMIC GAS

Reference State for Calculating ΔH_f°, ΔF_f°, and $\log_{10} K_p$: Solid Graphite from 298° to 3000°.

Gfw 12.011   GRAMS
($H^\circ_{298.15} - H^\circ_0$) = 1,559   CAL./GFW.

T TEMPERATURE °K	C_p° HEAT CAPACITY CAL./DEG./GFW.	$H^\circ_T - H^\circ_{298.15}$ HEAT CONTENT CAL./GFW.	S°_T ENTROPY CAL./DEG./GFW.	$-(F^\circ_T - H^\circ_{298.15})/T$ FREE ENERGY FUNCTION CAL./DEG./GFW.	HEAT ΔH_f° CAL./GFW.	FREE ENERGY ΔF_f° CAL./GFW.	LOG$_{10}$ K_p
298	4.98		37.76	37.76	170890	160040	-117.317
300	4.98	9	37.79	37.76	170895	159972	-116.549
400	4.97	507	39.22	37.96	171146	156294	-85.402
500	4.97	1004	40.33	38.33	171325	152560	-66.688
600	4.97	1501	41.23	38.73	171444	148800	-54.204
700	4.97	1998	42.01	39.16	171518	145009	-45.277
800	4.97	2495	42.67	39.56	171555	141219	-38.582
900	4.97	2992	43.26	39.94	171564	137427	-33.374
1000	4.97	3489	43.78	40.30	171556	133636	-29.208
1100	4.97	3986	44.25	40.63	171532	129842	-25.799
1200	4.97	4483	44.68	40.95	171499	126067	-22.961
1300	4.97	4980	45.08	41.25	171442	122276	-20.558
1400	4.97	5477	45.45	41.54	171377	118485	-18.497
1500	4.97	5975	45.79	41.81	171303	114723	-16.716
1600	4.98	6472	46.11	42.07	171220	110948	-15.154
1700	4.98	6970	46.41	42.31	171132	107195	-13.780
1800	4.99	7468	46.70	42.56	171038	103430	-12.557
1900	5.00	7968	46.97	42.78	170940	99671	-11.464
2000	5.01	8469	47.23	43.00	170839	95919	-10.481
2100	5.02	8971	47.47	43.20	170728	92188	-9.594
2200	5.03	9474	47.70	43.40	170619	88449	-8.786
2300	5.04	9977	47.93	43.60	170507	84717	-8.049
2400	5.06	10482	48.14	43.78	170392	80992	-7.375
2500	5.08	10989	48.35	43.96	170279	77254	-6.753
2600	5.09	11497	48.55	44.13	170157	73541	-6.181
2700	5.11	12007	48.74	44.30	170037	69840	-5.652
2800	5.13	12519	48.93	44.46	169909	66113	-5.160
2900	5.15	13033	49.11	44.62	169783	62396	-4.702
3000	5.17	13549	49.29	44.78	169659	58689	-4.275

M.P.   °K
ΔH_m CAL./GFW.

B.P.   °K
ΔH_v CAL./GFW.

S.P.   °K
ΔH_s CAL./GFW.

T.P.   °K
ΔH_t CAL./GFW.

T.P.   °K
ΔH_t CAL./GFW.

T_c =   °K
P_c =   ATM.

THERMODYNAMIC PROPERTIES OF THE ELEMENTS

CARBON

CARBON	C_2		Reference State for Calculating ΔH_f°, ΔF_f°, and
Gfw	24.022	GRAMS	$\log_{10} K_p$: Solid Graphite from 298° to 3000°.
$(H_{298.15}^\circ - H_0^\circ)$	2,096	CAL./GFW.	

IDEAL DIATOMIC GAS

	T TEMPERATURE °K	C_p° HEAT CAPACITY CAL./DEG./GFW.	$H_T^\circ - H_{298.15}^\circ$ HEAT CONTENT CAL./GFW.	S_T° ENTROPY CAL./DEG./GFW.	$-(F_T^\circ - H_{298.15}^\circ)$ FREE ENERGY FUNCTION CAL./DEG./GFW.	FORMATION FROM REFERENCE STATE		
						HEAT ΔH_f° CAL./GFW.	FREE ENERGY ΔF_f° CAL./GFW.	$\log_{10} K_p$
M.P. °K	298	7.00	0	47.91	47.91	200000	186532	-136.737
ΔH_m CAL./GFW.	300	7.00	12	47.95	47.91	200004	186447	-135.837
	400	7.15	720	49.99	48.19	200218	181894	-99.390
	500	7.36	1445	51.60	48.71	200307	177307	-77.506
	600	7.59	2192	52.97	49.32	200298	172704	-62.912
B.P. °K	700	7.80	2962	54.15	49.92	200222	168113	-52.491
ΔH_v CAL./GFW.	800	7.98	3750	55.20	50.52	200090	163530	-44.678
	900	8.14	4580	56.15	51.09	199924	158983	-38.609
	1000	8.27	5380	57.01	51.63	199734	154444	-33.756
	1100	8.37	6210	57.81	52.17	199522	149901	-29.785
S.P. °K	1200	8.45	7050	58.54	52.67	199302	145422	-26.487
ΔH_s CAL./GFW.	1300	8.52	7900	59.22	53.15	199044	140934	-23.695
	1400	8.59	8760	59.85	53.60	198780	136466	-21.305
	1500	8.64	9620	60.45	54.04	198496	132031	-19.238
	1600	8.69	10490	61.01	54.46	198206	127598	-17.428
	1700	8.73	11360	61.54	54.86	197904	123206	-15.839
T.P. °K	1800	8.76	12230	62.04	55.25	197590	118822	-14.426
ΔH_t CAL./GFW.	1900	8.79	13110	62.51	55.61	197274	114453	-13.164
	2000	8.81	13990	62.96	55.97	196950	110110	-12.031
	2100	8.83	14870	63.39	56.31	196604	105779	-11.008
	2200	8.85	15750	63.80	56.65	196250	101440	-10.077
T.P. °K	2300	8.86	16640	64.20	56.97	195923	97158	-9.231
ΔH_t CAL./GFW.	2400	8.88	17530	64.57	57.27	195570	92874	-8.457
	2500	8.90	18420	64.94	57.58	195220	88570	-7.742
	2600	8.92	19310	65.29	57.87	194850	84324	-7.088
$T_c =$ °K	2700	8.93	20200	65.62	58.14	194480	80108	-6.483
	2800	8.95	21090	65.95	58.42	194090	75846	-5.919
$P_c =$ ATM.	2900	8.96	21990	66.26	58.68	193710	71620	-5.397
	3000	8.98	22890	66.57	58.94	193330	67420	-4.911

CARBON

CARBON	C_3		Reference State for Calculating ΔH_f°, ΔF_f°, and
IDEAL TRIATOMIC GAS			$\log_{10} K_p$: Solid Graphite from 298° to 3000°.

Gfw	36,033	GRAMS
$(H°_{298.15} - H°_0) = $	2,541	CAL./GFW.

						FORMATION FROM REFERENCE STATE		
T TEMPERATURE °K	C_p° HEAT CAPACITY CAL./DEG./GFW.	$H°_T - H°_{298.15}$ HEAT CONTENT CAL./GFW.	$S°_T$ ENTROPY CAL./DEG./GFW.	$-(F°_T - H°_{298.15})/T$ FREE ENERGY FUNCTION CAL./DEG./GFW.		HEAT ΔH_f° CAL./GFW.	FREE ENERGY ΔF_f° CAL./GFW.	$\log_{10} K_p$
298	10.41	0	55.18	55.18		200000	184772	-135.447
300	10.43	19	55.24	55.18		200007	184677	-134.548
400	11.17	1101	58.35	55.60		200348	179516	-98.091
500	11.77	2249	60.91	56.42		200542	174287	-76.186
600	12.27	3452	63.10	57.35		200611	169033	-61.575
700	12.69	4700	65.03	58.32		200590	163763	-51.133
800	13.03	5980	66.74	59.27		200490	158498	-43.303
900	13.31	7310	68.29	60.17		200356	153286	-37.225
1000	13.53	8650	69.71	61.06		200181	148051	-32.359
1100	13.72	10010	71.01	61.91		199978	142822	-28.378
1200	13.87	11390	72.21	62.72		199768	137668	-25.074
1300	14.00	12790	73.32	63.49		199506	132504	-22.277
1400	14.11	14190	74.37	64.24		199220	127316	-19.876
1500	14.20	15610	75.34	64.94		198924	122229	-17.809
1600	14.27	17030	76.26	65.62		198604	117100	-15.994
1700	14.34	18460	77.13	66.28		198276	112035	-14.403
1800	14.39	19900	77.95	66.90		197940	106986	-12.989
1900	14.44	21340	78.73	67.50		197586	101921	-11.722
2000	14.48	22780	79.47	68.08		197220	96900	-10.588
2100	14.52	24240	80.18	68.64		196841	91904	-9.564
2200	14.55	25690	80.85	69.18		196455	86895	-8.632
2300	14.58	27140	81.50	69.70		196060	81957	-7.787
2400	14.60	28600	82.12	70.21		195660	76980	-7.009
2500	14.63	30070	82.72	70.70		195270	72020	-6.295
2600	14.65	31530	83.29	71.17		194840	67128	-5.642
2700	14.66	33000	83.85	71.63		194420	62228	-5.036
2800	14.68	34460	84.38	72.08		193960	57320	-4.473
2900	14.70	35930	84.89	72.51		193510	52425	-3.950
3000	14.71	37400	85.39	72.93		193060	47590	-3.466

M.P.	°K
ΔH_m	CAL./GFW.
B.P.	°K
ΔH_v	CAL./GFW.
S.P.	°K
ΔH_s	CAL./GFW.
T.P.	°K
ΔH_t	CAL./GFW.
T.P.	°K
ΔH_t	CAL./GFW.
$T_c =$	°K
$P_c =$	ATM.

THERMODYNAMIC PROPERTIES OF THE ELEMENTS

CERIUM

CERIUM   Ce

Gfw 140.13   GRAMS

REFERENCE STATE

$(H°_{298.15} - H°_0) = 1,742$ CAL./GFW.

Solid I from 298° to 1027°, Solid II from 1027° to 1077°, Liquid from 1077° to 3000°.

M.P. 1,077 °K
ΔH_m (2,200) CAL./GFW.

B.P. 3,200 °K
ΔH_v 75,000 CAL./GFW.

S.P.   °K
ΔH_s CAL./GFW.

T.P. 1,027 °K
ΔH_t (300) CAL./GFW.

T.P.   °K
ΔH_t CAL./GFW.

$T_c =$ °K
$P_c =$ ATM.

T TEMPERATURE °K	$C°_P$ HEAT CAPACITY CAL./DEG./GFW.	$H°_T - H°_{298.15}$ HEAT CONTENT CAL./GFW.	$S°_T$ ENTROPY CAL./DEG./GFW.	$-(F°_T - H°_{298.15})$ FREE ENERGY FUNCTION CAL./DEG./GFW.	HEAT $\Delta H°_f$ CAL./GFW.	FREE ENERGY $\Delta F°_f$ CAL./GFW.	LOG$_{10}$ K_P
298	6.89	0	16.64	16.64			
300	6.90	12	16.68	16.64			
400	7.30	722	18.72	16.92			
500	7.70	1472	20.39	17.45			
600	8.10	2262	21.83	18.06			
700	8.50	3092	23.11	18.70			
800	8.90	3962	24.27	19.32			
900	9.30	4872	25.34	19.93			
1000	9.70	5822	26.34	20.52			
1100	8.00	9170	29.48	21.15			
1200	8.00	9970	30.18	21.88			
1300	8.00	10770	30.82	22.54			
1400	8.00	11570	31.41	23.15			
1500	8.00	12370	31.96	23.72			
1600	8.00	13170	32.48	24.25			
1700	8.00	13970	32.97	24.76			
1800	8.00	14770	33.42	25.22			
1900	8.00	15570	33.86	25.67			
2000	8.00	16370	34.27	26.09			
2100	8.00	17170	34.66	26.49			
2200	8.00	17970	35.03	26.87			
2300	8.00	18770	35.38	27.22			
2400	8.00	19570	35.72	27.57			
2500	8.00	20370	36.05	27.91			
2600	8.00	21170	36.36	28.22			
2700	8.00	21970	36.67	28.54			
2800	8.00	22770	36.96	28.83			
2900	8.00	23570	37.24	29.12			
3000	8.00	24370	37.51	29.39			

CESIUM

CESIUM Cs — Solid from 298° to 301.8°, Liquid from 301.8° to 958°, Ideal Monatomic Gas from 958° to 3000°.

REFERENCE STATE

		GRAMS	CAL./GFW.
Gfw	132.91		
$(H°_{298.15} - H°_0)$ =		1,859	
M.P.	301.8	°K	
ΔH_m	510.		CAL./GFW.
B.P.	958.	°K	
ΔH_v	15,750.		CAL./GFW.
S.P.		°K	
ΔH_s			CAL./GFW.
T.P.		°K	
ΔH_t			CAL./GFW.
T.P.		°K	
ΔH_t			CAL./GFW.
T_c =		°K	
P_c =		ATM.	

T TEMPERATURE °K	$C°_p$ HEAT CAPACITY CAL./DEG./GFW.	$H°_T - H°_{298.15}$ HEAT CONTENT CAL./GFW.	$S°_T$ ENTROPY CAL./DEG./GFW.	$-(F°_T - H°_{298.15})/T$ FREE ENERGY FUNCTION CAL./DEG./GFW.	FORMATION FROM REFERENCE STATE		
					HEAT $\Delta H°_f$ CAL./GFW.	FREE ENERGY $\Delta F°_f$ CAL./GFW.	$LOG_{10} K_p$
298	7.50	0	20.16	20.16			
300	7.54	14	20.21	20.17			
400	7.60	1284	24.09	20.88			
500	7.60	2044	25.78	21.70			
600	7.60	2804	27.17	22.50			
700	7.60	3564	28.34	23.25			
800	7.60	4324	29.36	23.96			
900	7.60	5084	30.26	24.62			
1000	4.97	22157	47.96	25.81			
1100	4.97	22654	48.43	27.84			
1200	4.97	23150	48.86	29.57			
1300	4.97	23647	49.26	31.07			
1400	4.97	24145	49.63	32.39			
1500	4.98	24643	49.97	33.55			
1600	4.99	25141	50.29	34.58			
1700	5.01	25641	50.60	35.52			
1800	5.03	26143	50.88	36.36			
1900	5.06	26648	51.16	37.14			
2000	5.10	27156	51.42	37.85			
2100	5.15	27668	51.67	38.50			
2200	5.22	28187	51.91	39.10			
2300	5.29	28712	52.14	39.66			
2400	5.38	29246	52.37	40.19			
2500	5.49	29790	52.59	40.68			
2600	5.60	30345	52.81	41.14			
2700	5.73	30911	53.03	41.59			
2800	5.86	31491	53.24	42.00			
2900	6.02	32085	53.44	42.38			
3000	6.18	32695	53.65	42.76			

CESIUM

CESIUM   Cs

IDEAL MONATOMIC GAS

Gfw 132.91   GRAMS   CAL./GFW.

$(H°_{298.15} - H°_0) \mp 1,481$

Reference State for Calculating $\Delta H°_f$, $\Delta F°_f$, and $Log_{10} K_p$:
Solid from 298° to 301.8°, Liquid from 301.8° to 958°, Ideal Monatomic Gas from 958° to 3000°.

T TEMPERATURE °K	$C°_P$ HEAT CAPACITY CAL./DEG./GFW.	$H°_T - H°_{298.15}$ HEAT CONTENT CAL./GFW.	$S°_T$ ENTROPY CAL./DEG./GFW.	$-(F°_T - H°_{298.15})/T$ FREE ENERGY FUNCTION CAL./DEG./GFW.	FORMATION FROM REFERENCE STATE		
					HEAT $\Delta H°_f$ CAL./GFW.	FREE ENERGY $\Delta F°_f$ CAL./GFW.	$LOG_{10} K_P$
298	4.97	0	41.94	41.94	18670	12176	−8.925
300	4.97	9	41.97	41.94	18665	12137	−8.842
400	4.97	506	43.40	42.14	17892	10168	−5.555
500	4.97	1003	44.51	42.51	17629	8264	−3.612
600	4.97	1500	45.42	42.92	17366	6416	−2.337
700	4.97	1996	46.18	43.33	17102	4614	−1.440
800	4.97	2493	46.85	43.74	16839	2847	−.777
900	4.97	2990	47.43	44.11	16576	1123	−.272
1000	4.97	3487	47.96	44.48	0	0	0
1100	4.97	3984	48.43	44.81	0	0	0
1200	4.97	4480	48.86	45.13	0	0	0
1300	4.97	4977	49.26	45.44	0	0	0
1400	4.97	5475	49.63	45.72	0	0	0
1500	4.98	5973	49.97	45.99	0	0	0
1600	4.99	6471	50.29	46.25	0	0	0
1700	5.01	6971	50.60	46.50	0	0	0
1800	5.03	7473	50.88	46.73	0	0	0
1900	5.06	7978	51.16	46.97	0	0	0
2000	5.10	8486	51.42	47.18	0	0	0
2100	5.15	8998	51.67	47.39	0	0	0
2200	5.22	9517	51.91	47.59	0	0	0
2300	5.29	10042	52.14	47.78	0	0	0
2400	5.38	10576	52.37	47.97	0	0	0
2500	5.49	11120	52.59	48.15	0	0	0
2600	5.60	11675	52.81	48.32	0	0	0
2700	5.73	12241	53.03	48.50	0	0	0
2800	5.86	12821	53.24	48.67	0	0	0
2900	6.02	13415	53.44	48.82	0	0	0
3000	6.18	14025	53.65	48.98	0	0	0

M.P. °K ΔH_m CAL./GFW.

B.P. °K ΔH_v CAL./GFW.

S.P. °K ΔH_s CAL./GFW.

T.P. °K ΔH_t CAL./GFW.

T.P. °K ΔH_t CAL./GFW.

$T_c =$ °K

$P_c =$ ATM.

CESIUM

CESIUM Cs_2
IDEAL DIATOMIC GAS

Reference State for Calculating ΔH_f°, ΔF_f°, and $\log_{10} K_p$: Solid from 298° to 301.8°, Liquid from 301.8° to 958°, Ideal Monatomic Gas from 958° to 3000°.

Gfw 265.82 GRAMS
$(H°_{298.15} - H°_0) = 2,635$ CAL./GFW.

M.P. °K
ΔH_m CAL./GFW.

B.P. °K
ΔH_v CAL./GFW.

S.P. °K
ΔH_s CAL./GFW.

T.P. °K
ΔH_t CAL./GFW.

T.P. °K
ΔH_t CAL./GFW.

$T_c =$ °K
$P_c =$ ATM.

T TEMPERATURE °K	C_p° HEAT CAPACITY CAL./DEG./GFW.	$H°_T - H°_{298.15}$ HEAT CONTENT CAL./GFW.	$S°_T$ ENTROPY CAL./DEG./GFW.	$-(F°_T - H°_{298.15})/T$ FREE ENERGY FUNCTION CAL./DEG./GFW.	FORMATION FROM REFERENCE STATE		
					HEAT ΔH_f° CAL./GFW.	FREE ENERGY ΔF_f° CAL./GFW.	$\log_{10} K_p$
298	9.11	0	67.86	67.86	26630	18417	-13.500
300	9.11	17	67.91	67.86	26619	18372	-13.385
400	9.18	932	70.54	68.21	24994	16050	-8.770
500	9.24	1852	72.60	68.90	24394	13874	-6.064
600	9.30	2778	74.28	69.65	23800	11836	-4.311
700	9.36	3708	75.72	70.43	23210	9882	-3.085
800	9.42	4645	76.97	71.17	22627	8027	-2.193
900	9.48	5592	78.09	71.88	22054	6241	-1.515
1000	9.53	6546	79.09	72.55	-11138	5692	-1.244
1100	9.59	7497	80.00	73.19	-11181	7365	-1.463
1200	9.65	8466	80.84	73.79	-11204	9052	-1.648
1300	9.71	9443	81.62	74.36	-11221	10749	-1.807
1400	9.77	10414	82.34	74.91	-11246	12442	-1.942
1500	9.83	11391	83.01	75.42	-11265	14130	-2.058

THERMODYNAMIC PROPERTIES OF THE ELEMENTS

CHLORINE

Ideal Diatomic Gas from 298° to 3000°.

CHLORINE Cl_2		
Gfw 70.914	GRAMS	
$(H°_{298.15} - H°_0) = 2,194$	CAL./GFW.	
M.P. 172.16	°K	
ΔH_m 1,531.	CAL./GFW.	
B.P. 239.10	°K	
ΔH_v 4,878.	CAL./GFW.	
T_c = 417.	°K	
P_c = 76.1	ATM.	

T TEMPERATURE °K	$C°_P$ HEAT CAPACITY CAL./DEG./GFW.	$H°_T - H°_{298.15}$ HEAT CONTENT CAL./GFW.	$S°_T$ ENTROPY CAL./DEG./GFW.	$-(F°_T - H°_{298.15})/T$ FREE ENERGY FUNCTION CAL./DEG./GFW.	HEAT $\Delta H°_f$ CAL./GFW.	FREE ENERGY $\Delta F°_f$ CAL./GFW.	LOG$_{10}$ K_P
298	8.11	0	53.29	53.29			
300	8.12	15	53.34	53.29			
400	8.44	845	55.73	53.62			
500	8.62	1698	57.63	54.24			
600	8.74	2567	59.21	54.94			
700	8.82	3446	60.57	55.65			
800	8.88	4331	61.75	56.34			
900	8.92	5220	62.80	57.00			
1000	8.96	6115	63.74	57.63			
1100	8.99	7013	64.60	58.23			
1200	9.02	7913	65.38	58.79			
1300	9.04	8816	66.10	59.32			
1400	9.06	9721	66.77	59.83			
1500	9.08	10628	67.40	60.32			
1600	9.10	11536	67.99	60.78			
1700	9.11	12445	68.54	61.22			
1800	9.13	13358	69.06	61.64			
1900	9.14	14270	69.55	62.04			
2000	9.16	15186	70.02	62.43			
2100	9.17	16102	70.47	62.81			
2200	9.19	17021	70.90	63.17			
2300	9.20	17941	71.30	63.50			
2400	9.22	18862	71.70	63.85			
2500	9.23	19784	72.07	64.16			
2600	9.25	20708	72.43	64.47			
2700	9.26	21633	72.78	64.77			
2800	9.27	22560	73.12	65.07			
2900	9.28	23487	73.44	65.35			
3000	9.30	24416	73.76	65.63			

CHLORINE

CHLORINE	Cl		Reference State for Calculating ΔH_f°, ΔF_f°, and
IDEAL MONATOMIC GAS			$Log_{10} K_p$: Ideal Diatomic Gas from 298° to 3000°.

Gfw 35.457 GRAMS
$(H_{298.15}^\circ - H_0^\circ) = 1,499$ CAL./GFW.

						FORMATION FROM REFERENCE STATE		
T TEMPERATURE °K	C_p° HEAT CAPACITY CAL./DEG./GFW.	$H_T^\circ - H_{298.15}^\circ$ HEAT CONTENT CAL./GFW.	S_T° ENTROPY CAL./DEG./GFW.	$-(F^\circ - H_{298.15}^\circ)/T$ FREE ENERGY FUNCTION CAL./DEG./GFW.		HEAT ΔH_f° CAL./GFW.	FREE ENERGY ΔF_f° CAL./GFW.	$LOG_{10} K_p$
298	5.22	0	39.46	39.45		28942	25122	-18.415
300	5.22	10	39.49	39.46		28944	25098	-18.285
400	5.37	540	41.01	39.66		29059	23803	-13.006
500	5.44	1081	42.22	40.06		29174	22474	-9.824
600	5.44	1625	43.21	40.51		29283	21123	-7.694
700	5.42	2169	44.05	40.96		29388	19756	-6.168
800	5.39	2710	44.77	41.39		29486	18374	-5.019
900	5.35	3247	45.41	41.81		29579	16970	-4.121
1000	5.31	3780	45.97	42.19		29664	15564	-3.401
1100	5.28	4309	46.47	42.56		29744	14157	-2.812
1200	5.25	4836	46.93	42.90		29821	12733	-2.319
1300	5.22	5359	47.35	43.23		29893	11303	-1.900
1400	5.20	5880	47.74	43.54		29961	9871	-1.541
1500	5.17	6398	48.09	43.83		30026	8441	-1.229
1600	5.16	6915	48.43	44.11		30088	6984	-.953
1700	5.14	7430	48.74	44.37		30147	5548	-.713
1800	5.12	7943	49.03	44.62		30204	4104	-.498
1900	5.11	8454	49.31	44.87		30257	2650	-.304
2000	5.10	8965	49.57	45.09		30310	1190	-.130
2100	5.09	9475	49.82	45.31		30361	257	-.026
2200	5.08	9983	50.06	45.53		30410	-1732	.172
2300	5.07	10491	50.28	45.72		30457	-3192	.303
2400	5.07	10998	50.50	45.92		30503	-4657	.424
2500	5.06	11504	50.70	46.10		30547	-6103	.533
2600	5.05	12009	50.90	46.29		30590	-7578	.637
2700	5.05	12514	51.09	46.46		30631	-9032	.731
2800	5.04	13019	51.27	46.63		30672	-10516	.820
2900	5.04	13523	51.45	46.79		30712	-11976	.902
3000	5.03	14027	51.62	46.95		30750	-13440	.979

M.P. °K
ΔH_m CAL./GFW.

B.P. °K
ΔH_v CAL./GFW.

S.P. °K
ΔH_s CAL./GFW.

T.P. °K
ΔH_t CAL./GFW.

T.P. °K
ΔH_t CAL./GFW.

$T_c =$ °K
$P_c =$ ATM.

CHROMIUM

CHROMIUM Cr
Solid I from 298° to 2113°, Solid II from 2113° to 2176°, Liquid from 2176° to 2915°, Ideal Monatomic Gas from 2915° to 3000°.

REFERENCE STATE

Gfw 52.01
($H°_{298.15} - H°_0$) = 973 CAL./GFW. GRAMS

M.P.	2,176	°K
ΔH_m	3,300	CAL./GFW.
B.P.	2,915	°K
ΔH_v	83,360	CAL./GFW.
S.P.		°K
ΔH_s		CAL./GFW.
T.P.	2,113	°K
ΔH_t	(350)	CAL./GFW.
T.P.		°K
ΔH_t		CAL./GFW.
$T_c =$		°K
$P_c =$		ATM.

T TEMPERATURE °K	$C°_P$ HEAT CAPACITY CAL./DEG./GFW.	$H°_T - H°_{298.15}$ HEAT CONTENT CAL./GFW.	$S°_T$ ENTROPY CAL./DEG./GFW.	$-(F°_T - H°_{298.15})/T$ FREE ENERGY FUNCTION CAL./DEG./GFW.	FORMATION FROM REFERENCE STATE HEAT $\Delta H°_f$ CAL./GFW.	FORMATION FROM REFERENCE STATE FREE ENERGY $\Delta F°_f$ CAL./GFW.	LOG$_{10}$ K_P
298	5.56	0	5.70	5.70			
300	5.57	10	5.73	5.70			
400	6.08	594	7.41	5.93			
500	6.40	1220	8.80	6.36			
600	6.58	1870	9.99	6.88			
700	6.68	2530	11.01	7.40			
800	6.82	3210	11.91	7.90			
900	7.17	3900	12.73	8.40			
1000	7.54	4640	13.50	8.86			
1100	7.94	5410	14.24	9.33			
1200	8.35	6230	14.95	9.76			
1300	8.75	7080	15.63	10.19			
1400	8.99	7970	16.29	10.60			
1500	9.23	8880	16.92	11.00			
1600	9.47	9820	17.52	11.39			
1700	9.71	10780	18.10	11.76			
1800	9.95	11760	18.67	12.14			
1900	10.19	12770	19.21	12.49			
2000	10.43	13800	19.74	12.84			
2100	10.67	14850	20.25	13.18			
2200	9.70	19480	22.40	13.55			
2300	9.70	20450	22.83	13.94			
2400	9.70	21420	23.25	14.33			
2500	9.70	22390	23.64	14.69			
2600	9.70	23360	24.02	15.04			
2700	9.70	24330	24.39	15.38			
2800	9.70	25300	24.74	15.71			
2900	9.70	26270	25.08	16.03			
3000	7.35	110390	53.94	17.15			

CHROMIUM

CHROMIUM   **Cr**

IDEAL MONATOMIC GAS

Gfw 52.01 GRAMS
($H°_{298.15} - H°_0$) = 1,481 CAL./GFW.

Reference State for Calculating $\Delta H°_f$, $\Delta F°_f$, and $\log_{10} Kp$:
Solid I from 298° to 2113°, Solid II from 2113° to 2176°, Liquid from 2176° to 2915°, Ideal Monatomic Gas from 2915° to 3000°.

T TEMPERATURE °K	$C°_p$ HEAT CAPACITY CAL./DEG./GFW.	$H°_T - H°_{298.15}$ HEAT CONTENT CAL./GFW.	$S°_T$ ENTROPY CAL./DEG./GFW.	$-(F°_T - H°_{298.15})/T$ FREE ENERGY FUNCTION CAL./DEG./GFW.	HEAT $\Delta H°_f$ CAL./GFW.	FREE ENERGY $\Delta F°_f$ CAL./GFW.	$\log_{10} K_p$
298	4.97	0	41.64	41.64	95000	84284	−61.784
300	4.97	9	41.67	41.64	94999	84217	−61.357
400	4.97	506	43.10	41.84	94912	80636	−44.061
500	4.97	1003	44.20	42.20	94783	77083	−33.695
600	4.97	1500	45.11	42.61	94630	73558	−26.795
700	4.97	1996	45.88	43.03	94466	70057	−21.874
800	4.97	2493	46.54	43.43	94283	66579	−18.190
900	4.97	2990	47.13	43.81	94090	63130	−15.331
1000	4.98	3488	47.65	44.17	93848	59698	−13.048
1100	5.00	3987	48.12	44.50	93577	56309	−11.188
1200	5.02	4487	48.56	44.83	93257	52925	−9.639
1300	5.06	4992	48.96	45.12	92912	49583	−8.336
1400	5.12	5501	49.34	45.42	92531	46261	−7.222
1500	5.20	6017	49.70	45.69	92137	42967	−6.260
1600	5.30	6542	50.04	45.96	91722	39690	−5.421
1700	5.41	7078	50.36	46.20	91298	36456	−4.686
1800	5.54	7626	50.67	46.44	90866	33266	−4.038
1900	5.69	8187	50.98	46.68	90417	30054	−3.456
2000	5.84	8764	51.27	46.89	89964	26904	−2.939
2100	6.00	9356	51.56	47.11	89506	23755	−2.472
2200	6.17	9964	51.85	47.33	85484	20694	−2.055
2300	6.33	10589	52.12	47.52	85139	17772	−1.688
2400	6.49	11230	52.40	47.73	84810	14850	−1.352
2500	6.65	11887	52.66	47.91	84497	11947	−1.044
2600	6.80	12560	52.93	48.10	84200	9034	−.759
2700	6.95	13248	53.19	48.29	83918	6158	−.498
2800	7.09	13950	53.44	48.46	83650	3290	−.256
2900	7.22	14666	53.69	48.64	83396	427	−.032
3000	7.35	15394	53.94	48.81	0	0	0

M.P. °K
ΔH_m CAL./GFW.

B.P. °K
ΔH_v CAL./GFW.

S.P. °K
ΔH_s CAL./GFW.

T.P. °K
ΔH_t CAL./GFW.

T.P. °K
ΔH_t CAL./GFW.

T_c = °K
P_c = ATM.

COBALT

COBALT Co
Gfw 58.94 GRAMS
$(H°_{298.15} - H°_0) = 1,146.$ CAL./GFW.

Solid I from 298° to 720°, Solid II from 720° to 1768°, Liquid from 1768° to 3000°.

	T TEMPERATURE °K	$C°_P$ HEAT CAPACITY CAL./DEG./GFW.	$H°_T - H°_{298.15}$ HEAT CONTENT CAL./GFW.	$S°_T$ ENTROPY CAL./DEG./GFW.	$-(F°_T - H°_{298.15})/T$ FREE ENERGY FUNCTION CAL./DEG./GFW.	FORMATION FROM REFERENCE STATE		
						HEAT $\Delta H°_f$ CAL./GFW.	FREE ENERGY $\Delta F°_f$ CAL./GFW.	LOG$_{10}$ K_P
	298	5.89	0	7.18	7.18			
	300	5.90	10	7.21	7.18			
	400	6.35	623	8.97	7.42			
	500	6.80	1280	10.44	7.88			
	600	7.17	1980	11.71	8.41			
	700	7.35	2710	12.84	8.97			
	800	7.65	3510	13.91	9.53			
	900	8.20	4305	14.84	10.06			
	1000	8.90	5160	15.74	10.58			
	1100	9.64	6090	16.62	11.09			
	1200	10.50	7090	17.50	11.60			
	1300	11.50	8190	18.38	12.08			
	1400	9.60	9520	19.36	12.56			
	1500	9.60	10480	20.02	13.04			
	1600	9.60	11440	20.64	13.49			
	1700	9.60	12400	21.22	13.93			
	1800	8.30	16950	23.80	14.39			
	1900	8.30	17780	24.25	14.90			
	2000	8.30	18610	24.68	15.38			
	2100	8.30	19440	25.08	15.83			
	2200	8.30	20270	25.47	16.26			
	2300	8.30	21100	25.84	16.67			
	2400	8.30	21930	26.19	17.06			
	2500	8.30	22760	26.53	17.43			
	2600	8.30	23590	26.85	17.78			
	2700	8.30	24420	27.17	18.13			
	2800	8.30	25250	27.47	18.46			
	2900	8.30	26080	27.76	18.77			
	3000	8.30	26910	28.04	19.07			

REFERENCE STATE

M.P. 1,768. °K
ΔH_m 3,640. CAL./GFW.

B.P. 3,150. °K
ΔH_v 91,400. CAL./GFW.

S.P. °K
ΔH_s CAL./GFW.

T.P. 720. °K
ΔH_t 60. CAL./GFW.

T.P. 1,395. °K
ΔH_t 130. CAL./GFW.

T_c = °K
P_c = ATM.

COBALT

COBALT Co
IDEAL MONATOMIC GAS

G_{fw} 58.94 GRAMS
$(H°_{298.15} - H°_0) = 1,520$ CAL./GFW.

M.P. °K
ΔH_m CAL./GFW.

B.P. °K
ΔH_v CAL./GFW.

S.P. °K
ΔH_s CAL./GFW.

T.P. °K
ΔH_t CAL./GFW.

T.P. °K
ΔH_t CAL./GFW.

$T_c =$ °K
$P_c =$ ATM.

Reference State for Calculating $\Delta H°_f$, $\Delta F°_f$, and $\log_{10} K_p$: Solid I from 298° to 720°, Solid II from 720° to 1768°, Liquid from 1768° to 3000°.

TEMPERATURE °K	$C°_p$ HEAT CAPACITY CAL./DEG./GFW.	$H°_T - H°_{298.15}$ HEAT CONTENT CAL./GFW.	$S°_T$ ENTROPY CAL./DEG./GFW.	$-(F°_T - H°_{298.15})/T$ FREE ENERGY FUNCTION CAL./DEG./GFW.	FORMATION FROM REFERENCE STATE		
					HEAT $\Delta H°_f$ CAL./GFW.	FREE ENERGY $\Delta F°_f$ CAL./GFW.	$\log_{10} K_p$
298	5.50	0	42.88	42.88	101600	90955	-66.674
300	5.51	10	42.91	42.88	101600	90890	-66.218
400	5.86	580	44.55	43.10	101557	87325	-47.716
500	6.08	1177	45.88	43.53	101497	83777	-36.621
600	6.19	1791	47.00	44.02	101411	80237	-29.228
700	6.24	2413	47.96	44.52	101303	76719	-23.954
800	6.26	3038	48.79	45.00	101128	73224	-20.005
900	6.27	3664	49.53	45.46	100959	69738	-16.935
1000	6.29	4293	50.19	45.90	100733	66283	-14.487
1100	6.31	4923	50.79	46.32	100433	62846	-12.487
1200	6.33	5555	51.34	46.72	100065	59457	-10.829
1300	6.35	6188	51.85	47.09	99598	56087	-9.429
1400	6.36	6824	52.32	47.45	98904	52760	-8.236
1500	6.38	7461	52.76	47.79	98581	49471	-7.208
1600	6.38	8099	53.17	48.11	98259	46211	-6.311
1700	6.39	8737	53.56	48.43	97937	42959	-5.522
1800	6.39	9376	53.93	48.73	94026	39792	-4.831
1900	6.38	10014	54.27	49.00	93834	36796	-4.232
2000	6.37	10652	54.60	49.28	93642	33802	-3.693
2100	6.36	11289	54.91	49.54	93449	30806	-3.205
2200	6.35	11925	55.21	49.79	93255	27827	-2.764
2300	6.34	12559	55.49	50.03	93059	24864	-2.362
2400	6.33	13193	55.76	50.27	92863	21895	-1.993
2500	6.32	13825	56.01	50.48	92665	18965	-1.657
2600	6.30	14456	56.26	50.70	92466	16000	-1.344
2700	6.29	15086	56.50	50.92	92266	13075	-1.058
2800	6.28	15714	56.73	51.12	92064	10136	-.791
2900	6.28	16342	56.95	51.32	91862	7211	-.543
3000	6.27	16970	57.16	51.51	91660	4300	-.313

THERMODYNAMIC PROPERTIES OF THE ELEMENTS

COPPER

COPPER	Cu		Solid from 298° to 1356°, Liquid from 1356° to 2855°, Ideal Monatomic Gas from 2855° to 3000°.
Giw	63.54	GRAMS	
REFERENCE STATE			
$(H^°_{298.15} - H^°_0)$ =	1,201	CAL./GFW.	
M.P.	1,356	°K	
ΔH_m	3,120	CAL./GFW.	
B.P.	2,855	°K	
ΔH_v	72,800	CAL./GFW.	
S.P.		°K	
ΔH_s		CAL./GFW.	
T.P.		°K	
ΔH_t		CAL./GFW.	
T.P.		°K	
ΔH_t		CAL./GFW.	
T_c =		°K	
P_c =		ATM.	

T TEMPERATURE °K	$C^°_P$ HEAT CAPACITY CAL./DEG./GFW.	$H^°_T - H^°_{298.15}$ HEAT CONTENT CAL./GFW.	$S^°_T$ ENTROPY CAL./DEG./GFW.	$-(F^°_T - H^°_{298.15})$ FREE ENERGY FUNCTION CAL./DEG./GFW.	FORMATION FROM REFERENCE STATE		
					HEAT $\Delta H^°_f$ CAL./GFW.	FREE ENERGY $\Delta F^°_f$ CAL./GFW.	LOG$_{10}$ K$_P$
298	5.85	0	7.97	7.97			
300	5.85	10	8.01	7.98			
400	6.01	600	9.70	8.20			
500	6.16	1215	11.07	8.64			
600	6.31	1845	12.22	9.15			
700	6.46	2480	13.20	9.66			
800	6.61	3130	14.07	10.16			
900	6.76	3800	14.86	10.64			
1000	6.91	4490	15.58	11.09			
1100	7.06	5190	16.25	11.54			
1200	7.21	5895	16.87	11.96			
1300	7.36	6615	17.44	12.36			
1400	7.50	10480	20.29	12.81			
1500	7.50	11230	20.81	13.33			
1600	7.50	11980	21.29	13.81			
1700	7.50	12730	21.74	14.26			
1800	7.50	13480	22.17	14.69			
1900	7.50	14230	22.58	15.10			
2000	7.50	14980	22.96	15.47			
2100	7.50	15730	23.33	15.84			
2200	7.50	16480	23.68	16.19			
2300	7.50	17230	24.01	16.52			
2400	7.50	17980	24.33	16.84			
2500	7.50	18730	24.64	17.15			
2600	7.50	19480	24.93	17.44			
2700	7.50	20230	25.21	17.72			
2800	7.50	20980	25.49	18.00			
2900	5.89	94500	51.24	18.66			
3000	6.01	95090	51.44	19.75			

COPPER

COPPER   Cu
IDEAL MONATOMIC GAS

G_{fw} 63.54   GRAMS
$(H°_{298.15} - H°_0) = 1,481$ CAL./GFW.

Reference State for Calculating $\Delta H_f°$, $\Delta F_f°$, and $\text{Log}_{10} K_p$: Solid from 298° to 1356°, Liquid from 1356° to 2855°, Ideal Monatomic Gas from 2855° to 3000°.

T TEMPERATURE °K	$C_p°$ HEAT CAPACITY CAL./DEG./GFW.	$H°_T - H°_{298.15}$ HEAT CONTENT CAL./GFW.	$S°_T$ ENTROPY CAL./DEG./GFW.	$-(F°_T - H°_{298.15})/T$ FREE ENERGY FUNCTION CAL./DEG./GFW.	FORMATION FROM REFERENCE STATE		
					HEAT $\Delta H_f°$ CAL./GFW.	FREE ENERGY $\Delta F_f°$ CAL./GFW.	$\text{LOG}_{10} K_p$
298	4.97	0	39.74	39.74	81100	71628	-52.506
300	4.97	9	39.77	39.74	81099	71571	-52.143
400	4.97	506	41.20	39.94	81006	68406	-37.378
500	4.97	1003	42.31	40.31	80888	65268	-28.530
600	4.97	1500	43.22	40.72	80755	62155	-22.641
700	4.97	1996	43.98	41.13	80616	59070	-18.444
800	4.97	2493	44.65	41.54	80463	55999	-15.299
900	4.97	2990	45.23	41.91	80290	52957	-12.860
1000	4.97	3487	45.76	42.28	80097	49917	-10.910
1100	4.97	3984	46.26	42.61	79894	46916	-9.322
1200	4.97	4481	46.66	42.93	79686	43938	-8.002
1300	4.97	4978	47.06	43.24	79463	40957	-6.886
1400	4.98	5475	47.43	43.52	76095	38099	-5.948
1500	4.98	5973	47.77	43.79	75843	35403	-5.158
1600	5.00	6472	48.09	44.05	75592	32712	-4.468
1700	5.02	6973	48.40	44.30	75343	30021	-3.859
1800	5.04	7476	48.68	44.53	75096	27378	-3.323
1900	5.07	7981	48.96	44.76	74851	24729	-2.844
2000	5.12	8491	49.22	44.98	74611	22091	-2.413
2100	5.17	9005	49.47	45.19	74375	19481	-2.027
2200	5.23	9525	49.71	45.39	74145	16879	-1.676
2300	5.30	10051	49.95	45.58	73921	14259	-1.354
2400	5.38	10585	50.17	45.76	73705	11689	-1.064
2500	5.47	11127	50.39	45.94	73497	9122	-.797
2600	5.56	11679	50.62	46.12	73299	6531	-.548
2700	5.67	12241	50.82	46.29	73111	3964	-.320
2800	5.78	12813	51.03	46.46	72933	1421	-.110
2900	5.89	13396	51.24	46.63	0	0	0
3000	6.01	13991	51.44	46.78	0	0	0

M.P.	°K		
ΔH_m	CAL./GFW.		
B.P.	°K		
ΔH_v	CAL./GFW.		
S.P.	°K		
ΔH_s	CAL./GFW.		
T.P.	°K		
ΔH_t	CAL./GFW.		
T.P.	°K		
ΔH_t	CAL./GFW.		
$T_c =$	°K		
$P_c =$	ATM.		

THERMODYNAMIC PROPERTIES OF THE ELEMENTS

DYSPROSIUM

DYSPROSIUM Dy Solid from 298° to 1773°, Liquid from 1773° to 2600°, Ideal Monatomic Gas from 2600° to 3000°.

REFERENCE STATE

G/w	162.51	GRAMS
$(H°_{298.15} - H°_0)$ =	2,116	CAL./GFW.

M.P.	1,773	°K
ΔH_m	(4,100)	CAL./GFW.
B.P.	2,600	°K
ΔH_v	60,000	CAL./GFW.
S.P.		°K
ΔH_s		CAL./GFW.
T.P.		°K
ΔH_f		CAL./GFW.
T.P.		°K
ΔH_t		CAL./GFW.
$T_c =$		°K
$P_c =$		ATM.

T TEMPERATURE °K	$C°_P$ HEAT CAPACITY CAL./DEG./GFW.	$H°_T - H°_{298.15}$ HEAT CONTENT CAL./GFW.	$S°_T$ ENTROPY CAL./DEG./GFW.	$-(F°_T - H°_{298.15})/T$ FREE ENERGY FUNCTION CAL./DEG./GFW.	FORMATION FROM REFERENCE STATE		
					HEAT $\Delta H°_f$ CAL./GFW.	FREE ENERGY $\Delta F°_f$ CAL./GFW.	LOG$_{10}$ K$_P$
298	6.51	0	17.87	17.87			
300	6.51	11	17.91	17.88			
400	6.68	670	19.81	18.14			
500	6.85	1350	21.31	18.61			
600	7.02	2040	22.58	19.18			
700	7.19	2750	23.67	19.75			
800	7.36	3480	24.64	20.29			
900	7.53	4220	25.52	20.84			
1000	7.70	4990	26.32	21.33			
1100	7.87	5760	27.06	21.83			
1200	8.04	6560	27.76	22.30			
1300	8.21	7370	28.41	22.75			
1400	8.38	8200	29.02	23.17			
1500	8.55	9050	29.61	23.58			
1600	8.72	9911	30.16	23.97			
1700	8.89	10790	30.70	24.36			
1800	8.00	15760	33.51	24.76			
1900	8.00	16560	33.94	25.23			
2000	8.00	17360	34.36	25.68			
2100	8.00	18160	34.75	26.11			
2200	8.00	18960	35.12	26.51			
2300	8.00	19760	35.47	26.88			
2400	8.00	20560	35.81	27.25			
2500	8.00	21360	36.14	27.60			
2600	6.00	82160	59.53	27.93			
2700	6.00	82760	59.76	29.11			
2800	6.00	83360	59.98	30.21			
2900	6.00	83960	60.19	31.24			
3000	6.00	84560	60.39	32.21			

ERBIUM

ERBIUM Er
REFERENCE STATE Solid from 298° to 1800°, Liquid from 1800° to 2900°, Ideal Monatomic Gas from 2900° to 3000°.

Gfw 167.27 GRAMS
$(H°_{298.15} - H°_0) = 1,763$ CAL./GFW.

M.P. 1,800 °K
ΔH_m (4,100) CAL./GFW.

B.P. (2,900) °K
ΔH_v (70,000) CAL./GFW.

S.P.   °K
ΔH_s CAL./GFW.

T.P.   °K
ΔH_t CAL./GFW.

T.P.   °K
ΔH_t CAL./GFW.

T_c =   °K
P_c =   ATM.

T TEMPERATURE °K	$C_p°$ HEAT CAPACITY CAL./DEG./GFW.	$H°_T - H°_{298.15}$ HEAT CONTENT CAL./GFW.	$S°_T$ ENTROPY CAL./DEG./GFW.	$-(F°_T - H°_{298.15})/T$ FREE ENERGY FUNCTION CAL./DEG./GFW.	FORMATION FROM REFERENCE STATE		
					HEAT $\Delta H°_f$ CAL./GFW.	FREE ENERGY $\Delta F°_f$ CAL./GFW.	$\log_{10} K_p$
298	6.72	0	17.48	17.48			
300	6.72	12	17.52	17.48			
400	6.87	690	19.47	17.75			
500	7.02	1390	21.02	18.24			
600	7.17	2095	22.32	18.83			
700	7.32	2820	23.43	19.41			
800	7.47	3560	24.42	19.97			
900	7.62	4310	25.31	20.53			
1000	7.77	5080	26.12	21.04			
1100	7.92	5870	26.87	21.54			
1200	8.07	6670	27.56	22.01			
1300	8.22	7480	28.21	22.46			
1400	8.37	8310	28.83	22.90			
1500	8.52	9160	29.41	23.31			
1600	8.67	10020	29.97	23.71			
1700	8.82	10890	30.50	24.10			
1800	8.00	15880	33.29	24.47			
1900	8.00	16680	33.72	24.95			
2000	8.00	17480	34.13	25.39			
2100	8.00	18280	34.52	25.82			
2200	8.00	19080	34.89	26.22			
2300	8.00	19880	35.25	26.61			
2400	8.00	20680	35.59	26.98			
2500	8.00	21480	35.91	27.32			
2600	8.00	22280	36.23	27.67			
2700	8.00	23080	36.53	27.99			
2800	8.00	23880	36.82	28.30			
2900	7.00	94680	61.24	28.60			
3000	7.00	95380	61.47	29.68			

THERMODYNAMIC PROPERTIES OF THE ELEMENTS

EUROPIUM

EUROPIUM Eu
REFERENCE STATE Solid from 298° to 1100°, Liquid from 1100° to 1700°, Ideal Monatomic Gas from 1700° to 3000°.

Gfw 152.0 GRAMS CAL./GFW.
$(H°_{298.15} - H°_0) =$

						FORMATION FROM REFERENCE STATE		
T TEMPERATURE °K	$C°_p$ HEAT CAPACITY CAL./DEG./GFW.	$H°_T - H°_{298.15}$ HEAT CONTENT CAL./GFW.	$S°_T$ ENTROPY CAL./DEG./GFW.	$-(F°_T - H°_{298.15})/T$ FREE ENERGY FUNCTION CAL./DEG./GFW.		HEAT $\Delta H°_f$ CAL./GFW.	FREE ENERGY $\Delta F°_f$ CAL./GFW.	$LOG_{10} K_p$
298	6.40	0	17.00	17.00				
300	6.40	12	17.04	17.00				
400	6.60	660	18.91	17.26				
500	6.80	1330	20.40	17.74				
600	7.00	2020	21.66	18.30				
700	7.20	2730	22.76	18.86				
800	7.40	3460	23.73	19.41				
900	7.60	4210	24.62	19.95				
1000	7.80	4980	25.43	20.45				
1100	8.00	8270	28.45	20.94				
1200	8.00	9070	29.15	21.60				
1300	8.00	9870	29.79	22.20				
1400	8.00	10670	30.38	22.76				
1500	8.00	11470	30.93	23.29				
1600	8.00	12270	31.45	23.79				
1700	8.00	13070	31.94	24.26				
1800	5.02	50670	54.04	25.89				
1900	5.05	51170	54.31	27.38				
2000	5.10	51680	54.57	28.73				
2100	5.16	52190	54.82	29.97				
2200	5.24	52710	55.06	31.11				
2300	5.34	53240	55.29	32.15				
2400	5.46	53780	55.52	33.12				
2500	5.61	54330	55.75	34.02				
2600	5.78	54900	55.97	34.86				
2700	5.98	55490	56.20	35.65				
2800	6.21	56100	56.42	36.39				
2900	6.46	56730	56.64	37.08				
3000	6.74	57390	56.86	37.73				

M.P. (1,100) °K
ΔH_m (2,500) CAL./GFW.

B.P. (1,700) °K
ΔH_v (42,000) CAL./GFW.

S.P. °K
ΔH_s CAL./GFW.

T.P. °K
ΔH_t CAL./GFW.

T.P. °K
ΔH_t CAL./GFW.

$T_c =$ °K
$P_c =$ ATM.

EUROPIUM

EUROPIUM Eu
Gfw 152.0
$(H°_{298.15} - H°_0) = 1,481$
IDEAL MONATOMIC GAS

Reference State for Calculating $\Delta H°_f$, $\Delta F°_f$, and $\log_{10} K_p$: Solid from 298° to 1100°, Liquid from 1100° to 1700°, Ideal Monatomic Gas from 1700° to 3000°.

T TEMPERATURE °K	$C°_P$ HEAT CAPACITY CAL./DEG./GFW.	$H°_T - H°_{298.15}$ HEAT CONTENT CAL./GFW.	$S°_T$ ENTROPY CAL./DEG./GFW.	$-(F°_T - H°_{298.15})/T$ FREE ENERGY FUNCTION CAL./DEG./GFW.	FORMATION FROM REFERENCE STATE HEAT $\Delta H°_f$ CAL./GFW.	FORMATION FROM REFERENCE STATE FREE ENERGY $\Delta F°_f$ CAL./GFW.	$\log_{10} K_P$
298	4.97	0	45.10	45.10	43200	34821	−25.525
400	4.97	506	46.56	45.30	43046	31986	−17.477
500	4.97	1003	47.67	45.67	42873	29238	−12.780
600	4.97	1500	48.57	46.07	42680	26534	−9.665
700	4.97	1996	49.34	46.49	42466	23860	−7.450
800	4.97	2493	50.00	46.89	42233	21217	−5.796
900	4.97	2990	50.59	47.27	41980	18607	−4.518
1000	4.97	3487	51.11	47.63	41707	16027	−3.503
1100	4.97	3984	51.58	47.96	38914	13471	−2.676
1200	4.97	4480	52.02	48.29	38610	11166	−2.033
1300	4.97	4977	52.41	48.59	38307	8901	−1.496
1400	4.97	5474	52.78	48.87	38004	6644	−1.037
1500	4.98	5972	53.13	49.15	37702	4402	−.641
1600	4.98	6470	53.45	49.41	37400	2200	−.300
1700	5.00	6969	53.75	49.66	37099	22	−.002
1800	5.02	7469	54.04	49.90	0	0	0
1900	5.05	7973	54.31	50.12	0	0	0
2000	5.10	8480	54.57	50.33	0	0	0
2100	5.16	8992	54.82	50.54	0	0	0
2200	5.24	9512	55.06	50.74	0	0	0
2300	5.34	10040	55.29	50.93	0	0	0
2400	5.46	10580	55.52	51.12	0	0	0
2500	5.61	11133	55.75	51.30	0	0	0
2600	5.78	11702	55.97	51.47	0	0	0
2700	5.98	12290	56.20	51.65	0	0	0
2800	6.21	12900	56.42	51.82	0	0	0
2900	6.46	13534	56.64	51.98	0	0	0
3000	6.74	14194	56.86	52.13	0	0	0

M.P.	°K	
ΔH_m	CAL./GFW.	
B.P.	°K	
ΔH_v	CAL./GFW.	
S.P.	°K	
ΔH_s	CAL./GFW.	
T.P.	°K	
ΔH_t	CAL./GFW.	
T.P.	°K	
ΔH_t	CAL./GFW.	
$T_c =$	°K	
$P_c =$	ATM.	

FLUORINE

Ideal Diatomic Gas from 298° to 3000°.

FLUORINE	F_2		
Gfw	38.00	GRAMS	
$(H°_{298.15} - H°_0)$ =	2,110.	CAL./GFW.	
M.P.	53.54	°K	
ΔH_m	122.0	CAL./GFW.	
B.P.	85.02	°K	
ΔH_v	1,562.	CAL./GFW.	
S.P.		°K	
ΔH_s		CAL./GFW.	
T.P.	45.55	°K	
ΔH_f	173.9	CAL./GFW.	
T.P.		°K	
ΔH_t		CAL./GFW.	
T_c =	144.2	°K	
P_c =	55.	ATM.	

T TEMPERATURE °K	$C°_p$ HEAT CAPACITY CAL./DEG./GFW.	$H°_T - H°_{298.15}$ HEAT CONTENT CAL./GFW.	$S°_T$ ENTROPY CAL./DEG./GFW.	$-(F°_T - H°_{298.15})/T$ FREE ENERGY FUNCTION CAL./DEG./GFW.	FORMATION FROM REFERENCE STATE		
					HEAT $\Delta H°_f$ CAL./GFW.	FREE ENERGY $\Delta F°_f$ CAL./GFW.	$LOG_{10} K_p$
298	7.49	0	48.45	48.45			
300	7.49	14	48.49	48.45			
400	7.89	783	50.71	48.76			
500	8.19	1588	52.50	49.33			
600	8.41	2418	54.01	49.98			
700	8.56	3268	55.32	50.66			
800	8.68	4130	56.47	51.31			
900	8.77	5003	57.50	51.95			
1000	8.84	5882	58.43	52.55			
1100	8.90	6772	59.27	53.12			
1200	8.94	7663	60.05	53.67			
1300	8.99	8559	60.77	54.19			
1400	9.02	9461	61.44	54.69			
1500	9.06	10364	62.06	55.16			
1600	9.08	11269	62.64	55.60			
1700	9.11	12180	63.20	56.04			
1800	9.13	13093	63.72	56.45			
1900	9.15	14008	64.21	56.84			
2000	9.18	14924	64.68	57.22			
2100	9.20	15843	65.13	57.59			
2200	9.22	16764	65.56	57.94			
2300	9.23	17686	65.97	58.29			
2400	9.25	18612	66.36	58.61			
2500	9.27	19535	66.74	58.93			
2600	9.28	20463	67.10	59.23			
2700	9.30	21392	67.45	59.53			
2800	9.32	22323	67.79	59.82			
2900	9.33	23257	68.12	60.11			
3000	9.34	24191	68.44	60.38			

FLUORINE

FLUORINE F
IDEAL MONATOMIC GAS

Gfw 19.00
$(H°_{298.15} - H°_0) = 1,558$ CAL./GFW

Reference State for Calculating $\Delta H_f°$, $\Delta F_f°$, and $\text{Log}_{10} K_p$: Ideal Diatomic Gas from 298° to 3000°.

T TEMPERATURE °K	$C_p°$ HEAT CAPACITY CAL./DEG./GFW	$H°_T - H°_{298.15}$ HEAT CONTENT CAL./GFW	$S°_T$ ENTROPY CAL./DEG./GFW	$-(F°_T - H°_{298.15})$ FREE ENERGY FUNCTION CAL./DEG./GFW	HEAT $\Delta H_f°$ CAL./GFW	FREE ENERGY $\Delta F_f°$ CAL./GFW	$\text{LOG}_{10} K_p$
298	5.44	0	37.92	37.92	18903	14821	-10.864
300	5.44	10	37.95	37.92	18906	14796	-10.779
400	5.36	550	39.51	38.14	19061	13401	-7.322
500	5.28	1082	40.69	38.53	19191	11971	-5.232
600	5.22	1607	41.65	38.98	19301	10517	-3.831
700	5.17	2126	42.45	39.42	19395	9042	-2.823
800	5.13	2641	43.14	39.84	19479	7559	-2.065
900	5.10	3153	43.74	40.24	19554	6063	-1.472
1000	5.08	3662	44.28	40.62	19624	4564	-.997
1100	5.07	4170	44.76	40.97	19687	3055	-.607
1200	5.05	4676	45.20	41.31	19747	1543	-.281
1300	5.04	5180	45.61	41.63	19803	17	-.002
1400	5.03	5684	45.98	41.92	19856	1508	.235
1500	5.02	6187	46.33	42.21	19908	3042	.443
1600	5.02	6689	46.65	42.47	19957	4571	.624
1700	5.01	7190	46.95	42.73	20003	6092	.783
1800	5.01	7691	47.24	42.97	20047	7637	.927
1900	5.01	8192	47.51	43.20	20091	9169	1.054
2000	5.00	8692	47.77	43.43	20133	10727	1.172
2100	5.00	9192	48.01	43.64	20173	12251	1.274
2200	5.00	9692	48.24	43.84	20213	13799	1.370
2300	4.99	10191	48.47	44.04	20251	15353	1.458
2400	4.99	10691	48.68	44.23	20288	16912	1.540
2500	4.99	11190	48.88	44.41	20325	18450	1.612
2600	4.99	11689	49.08	44.59	20360	20018	1.682
2700	4.99	12188	49.26	44.75	20395	21536	1.743
2800	4.98	12687	49.45	44.92	20428	23112	1.803
2900	4.98	13185	49.62	45.08	20459	24665	1.858
3000	4.98	13683	49.79	45.23	20490	26220	1.910

M.P. °K
ΔH_m CAL./GFW

B.P. °K
ΔH_v CAL./GFW

S.P. °K
ΔH_s CAL./GFW

T.P. °K
ΔH_t CAL./GFW

T.P. °K
ΔH_t CAL./GFW

$T_c =$ °K
$P_c =$ ATM.

FRANCIUM

FRANCIUM	Fr		Solid from 298° to 300°, Liquid from 300° to 950°, Ideal Monatomic Gas from 950° to 3000°.
REFERENCE STATE			

Gfw	223.*	GRAMS	
$(H°_{298.15} - H°_0) =$		CAL./GFW.	
M.P.	(300)	°K	
ΔH_m	(500)	CAL./GFW.	
B.P.	(950)	°K	
ΔH_v	(15,200)	CAL./GFW.	
S.P.		°K	
ΔH_s		CAL./GFW.	
T.P.		°K	
ΔH_t		CAL./GFW.	
T.P.		°K	
ΔH_t		CAL./GFW.	
$T_c =$		°K	
$P_c =$		ATM.	

*Isotope of Longest Known Half Life.

T TEMPERATURE °K	$C°_P$ HEAT CAPACITY CAL./DEG./GFW.	$H°_T - H°_{298.15}$ HEAT CONTENT CAL./GFW.	$S°_T$ ENTROPY CAL./DEG./GFW.	$-(F°_T - H°_{298.15})/T$ FREE ENERGY FUNCTION CAL./DEG./GFW.	FORMATION FROM REFERENCE STATE		
					HEAT $\Delta H°_f$ CAL./GFW.	FREE ENERGY $\Delta F°_f$ CAL./GFW.	$LOG_{10} K_P$
298	7.60	0	22.50	22.50			
300	7.60	14	22.55	22.51			
400	7.60	1270	26.40	23.23			
500	7.60	2030	28.10	24.04			
600	7.60	2790	29.48	24.83			
700	7.60	3550	30.66	25.59			
800	7.60	4310	31.67	26.29			
900	7.60	5070	32.57	26.94			
1000	4.97	20890	49.49	28.60			
1100	4.97	21380	49.97	30.54			
1200	4.97	21880	50.40	32.17			
1300	4.98	22380	50.80	33.59			
1400	4.98	22880	51.17	34.83			
1500	4.99	23380	51.51	35.93			
1600	5.01	23880	51.83	36.91			
1700	5.03	24380	52.14	37.80			
1800	5.07	24880	52.43	38.61			
1900	5.12	25390	52.70	39.34			
2000	5.18	25910	52.97	40.02			
2100	5.25	25430	53.22	41.12			
2200	5.37	26960	53.47	41.22			
2300	5.47	27500	53.71	41.76			
2400	5.61	28050	53.94	42.26			
2500	5.78	28620	54.18	42.74			
2600	5.94	29210	54.41	43.18			
2700	6.14	29810	54.63	43.59			
2800	6.33	30430	54.86	44.00			
2900	6.57	31080	55.09	44.38			
3000	6.80	31750	55.31	44.73			

FRANCIUM

FRANCIUM   **Fr**

IDEAL MONATOMIC GAS

Reference State for Calculating ΔH_f°, ΔF_f°, and $\text{Log}_{10} K_p$: Solid from 298° to 300°, Liquid from 300° to 950°, Ideal Monatomic Gas from 950° to 3000°.

Gfw 223.*   GRAMS
$(H^\circ_{298.15} - H^\circ_0) = 1,481$ CAL./GFW.

	T TEMPERATURE °K	C_p° HEAT CAPACITY CAL./DEG./GFW.	$H^\circ_T - H^\circ_{298.15}$ HEAT CONTENT CAL./GFW.	S°_T ENTROPY CAL./DEG./GFW.	$-(F^\circ_T - H^\circ_{298.15})/T$ FREE ENERGY FUNCTION CAL./DEG./GFW.	FORMATION FROM REFERENCE STATE HEAT ΔH_f° CAL./GFW.	FORMATION FROM REFERENCE STATE FREE ENERGY ΔF_f° CAL./GFW.	$\text{LOG}_{10} K_p$
M.P. °K	298	4.97	0	43.48	43.48	17429	11145	8.169
ΔH_m CAL./GFW.	300	4.97	9	43.51	43.48	17395	11107	8.092
	400	4.97	506	44.94	43.68	16636	9220	5.037
	500	4.97	1003	46.05	44.05	16373	7398	3.233
	600	4.97	1500	46.96	44.46	16110	5622	2.047
	700	4.97	1996	47.72	44.87	15846	3904	1.218
B.P. °K	800	4.97	2493	48.38	45.27	15583	2215	.605
ΔH_v CAL./GFW.	900	4.97	2990	48.97	45.65	15320	560	.135
	1000	4.97	3487	49.49	46.01	0	0	0
	1100	4.97	3984	49.97	46.35	0	0	0
	1200	4.97	4481	50.40	46.67	0	0	0
S.P. °K	1300	4.98	4978	50.80	46.98	0	0	0
ΔH_s CAL./GFW.	1400	4.98	5476	51.17	47.26	0	0	0
	1500	4.99	5975	51.51	47.53	0	0	0
	1600	5.01	6475	51.83	47.79	0	0	0
	1700	5.03	6977	52.14	48.04	0	0	0
T.P. °K	1800	5.07	7481	52.43	48.28	0	0	0
ΔH_t CAL./GFW.	1900	5.12	7990	52.70	48.50	0	0	0
	2000	5.18	8505	52.97	48.72	0	0	0
	2100	5.25	9027	53.22	48.93	0	0	0
	2200	5.37	9556	53.47	49.13	0	0	0
T.P. °K	2300	5.47	10098	53.71	49.32	0	0	0
ΔH_t CAL./GFW.	2400	5.61	10652	53.94	49.51	0	0	0
	2500	5.78	11221	54.18	49.70	0	0	0
	2600	5.94	11807	54.41	49.87	0	0	0
	2700	6.14	12411	54.63	50.04	0	0	0
$T_c =$ °K	2800	6.33	13034	54.86	50.21	0	0	0
$P_c =$ ATM.	2900	6.57	13679	55.09	50.38	0	0	0
	3000	6.80	14347	55.31	50.53	0	0	0

THERMODYNAMIC PROPERTIES OF THE ELEMENTS

GADOLINIUM

GADOLINIUM Gd

REFERENCE STATE

Gfw 157.26 GRAMS
$(H°_{298.15} - H°_0) = 2,172$ CAL./GFW.

Solid from 298° to 1600°, Liquid from 1600° to 3000°.

M.P.	(1,600) °K
ΔH_m	(3,700) CAL./GFW.
B.P.	(3,000) °K
ΔH_v	(74,500) CAL./GFW.
S.P.	°K
ΔH_s	CAL./GFW.
T.P.	°K
ΔH_t	CAL./GFW.
T.P.	°K
ΔH_t	CAL./GFW.
$T_c =$	°K
$P_c =$	ATM.

T TEMPERATURE °K	$C°_p$ HEAT CAPACITY CAL./DEG./GFW.	$H°_T - H°_{298.15}$ HEAT CONTENT CAL./GFW.	$S°_T$ ENTROPY CAL./DEG./GFW.	$-(F°_T - H°_{298.15})$ FREE ENERGY FUNCTION CAL./DEG./GFW.	FORMATION FROM REFERENCE STATE		
					HEAT $\Delta H°_f$ CAL./GFW.	FREE ENERGY $\Delta F°_f$ CAL./GFW.	LOG$_{10}$ K_p
298	8.72	0	15.77	15.77			
300	8.67	16	15.83	15.78			
400	7.00	780	18.06	16.11			
500	7.17	1480	19.60	16.64			
600	7.33	2200	20.92	17.26			
700	7.50	2940	22.07	17.87			
800	7.67	3700	23.08	18.46			
900	7.83	4480	23.99	19.02			
1000	8.00	5270	24.83	19.56			
1100	8.17	6080	25.60	20.08			
1200	8.33	6900	26.31	20.56			
1300	8.50	7740	26.99	21.04			
1400	8.67	8600	27.62	21.48			
1500	8.84	9480	28.23	21.91			
1600	8.00	14070	31.11	22.32			
1700	8.00	14870	31.60	22.86			
1800	8.00	15670	32.06	23.36			
1900	8.00	16470	32.49	23.83			
2000	8.00	17270	32.90	24.27			
2100	8.00	18070	33.29	24.69			
2200	8.00	18870	33.66	25.09			
2300	8.00	19670	34.02	25.47			
2400	8.00	20470	34.36	25.84			
2500	8.00	21270	34.68	26.18			
2600	8.00	22070	35.00	26.52			
2700	8.00	22870	35.30	26.83			
2800	8.00	23670	35.59	27.14			
2900	8.00	24470	35.87	27.44			
3000	8.00	25270	36.14	27.72			

GADOLINIUM

GADOLINIUM Gd
IDEAL MONATOMIC GAS

Gfw 157.26 GRAMS
$(H°_{298.15} - H°_0) = 1,825$ CAL./GFW.

Reference State for Calculating $\Delta H°_f$, $\Delta F°_f$, and $Log_{10} Kp$:
Solid from 298° to 1600°, Liquid from 1600° to 3000°.

							FORMATION FROM REFERENCE STATE		
	T	$C°_p$	$H°_T - H°_{298.15}$	$S°_T$	$-(F°_T - H°_{298.15})/T$		HEAT $\Delta H°_f$	FREE ENERGY $\Delta F°_f$	$LOG_{10} K_p$
	TEMPERATURE °K	HEAT CAPACITY CAL./DEG./GFW.	HEAT CONTENT CAL./GFW.	ENTROPY CAL./DEG./GFW.	FREE ENERGY FUNCTION CAL./DEG./GFW.		CAL./GFW.	CAL./GFW.	
	298	6.58	0	46.42	46.42		82500	73361	-53.777
	300	6.58	12	46.46	46.42		82496	73307	-53.408
	400	6.52	668	48.34	46.67		82388	70276	-38.400
	500	6.43	1316	49.79	47.16		82336	67241	-29.393
	600	6.32	1953	50.95	47.70		82253	64235	-23.399
	700	6.20	2579	51.92	48.24		82139	61244	-19.122
	800	6.08	3193	52.74	48.75		81993	58265	-15.918
	900	5.97	3795	53.45	49.24		81815	55301	-13.429
	1000	5.89	4388	54.07	49.69		81618	52378	-11.448
	1100	5.83	4973	54.63	50.11		81393	49460	-9.827
	1200	5.79	5554	55.14	50.52		81154	46558	-8.480
	1300	5.79	6133	55.60	50.89		80893	43700	-7.347
	1400	5.81	6713	56.03	51.24		80613	40839	-6.375
	1500	5.86	7296	56.43	51.57		80316	38016	-5.539
	1600	5.92	7885	56.81	51.89		76315	35195	-4.807
	1700	6.01	8481	57.17	52.19		76111	32642	-4.196
	1800	6.11	9087	57.52	52.48		75917	30089	-3.653
	1900	6.22	9703	57.85	52.75		75733	27549	-3.168
	2000	6.33	10330	58.17	53.01		75560	25020	-2.733
	2100	6.46	10970	58.49	53.27		75400	22480	-2.339
	2200	6.58	11622	58.79	53.51		75252	19966	-1.983
	2300	6.71	12286	59.08	53.74		75116	17478	-1.660
	2400	6.83	12963	59.37	53.97		74993	14969	-1.363
	2500	6.95	13652	59.65	54.19		74882	12457	-1.088
	2600	7.07	14353	59.93	54.41		74783	9965	-.837
	2700	7.18	15066	60.20	54.62		74696	7466	-.604
	2800	7.29	15790	60.46	54.83		74620	4984	-.389
	2900	7.39	16524	60.72	55.03		74554	2489	-.187
	3000	7.49	17268	60.97	55.22		74498	8	-.000

M.P. °K
ΔH_m CAL./GFW.
B.P. °K
ΔH_v CAL./GFW.
S.P. °K
ΔH_s CAL./GFW.
T.P. °K
ΔH_t CAL./GFW.
T.P. °K
ΔH_t CAL./GFW.
T_c = °K
P_c = ATM.

GALLIUM

GALLIUM Ga

Solid from 298° to 303°, Liquid from 303° to 2,510°, Ideal Monatomic Gas from 2,510° to 3000°.

		Gfw 69.72 GRAMS		
		$(H°_{298.15} - H°_0) = 1,331$ CAL./GFW.		
M.P.	303 °K			
ΔH_m	1,335 CAL./GFW.			
B.P.	2,510 °K			
ΔH_v	61,200 CAL./GFW.			

REFERENCE STATE

T TEMPERATURE °K	$C_p°$ HEAT CAPACITY CAL./DEG./GFW.	$H°_T - H°_{298.15}$ HEAT CONTENT CAL./GFW.	$S°_T$ ENTROPY CAL./DEG./GFW.	$-(F°_T - H°_{298.15})/T$ FREE ENERGY FUNCTION CAL./DEG./GFW.	HEAT $\Delta H°_f$ CAL./GFW.	FREE ENERGY $\Delta F°_f$ CAL./GFW.	LOG$_{10}$ K_p
298	6.23	0	9.82	9.82			
300	6.24	11	9.86	9.83			
400	6.65	2010	16.18	11.16			
500	6.65	2675	17.66	12.31			
600	6.65	3340	18.87	13.31			
700	6.65	4005	19.90	14.18			
800	6.65	4670	20.79	14.96			
900	6.65	5335	21.57	15.65			
1000	6.65	6000	22.27	16.27			
1100	6.65	6665	22.90	16.85			
1200	6.65	7330	23.48	17.38			
1300	6.65	7995	24.01	17.86			
1400	6.65	8660	24.51	18.33			
1500	6.65	9325	24.97	18.76			
1600	6.65	9990	25.39	19.15			
1700	6.65	10655	25.80	19.54			
1800	6.65	11320	26.18	19.90			
1900	6.65	11985	26.54	20.24			
2000	6.65	12650	26.88	20.56			
2100	6.65	13315	27.20	20.86			
2200	6.65	13980	27.51	21.16			
2300	6.55	14645	27.81	21.45			
2400	6.65	15310	28.09	21.72			
2500	6.65	15975	28.36	21.97			
2600	5.07	77670	52.91	23.04			
2700	5.06	78180	53.10	24.15			
2800	5.06	78680	53.29	25.19			
2900	5.05	79190	53.47	26.17			
3000	5.05	79690	53.64	27.08			

S.P. °K
ΔH_s CAL./GFW.

T.P. °K
ΔH_t CAL./GFW.

T.P. °K
ΔH_t CAL./GFW.

T_c = °K
P_c = ATM.

GALLIUM

GALLIUM	Ga	Reference State for Calculating ΔH_f°, ΔF_f°, and $Log_{10}K_p$: Solid from 298° to 303°, Liquid from 303° to 2,510°, Ideal Monatomic Gas from 2,510° to 3000°.						
Gfw 69.72	GRAMS						FORMATION FROM REFERENCE STATE	
IDEAL MONATOMIC GAS		T TEMPERATURE °K	C_p° HEAT CAPACITY CAL./DEG./GFW.	$H_T^\circ - H_{298.15}^\circ$ HEAT CONTENT CAL./GFW.	S_T° ENTROPY CAL./DEG./GFW.	$-(F_T^\circ - H_{298.15}^\circ)/T$ FREE ENERGY FUNCTION CAL./DEG./GFW.	HEAT ΔH_f° CAL./GFW.	FREE ENERGY ΔF_f° CAL./GFW.	$LOG_{10} K_p$
$(H_{298.15}^\circ - H_0^\circ) = 1,566$	CAL./GFW.	298	6.06	0	40.38	40.38	65000	55888	−40.968
		300	6.07	11	40.41	40.38	65000	55835	−40.679
M.P.	°K	400	6.45	641	42.22	40.62	63631	53215	−29.077
ΔH_m	CAL./GFW.	500	6.45	1288	43.67	41.10	63613	50608	−22.122
		600	6.29	1926	44.83	41.62	63586	48010	−17.489
		700	6.09	2545	45.78	42.15	63540	45424	−14.183
B.P.	°K	800	5.91	3145	46.59	42.66	63475	42835	−11.702
ΔH_v	CAL./GFW.	900	5.75	3728	47.27	43.13	63393	40263	−9.777
		1000	5.63	4297	47.87	43.58	63297	37697	−8.239
		1100	5.53	4854	48.40	43.99	63189	35139	−6.982
		1200	5.44	5403	48.88	44.38	63073	32593	−5.936
S.P.	°K	1300	5.38	5944	49.31	44.74	62949	30059	−5.053
ΔH_s	CAL./GFW.	1400	5.32	6479	49.71	45.09	62819	27539	−4.299
		1500	5.28	7009	50.08	45.41	62684	25019	−3.645
		1600	5.24	7535	50.42	45.72	62545	22497	−3.072
		1700	5.21	8057	50.73	46.00	62402	20021	−2.573
T.P.	°K	1800	5.18	8577	51.03	46.27	62257	17527	−2.127
ΔH_t	CAL./GFW.	1900	5.16	9094	51.31	46.53	62109	15046	−1.730
		2000	5.14	9610	51.57	46.77	61960	12580	−1.374
		2100	5.13	10123	51.82	47.00	61808	10106	−1.051
		2200	5.11	10635	52.06	47.23	61655	7645	−.759
T.P.	°K	2300	5.10	11145	52.29	47.45	61500	5196	−.493
ΔH_t	CAL./GFW.	2400	5.09	11655	52.51	47.66	61345	2737	−.249
		2500	5.08	12163	52.71	47.85	61188	313	−.027
		2600	5.07	12671	52.91	48.04	0	0	0
$T_c =$	°K	2700	5.06	13177	53.10	48.22	0	0	0
		2800	5.06	13683	53.29	48.41	0	0	0
$P_c =$	ATM.	2900	5.05	14189	53.47	48.58	0	0	0
		3000	5.05	14693	53.64	48.75	0	0	0

GERMANIUM

GERMANIUM Ge Solid from 298° to 1210.4°, Liquid from 1210.4° to 3000°.

REFERENCE STATE

Gfw	72.60	GRAMS		
$(H°_{298.15} - H°_0)$ =	1,105	CAL./GFW.		
M.P.	1,210.4	°K		
ΔH_m	7,600.	CAL./GFW.		
B.P.	3,100.	°K		
ΔH_v	79,900.	CAL./GFW.		
S.P.		°K		
ΔH_s		CAL./GFW.		
T.P.		°K		
ΔH_t		CAL./GFW.		
T.P.		°K		
ΔH_t		CAL./GFW.		
T_c =		°K		
P_c =		ATM.		

T TEMPERATURE °K	$C°_P$ HEAT CAPACITY CAL./DEG./GFW.	$H°_T - H°_{298.15}$ HEAT CONTENT CAL./GFW.	$S°_T$ ENTROPY CAL./DEG./GFW.	$-(F°_T - H°_{298.15})$ FREE ENERGY FUNCTION CAL./DEG./GFW.	FORMATION FROM REFERENCE STATE		
					HEAT $\Delta H°_f$ CAL./GFW.	FREE ENERGY $\Delta F°_f$ CAL./GFW.	LOG$_{10}$ K$_p$
298	5.59	0	7.43	7.43			
300	5.60	10	7.46	7.43			
400	5.94	588	9.12	7.65			
500	6.16	1194	10.47	8.09			
600	6.32	1820	11.61	8.58			
700	6.45	2460	12.60	9.09			
800	6.56	3110	13.46	9.58			
900	6.66	3770	14.24	10.06			
1000	6.77	4440	14.95	10.51			
1100	6.87	5120	15.60	10.95			
1200	6.97	5810	16.20	11.36			
1300	7.00	14110	23.04	12.19			
1400	7.00	14810	23.56	12.99			
1500	7.00	15510	24.04	13.70			
1600	7.00	16210	24.50	14.37			
1700	7.00	16910	24.92	14.98			
1800	7.00	17610	25.32	15.54			
1900	7.00	18310	25.70	16.07			
2000	7.00	19010	26.06	16.55			
2100	7.00	19710	26.40	17.01			
2200	7.00	20410	26.73	17.45			
2300	7.00	21110	27.04	17.86			
2400	7.00	21810	27.33	18.24			
2500	7.00	22510	27.62	18.62			
2600	7.00	23210	27.89	18.97			
2700	7.00	23910	28.16	19.30			
2800	7.00	24610	28.41	19.62			
2900	7.00	25310	28.66	19.93			
3000	7.00	26010	28.90	20.23			

GERMANIUM

GERMANIUM Ge Reference State for Calculating ΔH_f°, ΔF_f°, and $\log_{10} K_P$:
IDEAL MONATOMIC GAS Solid from 298° to 1210.4°, Liquid from 1210.4° to 3000°.

G/W 72.60 GRAMS
$(H^\circ_{298.15} - H^\circ_0) = 1{,}768$ CAL./GFW

	T TEMPERATURE °K	C_P° HEAT CAPACITY CAL./DEG./GFW	$H_T^\circ - H_{298.15}^\circ$ HEAT CONTENT CAL./GFW	S_T° ENTROPY CAL./DEG./GFW	$-(F_T^\circ - H_{298.15}^\circ)/T$ FREE ENERGY FUNCTION CAL./DEG./GFW	FORMATION FROM REFERENCE STATE		
						HEAT ΔH_f° CAL./GFW	FREE ENERGY ΔF_f° CAL./GFW	$\log_{10} K_P$
	298	7.34	0	40.10	40.10	90000	80259	−58.833
	300	7.35	14	40.15	40.11	90004	80197	−58.428
	400	7.43	756	42.28	40.39	90168	76904	−42.021
	500	7.26	1491	43.93	40.95	90297	73567	−32.158
	600	6.99	2204	45.23	41.56	90384	70212	−25.576
	700	6.72	2889	46.28	42.16	90429	66853	−20.874
M.P.	800	6.46	3548	47.16	42.73	90438	63478	−17.342
ΔH_m	900	6.24	4183	47.91	43.27	90413	60110	−14.597
	1000	6.06	4797	48.56	43.77	90357	57747	−12.403
B.P.	1100	5.91	5395	49.13	44.23	90275	53392	−10.608
ΔH_v	1200	5.80	5981	49.64	44.66	90171	50043	−9.114
	1300	5.71	6556	50.10	45.06	82446	47268	−7.947
S.P.	1400	5.65	7124	50.52	45.44	82314	44570	−6.958
ΔH_s	1500	5.60	7686	50.91	45.79	82176	41871	−6.101
	1600	5.57	8244	51.27	46.12	82034	39202	−5.354
	1700	5.54	8800	51.60	46.43	81890	36534	−4.696
T.P.	1800	5.53	9353	51.92	46.73	81743	33863	−4.111
ΔH_t	1900	5.53	9906	52.22	47.01	81596	31208	−3.589
	2000	5.52	10458	52.50	47.28	81448	28568	−3.121
	2100	5.53	11011	52.77	47.53	81301	25924	−2.697
T.P.	2200	5.53	11564	53.03	47.78	81154	23294	−2.314
ΔH_t	2300	5.54	12117	53.28	48.02	81007	20655	−1.962
	2400	5.55	12671	53.51	48.24	80861	18029	−1.641
	2500	5.55	13226	53.74	48.45	80716	15416	−1.347
	2600	5.56	13782	53.96	48.66	80572	12790	−1.075
	2700	5.57	14338	54.17	48.86	80428	10201	−.825
$T_c =$ °K	2800	5.57	14896	54.37	49.05	80286	7598	−.593
$P_c =$ ATM.	2900	5.58	15453	54.56	49.24	80143	5033	−.379
	3000	5.59	16012	54.75	49.42	80002	2452	−.178

THERMODYNAMIC PROPERTIES OF THE ELEMENTS

GOLD

GOLD   **Au**

Gfw 197.0 GRAMS

Solid from 298° to 1336°, Liquid from 1336° to 2980°, Ideal Monatomic Gas from 2980° to 3000°.

REFERENCE STATE

$(H°_{298.15} - H°_0) = 1,434$ CAL./GFW.

M.P.	1,336	°K
ΔH$_m$	2,955	CAL./GFW.
B.P.	2,980	°K
ΔH$_v$	77,540	CAL./GFW.
S.P.		°K
ΔH$_s$		CAL./GFW.
T.P.		°K
ΔH$_t$		CAL./GFW.
T.P.		°K
ΔH$_t$		CAL./GFW.
T$_c$ =		°K
P$_c$ =		ATM.

T TEMPERATURE °K	C°$_P$ HEAT CAPACITY CAL./DEG./GFW.	H°$_T$ − H°$_{298.15}$ HEAT CONTENT CAL./GFW.	S°$_T$ ENTROPY CAL./DEG./GFW.	−(F°$_T$ − H°$_{298.15}$)/T FREE ENERGY FUNCTION CAL./DEG./GFW.	FORMATION FROM REFERENCE STATE		
					HEAT ΔH°$_f$ CAL./GFW.	FREE ENERGY ΔF°$_f$ CAL./GFW.	LOG$_{10}$ K$_P$
298	6.07	0	11.32	11.32			
300	6.07	11	11.36	11.33			
400	6.18	624	13.12	11.56			
500	6.28	1245	14.51	12.02			
600	6.40	1880	15.66	12.53			
700	6.52	2530	16.67	13.06			
800	6.65	3180	17.53	13.56			
900	6.78	3850	18.32	14.05			
1000	6.90	4530	19.04	14.51			
1100	7.02	5220	19.70	14.96			
1200	7.15	5930	20.32	15.38			
1300	7.27	6660	20.90	15.78			
1400	7.00	10330	23.64	16.27			
1500	7.00	11030	24.12	16.77			
1600	7.00	11730	24.57	17.24			
1700	7.00	12430	24.99	17.68			
1800	7.00	13130	25.39	18.10			
1900	7.00	13830	25.77	18.50			
2000	7.00	14530	26.13	18.87			
2100	7.00	15230	26.47	19.22			
2200	7.00	15930	26.80	19.56			
2300	7.00	16630	27.11	19.88			
2400	7.00	17330	27.41	20.19			
2500	7.00	18030	27.69	20.48			
2600	7.00	18730	27.97	20.77			
2700	7.00	19430	28.23	21.04			
2800	7.00	20130	28.49	21.31			
2900	7.00	20830	28.74	21.56			
3000	6.30	99060	54.97	21.95			

GOLD

GOLD Au
IDEAL MONATOMIC GAS

Gfw 197.0 GRAMS
$(H°_{298.15} - H°_0) = 1,481$ CAL./GFW.

Reference State for Calculating $\Delta H_f°$, $\Delta F_f°$, and $\text{Log}_{10} K_p$:
Solid from 298° to 1336°, Liquid from 1336° to 2980°, Ideal Monatomic Gas from 2980° to 3000°.

T TEMPERATURE °K	$C_p°$ HEAT CAPACITY CAL./DEG./GFW.	$H°_T - H°_{298.15}$ HEAT CONTENT CAL./GFW.	$S°_T$ ENTROPY CAL./DEG./GFW.	$-(F°_T - H°_{298.15})/T$ FREE ENERGY FUNCTION CAL./DEG./GFW.	FORMATION FROM REFERENCE STATE		
					HEAT $\Delta H_f°$ CAL./GFW.	FREE ENERGY $\Delta F_f°$ CAL./GFW.	$\text{LOG}_{10} K_p$
298	4.97	0	43.12	43.12	84700	75219	-55.139
300	4.97	9	43.15	43.12	84698	75161	-54.759
400	4.97	506	44.58	43.32	84582	71998	-39.341
500	4.97	1003	45.69	43.69	84458	68868	-30.104
600	4.97	1500	46.59	44.09	84320	65762	-23.955
700	4.97	1996	47.36	44.51	84166	62683	-19.572
800	4.97	2493	48.02	44.91	84013	59621	-16.289
900	4.97	2990	48.61	45.29	83840	56579	-13.740
1000	4.97	3487	49.13	45.65	83657	53567	-11.708
1100	4.97	3984	49.60	45.98	83464	50574	-10.049
1200	4.98	4482	50.04	46.31	83252	47588	-8.667
1300	4.99	4980	50.43	46.60	83020	44631	-7.503
1400	5.01	5480	50.81	46.90	79850	41812	-6.527
1500	5.04	5983	51.15	47.17	79653	39108	-5.698
1600	5.07	6488	51.48	47.43	79458	36402	-4.972
1700	5.12	6998	51.79	47.68	79268	33708	-4.333
1800	5.18	7513	52.08	47.91	79083	31041	-3.768
1900	5.25	8034	52.36	48.14	78904	28383	-3.264
2000	5.32	8562	52.63	48.35	78732	25732	-2.811
2100	5.41	9099	52.90	48.57	78569	23066	-2.400
2200	5.50	9644	53.15	48.77	78414	20444	-2.030
2300	5.59	10198	53.40	48.97	78268	17801	-1.691
2400	5.69	10762	53.64	49.16	78132	15180	-1.382
2500	5.80	11337	53.87	49.34	78007	12557	-1.097
2600	5.90	11922	54.10	49.52	77892	9954	-.836
2700	6.00	12517	54.32	49.69	77787	7344	-.594
2800	6.11	13122	54.54	49.86	77692	4752	-.370
2900	6.21	13738	54.76	50.03	77608	2150	-.162
3000	6.30	14364	54.97	50.19	0	0	0

M.P. °K
ΔH_m CAL./GFW.

B.P. °K
ΔH_v CAL./GFW.

S.P. °K
ΔH_s CAL./GFW.

T.P. °K
ΔH_t CAL./GFW.

T.P. °K
ΔH_t CAL./GFW.

$T_c =$ °K
$P_c =$ ATM.

THERMODYNAMIC PROPERTIES OF THE ELEMENTS

HAFNIUM

HAFNIUM Hf Solid from 298° to 2250°, Liquid from 2250° to 3000°.

Gfw 178.50 GRAMS
$(H°_{298.15} - H°_0) = 1,448$ CAL./GFW.

REFERENCE STATE

M.P.	2,250°	°K
ΔH_m	(5,200)	CAL./GFW.
B.P.	(5,500)	°K
ΔH_v	(158,000)	CAL./GFW.
S.P.		°K
ΔH_s		CAL./GFW.
T.P.		°K
ΔH_t		CAL./GFW.
T.P.		°K
ΔH_t		CAL./GFW.
$T_c =$		°K
$P_c =$		ATM.

T TEMPERATURE °K	$C°_P$ HEAT CAPACITY CAL./DEG./GFW.	$H°_T - H°_{298.15}$ HEAT CONTENT CAL./GFW.	$S°_T$ ENTROPY CAL./DEG./GFW.	$-(F°_T - H°_{298.15})/T$ FREE ENERGY FUNCTION CAL./DEG./GFW.	FORMATION FROM REFERENCE STATE		
					HEAT $\Delta H°_f$ CAL./GFW.	FREE ENERGY $\Delta F°_f$ CAL./GFW.	LOG$_{10}$ K$_P$
298	6.10	0	10.91	10.91			
300	6.10	11	10.95	10.92			
400	6.22	627	12.72	11.16			
500	6.34	1255	14.12	11.61			
600	6.46	1900	15.29	12.13			
700	6.58	2550	16.29	12.65			
800	6.70	3210	17.18	13.17			
900	6.82	3890	17.98	13.66			
1000	6.94	4575	18.70	14.13			
1100	7.06	5275	19.37	14.58			
1200	7.18	5990	19.99	15.00			
1300	7.30	6710	20.57	15.41			
1400	7.42	7450	21.11	15.79			
1500	7.54	8200	21.63	16.17			
1600	7.66	8960	22.12	16.52			
1700	7.78	9730	22.59	16.87			
1800	7.90	10510	23.04	17.21			
1900	8.02	11310	23.47	17.52			
2000	8.14	12120	23.88	17.82			
2100	8.26	12940	24.28	18.12			
2200	8.38	13770	24.67	18.42			
2300	8.00	19790	27.34	18.74			
2400	8.00	20590	27.68	19.11			
2500	8.00	21390	28.01	19.46			
2600	8.00	22190	28.33	19.80			
2700	8.00	22990	28.63	20.12			
2800	8.00	23790	28.92	20.43			
2900	8.00	24590	29.20	20.73			
3000	8.00	25390	29.47	21.01			

*D. K. Deardorf and E. T. Hayes, J. Metals 8, 509 (1956) have just reported what is probably the best determination of the melting point of hafnium as 2495° ± 30° K. The transition temperature remains uncertain.

HAFNIUM

HAFNIUM   Hf   Reference State for Calculating ΔH_f°, ΔF_f°, and $\log_{10} K_p$:
IDEAL MONATOMIC GAS   Solid from 298° to 2250°, Liquid from 2250° to 3000°.

Gfw 178.50   GRAMS   CAL./GFW.
$(H_{298.15}^\circ - H_0^\circ)$ = 1,481

T TEMPERATURE °K	C_p° HEAT CAPACITY CAL./DEG./GFW.	$H_T^\circ - H_{298.15}^\circ$ HEAT CONTENT CAL./GFW.	S_T° ENTROPY CAL./DEG./GFW.	$-(F_T^\circ - H_{298.15}^\circ)/T$ FREE ENERGY FUNCTION CAL./DEG./GFW.	FORMATION FROM REFERENCE STATE		
					HEAT ΔH_f° CAL./GFW.	FREE ENERGY ΔF_f° CAL./GFW.	LOG$_{10}$ K$_p$
298	4.97	0	44.64	44.64	168000	157943	-115.780
300	4.97	9	44.67	44.64	167998	157882	-115.026
400	5.01	508	46.11	44.84	167881	154525	-84.435
500	5.11	1013	47.24	45.22	167758	151198	-66.093
600	5.29	1533	48.18	45.63	167633	147899	-53.876
700	5.50	2072	49.01	46.05	167522	144618	-45.155
800	5.73	2634	49.76	46.47	167424	141360	-38.620
900	5.97	3219	50.45	46.88	167329	138106	-33.539
1000	6.20	3827	51.09	47.27	167252	134862	-29.476
1100	6.41	4458	51.69	47.64	167183	131631	-26.155
1200	6.59	5108	52.26	48.01	167118	128394	-23.385
1300	6.76	5776	52.79	48.35	167066	125180	-21.046
1400	6.90	6459	53.30	48.69	167009	121943	-19.037
1500	7.01	7155	53.78	49.01	166955	118730	-17.300
1600	7.10	7861	54.24	49.33	166901	115509	-15.777
1700	7.17	8574	54.67	49.63	166844	112308	-14.438
1800	7.21	9293	55.08	49.92	166783	109111	-13.247
1900	7.24	10016	55.47	50.20	166706	105906	-12.181
2000	7.24	10740	55.84	50.47	166620	102700	-11.222
2100	7.24	11464	56.19	50.74	166524	99513	-10.356
2200	7.22	12187	56.53	51.00	166417	96325	-9.568
2300	7.20	12908	56.85	51.24	161118	93245	-8.860
2400	7.17	13627	57.16	51.49	161037	90285	-8.221
2500	7.13	14341	57.45	51.72	160951	87351	-7.636
2600	7.09	15052	57.73	51.95	160862	84422	-7.096
2700	7.05	15759	57.99	52.16	160769	81497	-6.596
2800	7.01	16462	58.25	52.38	160672	78548	-6.130
2900	6.97	17161	58.50	52.59	160571	75601	-5.697
3000	6.93	17855	58.73	52.78	160465	72685	-5.295

M.P. °K ΔH_m CAL./GFW.
B.P. °K ΔH_v CAL./GFW.
S.P. °K ΔH_s CAL./GFW.
T.P. °K ΔH_t CAL./GFW.
T.P. °K ΔH_t CAL./GFW.
T_c =   °K
P_c =   ATM.

THERMODYNAMIC PROPERTIES OF THE ELEMENTS

HELIUM

HELIUM He
Ideal Monatomic Gas from 298° to 3000°.

REFERENCE STATE

Gfw	4.003	GRAMS
$(H°_{298.15} - H°_0)$ =	1,481	CAL./GFW.
M.P. $\begin{pmatrix} 103 \\ ATM \end{pmatrix}$	3.5	°K
ΔH_m	5.	CAL./GFW.
B.P.	4.22	°K
ΔH_v	20.	CAL./GFW.
S.P.		°K
ΔH_s		CAL./GFW.
T.P.	2.19	°K
ΔH_t	0.	CAL./GFW.
T.P.		°K
ΔH_t		CAL./GFW.
T_c =	5.3	°K
P_c =	2.26	ATM.

T TEMPERATURE °K	$C°_p$ HEAT CAPACITY CAL./DEG./GFW.	$H°_T - H°_{298.15}$ HEAT CONTENT CAL./GFW.	$S°_T$ ENTROPY CAL./DEG./GFW.	$-(F°_T - H°_{298.15})$ FREE ENERGY FUNCTION CAL./DEG./GFW.	HEAT $\Delta H°_f$ CAL./GFW.	FREE ENERGY $\Delta F°_f$ CAL./GFW.	LOG$_{10}$ K_p
298	4.97	0	30.13	30.13			
300	4.97	9	30.16	30.13			
400	4.97	506	31.59	30.33			
500	4.97	1003	32.69	30.69			
600	4.97	1500	33.60	31.10			
700	4.97	1996	34.36	31.51			
800	4.97	2493	35.03	31.92			
900	4.97	2990	35.61	32.29			
1000	4.97	3487	36.14	32.66			
1100	4.97	3984	36.61	32.99			
1200	4.97	4480	37.04	33.31			
1300	4.97	4977	37.44	33.62			
1400	4.97	5474	37.81	33.90			
1500	4.97	5971	38.15	34.17			
1600	4.97	6468	38.47	34.43			
1700	4.97	6964	38.77	34.68			
1800	4.97	7461	39.06	34.92			
1900	4.97	7958	39.33	35.15			
2000	4.97	8455	39.58	35.36			
2100	4.97	8952	39.82	35.56			
2200	4.97	9448	40.06	35.77			
2300	4.97	9945	40.28	35.96			
2400	4.97	10442	40.49	36.14			
2500	4.97	10939	40.69	36.32			
2600	4.97	11436	40.88	36.49			
2700	4.97	11932	41.07	36.66			
2800	4.97	12429	41.25	36.82			
2900	4.97	12926	41.43	36.98			
3000	4.97	13423	41.60	37.13			

HOLMIUM

HOLMIUM	Ho							
Gfw 164.94	GRAMS							
REFERENCE STATE		Solid from 298° to 1773°, Liquid from 1773° to 3000°.						

$(H°_{298.15} - H°_0)$ =	CAL./GFW.							
M.P. (1,773)	°K							
ΔH_m (4,100)	CAL./GFW.							
B.P. (2,600)	°K							
ΔH_v (60,000)	CAL./GFW.							
S.P.	°K							
ΔH_s	CAL./GFW.							
T.P.	°K							
ΔH_t	CAL./GFW.							
T.P.	°K							
ΔH_t	CAL./GFW.							
T_c =	°K							
P_c =	ATM.							

					FORMATION FROM REFERENCE STATE		
T TEMPERATURE °K	$C°_P$ HEAT CAPACITY CAL./DEG./GFW.	$H°_T - H°_{298.15}$ HEAT CONTENT CAL./GFW.	$S°_T$ ENTROPY CAL./DEG./GFW.	$-(F°_T - H°_{298.15})/T$ FREE ENERGY FUNCTION CAL./DEG./GFW.	HEAT $\Delta H°_f$ CAL./GFW.	FREE ENERGY $\Delta F°_f$ CAL./GFW.	$LOG_{10} K_p$
298	6.51	0	17.77	17.77			
300	6.51	11	17.81	17.78			
400	6.68	670	19.71	18.04			
500	6.85	1350	21.21	18.51			
600	7.02	2040	22.48	19.08			
700	7.19	2750	23.57	19.65			
800	7.36	3480	24.54	20.19			
900	7.53	4220	25.42	20.74			
1000	7.70	4985	26.22	21.24			
1100	7.87	5760	26.96	21.73			
1200	8.04	6560	27.66	22.20			
1300	8.21	7370	28.31	22.65			
1400	8.38	8200	28.92	23.07			
1500	8.55	9050	29.51	23.48			
1600	8.72	9910	30.06	23.87			
1700	8.89	10790	30.60	24.26			
1800	8.00	15760	33.41	24.66			
1900	8.00	16560	33.84	25.13			
2000	8.00	17360	34.26	25.58			
2100	8.00	18160	34.65	26.01			
2200	8.00	18960	35.02	26.41			
2300	8.00	19760	35.37	26.78			
2400	8.00	20560	35.71	27.15			
2500	8.00	21360	36.04	27.50			
2600	8.00	22160	36.35	27.83			
2700	6.00	82760	59.66	29.01			
2800	6.00	83360	59.88	30.11			
2900	6.00	83960	60.09	31.14			
3000	6.00	84560	60.29	32.11			

HYDROGEN

HYDROGEN H₂   Ideal Diatomic Gas from 298° to 3000°.

HYDROGEN	H₂		
Gfw	2.0160	GRAMS	
REFERENCE STATE			
$(H^\circ_{298.15} - H^\circ_0) =$	2,024	CAL./GFW.	
M.P.	13.96	°K	
ΔH_m	28.0	CAL./GFW.	
B.P.	20.39	°K	
ΔH_v	215.8	CAL./GFW.	
S.P.		°K	
ΔH_s		CAL./GFW.	
T.P.		°K	
ΔH_t		CAL./GFW.	
T.P.		°K	
ΔH_t		CAL./GFW.	
T_c =	33.24	°K	
P_c =	12.80	ATM.	

T TEMPERATURE °K	C°_p HEAT CAPACITY CAL./DEG./GFW.	H°_T − H°_298.15 HEAT CONTENT CAL./GFW.	S°_T ENTROPY CAL./DEG./GFW.	−(F°_T − H°_298.15) FREE ENERGY FUNCTION CAL./DEG./GFW.	FORMATION FROM REFERENCE STATE		
					HEAT ΔH°_f CAL./GFW.	FREE ENERGY ΔF°_f CAL./GFW.	LOG₁₀ K_p
298	6.89	0	31.21	31.21			
300	6.89	12	31.25	31.21			
400	6.98	706	33.25	31.49			
500	6.99	1406	34.81	32.00			
600	7.01	2105	36.08	32.58			
700	7.04	2808	37.17	33.16			
800	7.08	3514	38.11	33.72			
900	7.14	4224	38.95	34.26			
1000	7.22	4942	39.70	34.76			
1100	7.32	5669	40.40	35.25			
1200	7.43	6407	41.04	35.71			
1300	7.52	7154	41.64	36.14			
1400	7.62	7911	42.20	36.55			
1500	7.72	8678	42.73	36.95			
1600	7.82	9456	43.23	37.32			
1700	7.91	10242	43.70	37.68			
1800	8.00	11038	44.16	38.03			
1900	8.10	11842	44.59	38.36			
2000	8.20	12658	45.01	38.69			
2100	8.29	13482	45.41	38.99			
2200	8.38	14316	45.80	39.30			
2300	8.46	15158	46.18	39.59			
2400	8.53	16007	46.54	39.88			
2500	8.60	16864	46.89	40.15			
2600	8.67	17727	47.23	40.42			
2700	8.73	18598	47.55	40.67			
2800	8.78	19473	47.87	40.92			
2900	8.82	20353	48.18	41.17			
3000	8.86	21237	48.48	41.41			

HYDROGEN

HYDROGEN H
IDEAL MONATOMIC GAS

Gfw 1.0080 GRAMS
$(H^°_{298.15} - H^°_0) = 1,481$ CAL./GFW.

Reference State for Calculating $\Delta H_f^°$, $\Delta F_f^°$, and $\text{Log}_{10} K_p$: Ideal Diatomic Gas from 298° to 3000°.

T TEMPERATURE °K	$C_p^°$ HEAT CAPACITY CAL./DEG./GFW.	$H_T^° - H_{298.15}^°$ HEAT CONTENT CAL./GFW.	$S_T^°$ ENTROPY CAL./DEG./GFW.	$-(F_T^° - H_{298.15}^°)/T$ FREE ENERGY FUNCTION CAL./DEG./GFW.	FORMATION FROM REFERENCE STATE		
					HEAT $\Delta H_f^°$ CAL./GFW.	FREE ENERGY $\Delta F_f^°$ CAL./GFW.	$\text{LOG}_{10} K_p$
298	4.97	0	27.39	27.39	52090	48578	-35.610
300	4.97	9	27.42	27.39	52093	48556	-35.375
400	4.97	506	28.85	27.59	52243	47355	-25.875
500	4.97	1003	29.96	27.96	52390	46115	-20.158
600	4.97	1500	30.87	28.37	52537	44839	-16.333
700	4.97	1996	31.63	28.78	52682	43554	-13.599
800	4.97	2493	32.30	29.19	52826	42234	-11.538
900	4.97	2990	32.88	29.56	52968	40908	-9.934
1000	4.97	3487	33.40	29.92	53106	39556	-8.645
1100	4.97	3983	33.88	30.26	53238	38190	-7.588
1200	4.97	4480	34.31	30.58	53366	36818	-6.706
1300	4.97	4977	34.71	30.89	53490	35433	-5.957
1400	4.97	5474	35.08	31.17	53608	34036	-5.313
1500	4.97	5971	35.42	31.44	53722	32647	-4.756
1600	4.97	6468	35.74	31.70	53830	31238	-4.266
1700	4.97	6964	36.04	31.95	53933	29810	-3.832
1800	4.97	7461	36.32	32.18	54032	28400	-3.448
1900	4.97	7958	36.59	32.41	54127	26976	-3.102
2000	4.97	8455	36.85	32.63	54216	25536	-2.790
2100	4.97	8952	37.09	32.83	54301	24103	-2.508
2200	4.97	9449	37.32	33.03	54381	22657	-2.250
2300	4.97	9945	37.54	33.22	54506	21271	-2.021
2400	4.97	10442	37.75	33.40	54528	19776	-1.800
2500	4.97	10939	37.96	33.59	54597	18322	-1.601
2600	4.97	11436	38.15	33.76	54662	16884	-1.419
2700	4.97	11933	38.34	33.93	54724	15412	-1.247
2800	4.97	12429	38.52	34.09	54782	13958	-1.089
2900	4.97	12926	38.70	34.25	54839	12470	-.940
3000	4.97	13423	38.86	34.39	54895	11035	-.803

M.P. °K
ΔH_m CAL./GFW.

B.P. °K
ΔH_v CAL./GFW.

S.P. °K
ΔH_s CAL./GFW.

T.P. °K
ΔH_t CAL./GFW.

T.P. °K
ΔH_t CAL./GFW.

$T_c =$ °K
$P_c =$ ATM.

THERMODYNAMIC PROPERTIES OF THE ELEMENTS

INDIUM

INDIUM	In	Solid from 298° to 429.32°, Liquid from 429.32° to 2320°, Ideal Monatomic Gas from 2320° to 3000°.

REFERENCE STATE

Gfw 114.82 GRAMS
$(H°_{298.15} - H°_0) = 1,578$ CAL./GFW.

M.P. 429.32 °K
ΔH_m 780. CAL./GFW.

B.P. 2,320. °K
ΔH_v 54,100. CAL./GFW.

S.P. °K
ΔH_s CAL./GFW.

T.P. °K
ΔH_t CAL./GFW.

T.P. °K
ΔH_t CAL./GFW.

$T_c =$ °K
$P_c =$ ATM.

T TEMPERATURE °K	$C°_P$ HEAT CAPACITY CAL./DEG./GFW.	$H°_T - H°_{298.15}$ HEAT CONTENT CAL./GFW.	$S°_T$ ENTROPY CAL./DEG./GFW.	$-(F°_T - H°_{298.15})$ FREE ENERGY FUNCTION CAL./DEG./GFW.	HEAT $\Delta H°_f$ CAL./GFW.	FREE ENERGY $\Delta F°_f$ CAL./GFW.	LOG$_{10}$ K_P
298	6.39	0	13.82	13.82			
300	6.40	12	13.86	13.82			
400	6.90	680	15.81	14.11			
500	7.10	2170	19.20	14.86			
600	7.10	2880	20.49	15.69			
700	7.10	3590	21.59	16.47			
800	7.10	4300	22.54	17.17			
900	7.10	5010	23.37	17.81			
1000	7.10	5720	24.12	18.40			
1100	7.10	6430	24.80	18.96			
1200	7.10	7140	25.42	19.47			
1300	7.10	7850	25.98	19.95			
1400	7.10	8560	26.51	20.40			
1500	7.10	9270	27.00	20.82			
1600	7.10	9980	27.46	21.23			
1700	7.10	10690	27.89	21.61			
1800	7.10	11400	28.29	21.96			
1900	7.10	12110	28.68	22.31			
2000	7.10	12820	29.04	22.63			
2100	7.10	13530	29.39	22.95			
2200	7.10	14240	29.72	23.25			
2300	7.10	14950	30.04	23.54			
2400	5.76	69640	53.63	24.62			
2500	5.71	70210	53.86	25.78			
2600	5.66	70780	54.09	26.87			
2700	5.62	71340	54.30	27.88			
2800	5.58	71900	54.50	28.83			
2900	5.54	72460	54.70	29.72			
3000	5.51	73010	54.89	30.56			

INDIUM

INDIUM In Reference State for Calculating ΔH_f°, ΔF_f°, and $\log_{10} K_p$: Solid from 298° to 429.32°, Liquid from 429.32° to 2320°, Ideal Monatomic Gas from 2320° to 3000°.

IDEAL MONATOMIC GAS

G_{fw} 114.82 GRAMS
$(H_{298.15}^\circ - H_0^\circ) = 1,482$ CAL./GFW.

					FORMATION FROM REFERENCE STATE		
T TEMPERATURE °K	C_p° HEAT CAPACITY CAL./DEG./GFW.	$H_T^\circ - H_{298.15}^\circ$ HEAT CONTENT CAL./GFW.	S_T° ENTROPY CAL./DEG./GFW.	$-(F^\circ - H_T^\circ)/T$ (298.15) FREE ENERGY FUNCTION CAL./DEG./GFW.	HEAT ΔH_f° CAL./GFW.	FREE ENERGY ΔF_f° CAL./GFW.	$\log_{10} K_p$
298	4.98	0	41.51	41.51	57000	48744	-35.731
300	4.98	9	41.54	41.51	56997	48693	-35.475
400	5.06	510	42.98	41.71	56830	45962	-25.114
500	5.24	1024	44.13	42.09	55854	43389	-18.966
600	5.51	1562	45.11	42.51	55682	40910	-14.902
700	5.80	2127	45.98	42.95	55537	38464	-12.009
800	6.06	2721	46.77	43.37	55421	36037	-9.845
900	6.26	3338	47.50	43.80	55328	33611	-8.162
1000	6.39	3971	48.16	44.19	55251	31211	-6.821
1100	6.46	4614	48.78	44.59	55184	28806	-5.723
1200	6.48	5261	49.34	44.96	55121	26417	-4.811
1300	6.47	5909	49.86	45.32	55059	24015	-4.037
1400	6.42	6554	50.33	45.65	54994	21646	-3.379
1500	6.36	7192	50.78	45.99	54922	19252	-2.805
1600	6.29	7826	51.18	46.29	54846	16894	-2.307
1700	6.22	8452	51.56	46.59	54762	14523	-1.867
1800	6.15	9070	51.92	46.89	54670	12136	-1.473
1900	6.07	9681	52.25	47.16	54571	9788	-1.125
2000	6.00	10285	52.56	47.42	54465	7425	-.811
2100	5.94	10882	52.85	47.67	54352	5086	-.529
2200	5.87	11472	53.12	47.91	54232	2752	-.273
2300	5.81	12057	53.38	48.14	54107	425	-.040
2400	5.76	12635	53.63	48.37	0	0	0
2500	5.71	13209	53.86	48.58	0	0	0
2600	5.66	13777	54.09	48.80	0	0	0
2700	5.62	14342	54.30	48.99	0	0	0
2800	5.58	14902	54.50	49.18	0	0	0
2900	5.54	15458	54.70	49.37	0	0	0
3000	5.51	16010	54.89	49.56	0	0	0

M.P. °K
ΔH_m CAL./GFW.

B.P. °K
ΔH_v CAL./GFW.

S.P. °K
ΔH_s CAL./GFW.

T.P. °K
ΔH_t CAL./GFW.

T.P. °K
ΔH_t CAL./GFW.

$T_c =$ °K
$P_c =$ ATM.

THERMODYNAMIC PROPERTIES OF THE ELEMENTS

IODINE

IODINE I_2

Solid from 298° to 386.8°, Liquid from 386.8° to 456°, Ideal Diatomic Gas from 456° to 3000°.

Gfw	253.82	GRAMS
REFERENCE STATE		
$(H°_{298.15} - H°_0) =$	3,178	CAL./GFW.
M.P.	386.8	°K
ΔH_m	3,770.	CAL./GFW.
B.P.	456.	°K
ΔH_v	9,970.	CAL./GFW.
S.P.		°K
ΔH_s		CAL./GFW.
T.P.		°K
ΔH_t		CAL./GFW.
T.P.		°K
ΔH_t		CAL./GFW.
$T_c =$		°K
$P_c =$		ATM.

T TEMPERATURE °K	$C°_P$ HEAT CAPACITY CAL./DEG./GFW.	$H°_T - H°_{298.15}$ HEAT CONTENT CAL./GFW.	$S°_T$ ENTROPY CAL./DEG./GFW.	$-(F°_T - H°_{298.15})/T$ FREE ENERGY FUNCTION CAL./DEG./GFW.	FORMATION FROM REFERENCE STATE		
					HEAT $\Delta H°_f$ CAL./GFW.	FREE ENERGY $\Delta F°_f$ CAL./GFW.	LOG$_{10}$ K_P
298	13.14	0	27.90	27.90			
300	13.16	24	27.98	27.90			
400	19.20	5234	41.85	28.77			
500	8.95	16676	66.88	33.53			
600	8.98	17573	68.51	39.23			
700	9.00	18472	69.90	43.52			
800	9.02	19373	71.10	46.89			
900	9.04	20277	72.17	49.64			
1000	9.06	21182	73.12	51.94			
1100	9.08	22088	73.98	53.90			
1200	9.09	22997	74.77	55.61			
1300	9.11	23907	75.50	57.11			
1400	9.12	24818	76.18	58.46			
1500	9.14	25731	76.81	59.66			
1600	9.15	26645	77.40	60.75			
1700	9.16	27561	77.96	61.75			
1800	9.18	28478	78.48	62.66			
1900	9.19	29396	78.98	63.51			
2000	9.21	30317	79.45	64.30			
2100	9.22	31237	79.90	65.03			
2200	9.23	32160	80.33	65.72			
2300	9.25	33085	80.74	66.36			
2400	9.26	34010	81.13	66.96			
2500	9.27	34936	81.51	67.54			
2600	9.29	35864	81.87	68.08			
2700	9.30	36794	82.22	68.60			
2800	9.31	37724	82.56	69.09			
2900	9.33	38655	82.89	69.57			
3000	9.34	39586	83.20	70.01			

IODINE

IODINE I_2
IDEAL DIATOMIC GAS

Gfw 253.82 GRAMS
$(H°_{298.15} - H°_0) = 2,418$ CAL./GFW.

Reference State for Calculating $\Delta H°_f$, $\Delta F°_f$, and $\log_{10} K_p$:
Solid from 298° to 386.8°, Liquid from 386.8° to 456°, Ideal Diatomic Gas from 456° to 3000°.

T TEMPERATURE °K	$C°_p$ HEAT CAPACITY CAL./DEG./GFW.	$H°_T - H°_{298.15}$ HEAT CONTENT CAL./GFW.	$S°_T$ ENTROPY CAL./DEG./GFW.	$-(F°_T - H°_{298.15})/T$ FREE ENERGY FUNCTION CAL./DEG./GFW.	FORMATION FROM REFERENCE STATE		
					HEAT $\Delta H°_f$ CAL./GFW.	FREE ENERGY $\Delta F°_f$ CAL./GFW.	$\log_{10} K_p$
298	8.81	0	62.28	62.28	14880	4629	-3.393
300	8.82	16	62.34	62.29	14872	4564	-3.325
400	8.90	903	64.89	62.64	10549	1337	-.730
500	8.95	1795	66.88	63.29	0	0	0
600	8.98	2692	68.51	64.03	0	0	0
700	9.00	3591	69.90	64.77	0	0	0
800	9.02	4493	71.10	65.49	0	0	0
900	9.04	5396	72.17	66.18	0	0	0
1000	9.06	6301	73.12	66.82	0	0	0
1100	9.08	7207	73.98	67.43	0	0	0
1200	9.09	8116	74.77	68.01	0	0	0
1300	9.11	9026	75.50	68.56	0	0	0
1400	9.12	9937	76.18	69.09	0	0	0
1500	9.14	10850	76.81	69.58	0	0	0
1600	9.15	11765	77.40	70.05	0	0	0
1700	9.16	12680	77.96	70.51	0	0	0
1800	9.18	13597	78.48	70.93	0	0	0
1900	9.19	14515	78.98	71.35	0	0	0
2000	9.21	15436	79.45	71.74	0	0	0
2100	9.22	16356	79.90	72.12	0	0	0
2200	9.23	17280	80.33	72.48	0	0	0
2300	9.25	18204	80.74	72.83	0	0	0
2400	9.26	19129	81.13	73.16	0	0	0
2500	9.27	20055	81.51	73.49	0	0	0
2600	9.29	20983	81.87	73.80	0	0	0
2700	9.30	21912	82.22	74.11	0	0	0
2800	9.31	22842	82.56	74.41	0	0	0
2900	9.33	23772	82.89	74.70	0	0	0
3000	9.34	24705	83.20	74.97	0	0	0

M.P. °K
ΔH_m CAL./GFW.

B.P. °K
ΔH_v CAL./GFW.

S.P. °K
ΔH_s CAL./GFW.

T.P. °K
ΔH_t CAL./GFW.

T.P. °K
ΔH_t CAL./GFW.

$T_c =$ °K
$P_c =$ ATM.

IODINE

IODINE I

IDEAL MONATOMIC GAS

Reference State for Calculating ΔH_f°, ΔF_f°, and $Log_{10} K_p$:
Solid from 298° to 386.8°, Liquid from 386.8° to 456°, Ideal Diatomic Gas from 456° to 3000°.

Gfw 126.91 GRAMS
$(H^\circ_{298.15} - H^\circ_0) = 1,481$ CAL./GFW.

M.P.	°K	
ΔH_m	CAL./GFW.	
B.P.	°K	
ΔH_v	CAL./GFW.	
S.P.	°K	
ΔH_s	CAL./GFW.	
T.P.	°K	
ΔH_t	CAL./GFW.	
T.P.	°K	
ΔH_t	CAL./GFW.	
$T_c =$	°K	
$P_c =$	ATM.	

TEMPERATURE °K	C_p° HEAT CAPACITY CAL./DEG./GFW.	$H^\circ_T - H^\circ_{298.15}$ HEAT CONTENT CAL./GFW.	S°_T ENTROPY CAL./DEG./GFW.	$-(F^\circ_T - H^\circ_{298.15})/T$ FREE ENERGY FUNCTION CAL./DEG./GFW.	HEAT ΔH°_f CAL./GFW.	FREE ENERGY ΔF°_f CAL./GFW.	LOG$_{10}$ K_p
298	4.97	0	43.18	43.18	25482	16767	-12.291
300	4.97	9	43.21	43.18	25479	16713	-12.176
400	4.97	506	44.64	43.38	23371	13887	-7.588
500	4.97	1003	45.75	43.75	18147	11992	-5.242
600	4.97	1500	46.66	44.16	18195	10755	-3.917
700	4.97	1996	47.42	44.57	18242	9513	-2.970
800	4.97	2493	48.09	44.98	18288	8256	-2.255
900	4.97	2990	48.67	45.35	18333	7011	-1.702
1000	4.97	3487	49.20	45.72	18378	5738	-1.254
1100	4.97	3984	49.67	46.05	18422	4474	-.888
1200	4.98	4482	50.10	46.37	18465	3213	-.585
1300	4.98	4979	50.50	46.67	18507	1932	-.324
1400	4.99	5478	50.87	46.96	18551	659	-.102
1500	5.00	5978	51.22	47.24	18594	621	.090
1600	5.02	6479	51.54	47.50	18638	1906	.260
1700	5.03	6982	51.84	47.74	18683	3179	.408
1800	5.05	7486	52.13	47.98	18729	4473	.543
1900	5.07	7992	52.41	48.21	18776	5772	.663
2000	5.09	8500	52.67	48.42	18823	7057	.771
2100	5.11	9011	52.92	48.63	18874	8363	.870
2200	5.14	9523	53.15	48.83	18925	9631	.956
2300	5.16	10038	53.38	49.02	18977	10946	1.040
2400	5.18	10555	53.60	49.21	19032	12240	1.114
2500	5.20	11075	53.81	49.38	19089	13536	1.183
2600	5.22	11596	54.01	49.55	19146	14836	1.247
2700	5.24	12119	54.21	49.73	19204	16166	1.308
2800	5.26	12644	54.40	49.89	19264	17472	1.363
2900	5.28	13171	54.59	50.05	19325	18752	1.413
3000	5.30	13703	54.77	50.21	19392	20118	1.465

IRIDIUM

IRIDIUM   Ir   Solid from 298° to 2727°   Liquid from 2727° to 3000°.
REFERENCE STATE

Gfw	192.2	GRAMS							FORMATION FROM REFERENCE STATE		
$(H°_{298.15} - H°_0)$ =		CAL./GFW.							HEAT $\Delta H°_f$ CAL./GFW.	FREE ENERGY $\Delta F°_f$ CAL./GFW.	LOG$_{10}$ K$_p$
M.P.	2,727	°K	T TEMPERATURE °K	C°$_p$ HEAT CAPACITY CAL./DEG./GFW.	$H°_T - H°_{298.15}$ HEAT CONTENT CAL./GFW.	S°$_T$ ENTROPY CAL./DEG./GFW.	$-(F°_T - H°_{298.15})/T$ FREE ENERGY FUNCTION CAL./DEG./GFW.				
ΔH_m	(6,300)	CAL./GFW.	298	6.00	0	8.70	8.70				
			300	6.00	11	8.74	8.71				
			400	6.13	620	10.49	8.94				
			500	6.27	1235	11.86	9.39				
			600	6.41	1870	13.02	9.91				
			700	6.55	2525	14.03	10.43				
B.P.	(4,400)	°K	800	6.70	3190	14.92	10.94				
			900	6.84	3860	15.70	11.42				
ΔH_v	(134,700)	CAL./GFW.	1000	6.98	4545	16.43	11.89				
			1100	7.12	5250	17.10	12.33				
			1200	7.26	5970	17.73	12.76				
S.P.		°K	1300	7.41	6700	18.31	13.16				
			1400	7.55	7440	18.86	13.55				
ΔH_s		CAL./GFW.	1500	7.69	8200	19.38	13.92				
			1600	7.83	8980	19.89	14.28				
			1700	7.97	9780	20.37	14.62				
T.P.		°K	1800	8.12	10590	20.83	14.95				
			1900	8.26	11410	21.27	15.27				
ΔH_t		CAL./GFW.	2000	8.40	12240	21.70	15.58				
			2100	8.54	13090	22.11	15.88				
			2200	8.68	13950	22.51	16.17				
T.P.		°K	2300	8.83	14820	22.90	16.46				
			2400	8.97	15710	23.28	16.74				
ΔH_t		CAL./GFW.	2500	9.11	16620	23.65	17.01				
			2600	9.25	17540	24.01	17.27				
			2700	9.39	18470	24.36	17.52				
T_c =		°K	2800	9.50	25710	27.02	17.84				
			2900	9.50	26660	27.35	18.16				
P_c =		ATM.	3000	9.50	27610	27.67	18.47				

THERMODYNAMIC PROPERTIES OF THE ELEMENTS

IRIDIUM

IRIDIUM   Ir

IDEAL MONATOMIC GAS

Gfw 192.2 GRAMS

$(H°_{298.15} - H°_0) = 1,481$ CAL./GFW.

Reference State for Calculating $\Delta H°_f$, $\Delta F°_f$, and $\log_{10} K_p$:
Solid from 298° to 2727°, Liquid from 2727° to 3000°.

T TEMPERATURE °K	$C°_P$ HEAT CAPACITY CAL./DEG./GFW.	$H°_T - H°_{298.15}$ HEAT CONTENT CAL./GFW.	$S°_T$ ENTROPY CAL./DEG./GFW.	$-(F°_T - H°_{298.15})/T$ FREE ENERGY FUNCTION CAL./DEG./GFW.	FORMATION FROM REFERENCE STATE		
					HEAT $\Delta H°_f$ CAL./GFW.	FREE ENERGY $\Delta F°_f$ CAL./GFW.	$\log_{10} K_p$
298	4.97	0	46.24	46.24	150000	138807	-101.752
300	4.97	9	46.27	46.24	149998	138739	-101.079
400	4.98	506	47.70	46.44	149886	135002	-73.767
500	5.01	1005	48.82	46.81	149770	131290	-57.390
600	5.07	1509	49.73	47.22	149639	127613	-46.486
700	5.18	2021	50.52	47.64	149496	123953	-38.703
800	5.31	2546	51.22	48.04	149356	120316	-32.871
900	5.46	3085	51.86	48.44	149225	116681	-28.335
1000	5.62	3639	52.44	48.81	149094	113084	-24.716
1100	5.78	4209	52.99	49.17	148959	109480	-21.753
1200	5.93	4795	53.49	49.50	148825	105913	-19.290
1300	6.08	5396	53.98	49.83	148696	102325	-17.203
1400	6.21	6010	54.43	50.14	148570	98772	-15.420
1500	6.33	6637	54.86	50.44	148437	95217	-13.874
1600	6.44	7276	55.28	50.74	148296	91672	-12.521
1700	6.54	7925	55.67	51.01	148145	88135	-11.330
1800	6.62	8584	56.05	51.29	147994	84598	-10.271
1900	6.69	9249	56.41	51.55	147839	81073	-9.325
2000	6.75	9921	56.75	51.79	147681	77581	-8.477
2100	6.79	10599	57.08	52.04	147509	74072	-7.708
2200	6.83	11280	57.40	52.28	147330	70572	-7.010
2300	6.85	11974	57.70	52.50	147154	67114	-6.377
2400	6.87	12650	57.99	52.72	146940	63636	-5.794
2500	6.87	13337	58.27	52.94	146717	60167	-5.259
2600	6.87	14024	58.54	53.15	146484	56706	-4.766
2700	6.86	14710	58.80	53.36	146240	53252	-4.310
2800	6.85	15396	59.05	53.56	139686	50002	-3.902
2900	6.83	16079	59.29	53.75	139419	46793	-3.526
3000	6.81	16761	59.52	53.94	139151	43601	-3.176

M.P. °K
ΔH_m CAL./GFW.

B.P. °K
ΔH_v CAL./GFW.

S.P. °K
ΔH_s CAL./GFW.

T.P. °K
ΔH_t CAL./GFW.

T.P. °K
ΔH_t CAL./GFW.

$T_c =$ °K
$P_c =$ ATM.

IRON

IRON   **Fe**

Solid I from 298° to 1183°, Solid II from 1183° to 1673°,
Solid III from 1673° to 1812°, Liquid from 1812° to 3000°.

REFERENCE STATE

Gfw	55.85	GRAMS
$(H°_{298.15} - H°_0)$ =	1,070	CAL./GFW.
M.P.	1,812	°K
ΔH_m	3,670	CAL./GFW.
B.P.	3,160	°K
ΔH_v	83,900	CAL./GFW.
S.P.		°K
ΔH_s		CAL./GFW.
T.P.	1,033	°K
ΔH_t	0	CAL./GFW.
T.P.	1,183	°K
ΔH_t	215	CAL./GFW.
T_c =	1,673	°K
P_c =	165	ATM.

T TEMPERATURE °K	$C°_p$ HEAT CAPACITY CAL./DEG./GFW.	$H°_T - H°_{298.15}$ HEAT CONTENT CAL./GFW.	$S°_T$ ENTROPY CAL./DEG./GFW.	$-(F°_T - H°_{298.15})$ FREE ENERGY FUNCTION CAL./DEG./GFW.	FORMATION FROM REFERENCE STATE		
					HEAT $\Delta H°_f$ CAL./GFW.	FREE ENERGY $\Delta F°_f$ CAL./GFW.	LOG$_{10}$ K_p
298	5.99	0	6.49	6.49			
300	6.00	11	6.52	6.49			
400	6.55	639	8.33	6.74			
500	6.99	1317	9.84	7.21			
600	7.54	2042	11.16	7.76			
700	8.25	2830	12.37	8.33			
800	9.23	3700	13.53	8.91			
900	10.74	4694	14.70	9.49			
1000	13.80	5882	15.95	10.07			
1100	10.83	7229	17.25	10.68			
1200	8.17	8438	18.30	11.27			
1300	8.38	9266	18.96	11.84			
1400	8.58	10114	19.59	12.37			
1500	8.77	10981	20.19	12.87			
1600	8.97	11868	20.76	13.35			
1700	9.48	12949	21.41	13.80			
1800	9.58	13902	21.96	14.24			
1900	10.54	18613	24.55	14.76			
2000	10.58	19669	25.09	15.26			
2100	10.62	20729	25.61	15.74			
2200	10.66	21793	26.11	16.21			
2300	10.70	22861	26.58	16.65			
2400	10.74	23933	27.04	17.07			
2500	10.78	25009	27.48	17.48			
2600	10.82	26089	27.90	17.87			
2700	10.86	27173	28.31	18.25			
2800	10.90	28261	28.70	18.61			
2900	10.94	29353	29.09	18.97			
3000	10.98	30449	29.46	19.32			

THERMODYNAMIC PROPERTIES OF THE ELEMENTS

IRON

IRON   **Fe**

G/W 55.85   **GRAMS**

$(H°_{298.15} - H°_0) = 1,637$ CAL./GFW.

IDEAL MONATOMIC GAS

Reference State for Calculating $\Delta H°_f$, $\Delta F°_f$, and $\log_{10} K_p$:
Solid I from 298° to 1183°, Solid II from 1183° to 1673°,
Solid III from 1673° to 1812°, Liquid from 1812° to 3000°.

TEMPERATURE °K	$C°_p$ HEAT CAPACITY CAL./DEG./GFW.	$H°_T - H°_{298.15}$ HEAT CONTENT CAL./GFW.	$S°_T$ ENTROPY CAL./DEG./GFW.	$-(F°_T - H°_{298.15})/T$ FREE ENERGY FUNCTION CAL./DEG./GFW.	FORMATION FROM REFERENCE STATE		
					HEAT $\Delta H°_f$ CAL./GFW.	FREE ENERGY $\Delta F°_f$ CAL./GFW.	$\log_{10} K_p$
298	6.14	0	43.11	43.11	99830	88912	−65.176
300	6.14	11	43.15	43.12	99830	88841	−64.725
400	6.10	625	44.92	43.36	99816	85180	−46.544
500	5.95	1228	46.26	43.81	99741	81531	−35.639
600	5.78	1815	47.33	44.31	99603	77901	−28.377
700	5.64	2386	48.21	44.81	99386	74298	−23.198
800	5.53	2944	48.96	45.28	99074	70730	−19.324
900	5.44	3492	49.60	45.72	98628	67218	−16.323
1000	5.37	4033	50.17	46.14	97981	63761	−13.936
1100	5.33	4568	50.68	46.53	97169	60396	−12.000
1200	5.30	5099	51.15	46.91	96491	57071	−10.394
1300	5.29	5629	51.57	47.24	96193	53800	−9.045
1400	5.30	6158	51.96	47.57	95874	50556	−7.892
1500	5.31	6688	52.33	47.88	95537	47327	−6.896
1600	5.34	7221	52.67	48.16	95183	44127	−6.027
1700	5.38	7757	53.00	48.44	94638	40935	−5.262
1800	5.43	8298	53.31	48.70	94226	37796	−4.588
1900	5.49	8844	53.60	48.95	90061	34866	−4.010
2000	5.55	9396	53.88	49.19	89557	31977	−3.494
2100	5.61	9954	54.16	49.42	89055	29100	−3.028
2200	5.68	10519	54.42	49.64	88556	26274	−2.610
2300	5.75	11090	54.67	49.85	88059	23452	−2.228
2400	5.82	11669	54.92	50.06	87566	20654	−1.880
2500	5.90	12255	55.16	50.26	87076	17876	−1.562
2600	5.97	12848	55.39	50.45	86589	15115	−1.270
2700	6.04	13449	55.62	50.64	86106	12369	−1.001
2800	6.12	14056	55.84	50.82	85625	9633	−.751
2900	6.19	14673	56.06	51.01	85150	6937	−.522
3000	6.26	15294	56.27	51.18	84675	4245	−.309

M.P.		°K	
ΔH_m		CAL./GFW.	
B.P.		°K	
ΔH_v		CAL./GFW.	
S.P.		°K	
ΔH_s		CAL./GFW.	
T.P.		°K	
ΔH_t		CAL./GFW.	
T.P.		°K	
ΔH_t		CAL./GFW.	
$T_c =$		°K	
$P_c =$		ATM.	

KRYPTON

Ideal Monatomic Gas from 298° to 3000°.

KRYPTON	Kr		
Gfw	83.80	GRAMS	
$(H°_{298.15} - H°_0)$ =	1,481	CAL./GFW.	
M.P.	115.9	°K	
ΔH_m	391.	CAL./GFW.	
B.P.	119.75	°K	
ΔH_v	2,158.	CAL./GFW.	
S.P.		°K	
ΔH_s		CAL./GFW.	
T.P.		°K	
ΔH_t		CAL./GFW.	
T.P.		°K	
ΔH_t		CAL./GFW.	
T_c =	209.4	°K	
P_c =	54.3	ATM.	

REFERENCE STATE

T TEMPERATURE °K	$C°_P$ HEAT CAPACITY CAL./DEG./GFW.	$H°_T - H°_{298.15}$ HEAT CONTENT CAL./GFW.	$S°_T$ ENTROPY CAL./DEG./GFW.	$-(F°_T - H°_{298.15})/T$ FREE ENERGY FUNCTION CAL./DEG./GFW.	FORMATION FROM REFERENCE STATE		
					HEAT $\Delta H°_f$ CAL./GFW.	FREE ENERGY $\Delta F°_f$ CAL./GFW.	$LOG_{10} K_P$
298	4.97	0	39.19	39.19			
300	4.97	9	39.22	39.19			
400	4.97	506	40.65	39.39			
500	4.97	1003	41.76	39.76			
600	4.97	1500	42.67	40.17			
700	4.97	1996	43.43	40.58			
800	4.97	2493	44.10	40.99			
900	4.97	2990	44.68	41.36			
1000	4.97	3487	45.20	41.72			
1100	4.97	3984	45.68	42.06			
1200	4.97	4480	46.11	42.38			
1300	4.97	4977	46.51	42.69			
1400	4.97	5474	46.88	42.97			
1500	4.97	5971	47.22	43.24			
1600	4.97	6468	47.54	43.50			
1700	4.97	6964	47.84	43.75			
1800	4.97	7461	48.12	43.98			
1900	4.97	7958	48.39	44.21			
2000	4.97	8455	48.65	44.43			
2100	4.97	8952	48.89	44.63			
2200	4.97	9448	49.12	44.83			
2300	4.97	9945	49.34	45.02			
2400	4.97	10442	49.55	45.20			
2500	4.97	10939	49.76	45.39			
2600	4.97	11436	49.95	45.56			
2700	4.97	11932	50.14	45.73			
2800	4.97	12429	50.32	45.89			
2900	4.97	12926	50.49	46.04			
3000	4.97	13423	50.66	46.19			

THERMODYNAMIC PROPERTIES OF THE ELEMENTS

LANTHANUM

LANTHANUM La Solid from 298° to 1193°, Liquid from 1193° to 3000°.

REFERENCE STATE

Gfw	138.92	GRAMS
$(H°_{298.15} - H°_0)$ =	1,569	CAL./GFW.
M.P.	1,193	°K
ΔH_m	(2,700)	CAL./GFW.
B.P.	3,640	°K
ΔH_v	95,500	CAL./GFW.
S.P.		°K
ΔH_s		CAL./GFW.
T.P.		°K
ΔH_t		CAL./GFW.
T.P.		°K
ΔH_t		CAL./GFW.
T_c =		°K
P_c =		ATM.

T TEMPERATURE °K	$C°_p$ HEAT CAPACITY CAL./DEG./GFW.	$H°_T - H°_{298.15}$ HEAT CONTENT CAL./GFW.	$S°_T$ ENTROPY CAL./DEG./GFW.	$-(F°_T - H°_{298.15})/T$ FREE ENERGY FUNCTION CAL./DEG./GFW.	HEAT $\Delta H°_f$ CAL./GFW.	FREE ENERGY $\Delta F°_f$ CAL./GFW.	$LOG_{10} K_p$
298	6.65	0	13.60	13.60			
300	6.65	12	13.64	13.60			
400	6.81	685	15.57	13.86			
500	6.97	1370	17.11	14.37			
600	7.13	2080	18.40	14.94			
700	7.29	2800	19.51	15.51			
800	7.45	3540	20.49	16.07			
900	7.61	4290	21.38	16.62			
1000	7.77	5060	22.19	17.13			
1100	7.93	5840	22.94	17.64			
1200	8.00	9380	25.94	18.13			
1300	8.00	10180	26.58	18.75			
1400	8.00	10980	27.17	19.33			
1500	8.00	11780	27.73	19.88			
1600	8.00	12580	28.24	20.38			
1700	8.00	13380	28.73	20.86			
1800	8.00	14180	29.18	21.31			
1900	8.00	14980	29.62	21.74			
2000	8.00	15780	30.03	22.14			
2100	8.00	16580	30.42	22.53			
2200	8.00	17380	30.79	22.84			
2300	8.00	18180	31.15	23.25			
2400	8.00	18980	31.49	23.59			
2500	8.00	19780	31.81	23.90			
2600	8.00	20580	32.13	24.22			
2700	8.00	21380	32.43	24.52			
2800	8.00	22180	32.72	24.80			
2900	8.00	22980	33.00	25.08			
3000	8.00	23780	33.27	25.35			

LANTHANUM

LANTHANUM La
Gfw 138.92 **GRAMS**
$(H°_{298.15} - H°_0) = 1,509$ **CAL./GFW.**

IDEAL MONATOMIC GAS

Reference State for Calculating $\Delta H_f°$, $\Delta F_f°$, and $Log_{10}K_p$:
Solid from 298° to 1193°, Liquid from 1193° to 3000°.

TEMPERATURE °K	$C_p°$ HEAT CAPACITY CAL./DEG./GFW.	$H°_T - H°_{298.15}$ HEAT CONTENT CAL./GFW.	$S°_T$ ENTROPY CAL./DEG./GFW.	$-(F°_T - H°_{298.15})/T$ FREE ENERGY FUNCTION CAL./DEG./GFW.	HEAT $\Delta H_f°$ CAL./GFW.	FREE ENERGY $\Delta F_f°$ CAL./GFW.	$LOG_{10} K_p$
298	5.44	0	43.56	43.56	99600	90667	-66.463
300	5.45	10	43.60	43.57	99598	90610	-66.014
400	5.89	578	45.23	43.79	99493	87629	-47.882
500	6.22	1185	46.58	44.21	99415	84680	-37.016
600	6.46	1819	47.74	44.71	99339	81735	-29.774
700	6.67	2475	48.75	45.22	99275	78807	-24.606
800	6.87	3152	49.65	45.71	99212	75884	-20.732
900	7.07	3849	50.47	46.20	99159	72978	-17.722
1000	7.25	4566	51.23	46.67	99106	70066	-15.314
1100	7.39	5298	51.92	47.11	99058	67180	-13.348
1200	7.50	6043	52.57	47.54	96263	64307	-11.712
1300	7.57	6796	53.18	47.96	96216	61636	-10.362
1400	7.62	7556	53.74	48.35	96176	58978	-9.207
1500	7.64	8319	54.27	48.73	96139	56329	-8.207
1600	7.66	9084	54.76	49.09	96104	53672	-7.331
1700	7.66	9850	55.22	49.43	96070	51037	-6.561
1800	7.66	10616	55.66	49.77	96036	48372	-5.872
1900	7.65	11382	56.08	50.09	96002	45728	-5.259
2000	7.65	12147	56.47	50.40	95967	43087	-4.708
2100	7.64	12911	56.84	50.70	95931	40449	-4.209
2200	7.64	13675	57.20	50.99	95895	37793	-3.754
2300	7.64	14439	57.54	51.27	95859	35162	-3.341
2400	7.64	15203	57.86	51.53	95823	32535	-2.962
2500	7.64	15967	58.17	51.79	95787	29887	-2.612
2600	7.65	16732	58.47	52.04	95752	27268	-2.292
2700	7.66	17498	58.76	52.28	95718	24627	-1.993
2800	7.67	18264	59.04	52.52	95684	21988	-1.716
2900	7.69	19032	59.31	52.75	95652	19353	-1.458
3000	7.70	19802	59.57	52.97	95622	16722	-1.218

M.P. °K
ΔH_m CAL./GFW.
B.P. °K
ΔH_v CAL./GFW.
S.P. °K
ΔH_s CAL./GFW.
T.P. °K
ΔH_t CAL./GFW.
T.P. °K
ΔH_t CAL./GFW.
$T_c =$ °K
$P_c =$ ATM.

THERMODYNAMIC PROPERTIES OF THE ELEMENTS

LEAD

LEAD	Pb		Solid from 298° to 600.6°, Liquid from 600.6° to 2024°, Ideal Monatomic Gas from 2024° to 3000°.
Gw	207.21	GRAMS	
$(H°_{298.15} - H°_0)$ =	1,644	CAL./GFW.	
M.P.	600.6	°K	
ΔH_m	1,141.	CAL./GFW.	
B.P.	2,024.	°K	
ΔH_v	42,880.	CAL./GFW.	
S.P.		°K	
ΔH_s		CAL./GFW.	
T.P.		°K	
ΔH_t		CAL./GFW.	
T.P.		°K	
ΔH_t		CAL./GFW.	
T_c =		°K	
P_c =		ATM.	

T TEMPERATURE °K	$C_p°$ HEAT CAPACITY CAL./DEG./GFW.	$H°_T - H°_{298.15}$ HEAT CONTENT CAL./GFW.	$S°_T$ ENTROPY CAL./DEG./GFW.	$-(F°_T - H°_{298.15})$ FREE ENERGY FUNCTION CAL./DEG./GFW.	FORMATION FROM REFERENCE STATE		
					HEAT $\Delta H°_f$ CAL./GFW.	FREE ENERGY $\Delta F°_f$ CAL./GFW.	LOG$_{10}$ K$_p$
298	6.32	0	15.49	15.49			
300	6.33	12	15.53	15.49			
400	6.56	656	17.38	15.74			
500	6.79	1324	18.87	16.23			
600	7.02	2014	20.13	16.78			
700	7.25	3884	23.15	17.61			
800	7.17	4605	24.11	18.36			
900	7.10	5318	25.00	19.10			
1000	7.03	6024	25.70	19.68			
1100	6.95	6723	26.37	20.76			
1200	6.88	7415	26.97	20.80			
1300	6.80	8100	27.51	21.28			
1400	6.73	8780	28.02	21.75			
1500	6.66	9450	28.48	22.18			
1600	6.58	10110	28.91	22.60			
1700	6.51	10760	29.30	22.98			
1800	6.43	11410	29.67	23.34			
1900	6.36	12050	30.02	23.68			
2000	6.29	12680	30.34	24.00			
2100	6.10	56165	51.82	25.08			
2200	6.30	56785	52.11	26.30			
2300	6.52	57430	52.39	27.43			
2400	6.74	58090	52.67	28.47			
2500	6.95	58770	52.95	29.45			
2600	7.17	59480	53.23	30.36			
2700	7.37	60210	53.51	31.21			
2800	7.57	60950	53.78	32.02			
2900	7.76	61720	54.05	32.77			
3000	7.93	62500	54.31	33.48			

LEAD

LEAD **Pb**
IDEAL MONATOMIC GAS

Reference State for Calculating ΔH_f°, ΔF_f°, and $\log_{10} K_p$:
Solid from 298° to 600.6°, Liquid from 600.6° to 2024°, Ideal Monatomic Gas from 2024° to 3000°.

Gfw 207.21 GRAMS
$(H_{298.15}^\circ - H_0^\circ) = 1,481$ CAL./GFW.

TEMPERATURE °K	C_p° HEAT CAPACITY CAL./DEG./GFW.	$H_T^\circ - H_{298.15}^\circ$ HEAT CONTENT CAL./GFW.	S_T° ENTROPY CAL./DEG./GFW.	$-(F_T^\circ - H_{298.15}^\circ)/T$ FREE ENERGY FUNCTION CAL./DEG./GFW.	FORMATION FROM REFERENCE STATE		
					HEAT ΔH_f° CAL./GFW.	FREE ENERGY ΔF_f° CAL./GFW.	$\log_{10} K_p$
298	4.97	0	41.89	41.89	46800	38929	-28.536
300	4.97	9	41.92	41.89	46797	38880	-28.326
400	4.97	506	43.35	42.09	46650	36262	-19.814
500	4.97	1003	44.46	42.46	46479	33684	-14.724
600	4.97	1500	45.36	42.86	46286	31148	-11.346
700	4.97	1996	46.13	43.28	44912	28826	-9.000
800	4.97	2493	46.79	43.68	44688	26544	-7.252
900	4.97	2990	47.38	44.06	44472	24330	-5.908
1000	4.98	3488	47.90	44.42	44264	22064	-4.822
1100	4.99	3986	48.38	44.76	44063	19852	-3.944
1200	5.02	4487	48.81	45.08	43872	17664	-3.217
1300	5.06	4990	49.22	45.39	43690	15467	-2.600
1400	5.11	5498	49.59	45.67	43518	13320	-2.079
1500	5.19	6014	49.95	45.95	43364	11159	-1.625
1600	5.29	6537	50.29	46.21	43227	9019	-1.231
1700	5.41	7072	50.61	46.45	43112	6885	-.885
1800	5.56	7621	50.92	46.69	43011	4761	-.578
1900	5.72	8184	51.23	46.93	42934	2635	-.303
2000	5.90	8765	51.53	47.15	42885	505	-.055
2100	6.10	9365	51.82	47.37	0	0	0
2200	6.30	9985	52.11	47.58	0	0	0
2300	6.52	10626	52.39	47.77	0	0	0
2400	6.74	11288	52.67	47.97	0	0	0
2500	6.95	11973	52.95	48.17	0	0	0
2600	7.17	12679	53.23	48.36	0	0	0
2700	7.37	13406	53.51	48.55	0	0	0
2800	7.57	14153	53.78	48.73	0	0	0
2900	7.76	14920	54.05	48.91	0	0	0
3000	7.93	15704	54.31	49.08	0	0	0

M.P. °K
ΔH_m CAL./GFW.
B.P. °K
ΔH_v CAL./GFW.
S.P. °K
ΔH_s CAL./GFW.
T.P. °K
ΔH_t CAL./GFW.
T.P. °K
ΔH_t CAL./GFW.
T_c = °K
P_c = ATM.

THERMODYNAMIC PROPERTIES OF THE ELEMENTS

LITHIUM

LITHIUM   Li

Gfw 6.940 **GRAMS**

$(H°_{298.15} - H°_0)$ = 1,092 CAL./GFW.

REFERENCE STATE Solid from 298° to 453.7°, Liquid from 453.7° to 1604°, Ideal Monatomic Gas from 1604° to 3000°.

M.P.	453.70	°K
ΔH_m	722.8	CAL./GFW.
B.P.	1,604.	°K
ΔH_v	32,190.	CAL./GFW.
S.P.		°K
ΔH_s		CAL./GFW.
T.P.		°K
ΔH_t		CAL./GFW.
T.P.		°K
ΔH_t		CAL./GFW.
$T_c =$		°K
$P_c =$		ATM.

T TEMPERATURE °K	$C°_P$ HEAT CAPACITY CAL./DEG./GFW.	$H°_T - H°_{298.15}$ HEAT CONTENT CAL./GFW.	$S°_T$ ENTROPY CAL./DEG./GFW.	$-(F°_T - H°_{298.15})$ FREE ENERGY FUNCTION CAL./DEG./GFW.	FORMATION FROM REFERENCE STATE		
					HEAT $\Delta H°_f$ CAL./GFW.	FREE ENERGY $\Delta F°_f$ CAL./GFW.	LOG$_{10}$ K$_P$
298	5.91	0	6.75	6.75			
300	5.92	11	6.79	6.76			
400	6.50	630	8.57	7.00			
500	7.20	2049	11.71	7.62			
600	7.06	2763	13.01	8.41			
700	6.95	3462	14.09	9.15			
800	6.91	4155	15.01	9.82			
900	6.90	4846	15.83	10.45			
1000	6.89	5536	16.55	11.02			
1100	6.88	6224	17.21	11.56			
1200	6.87	6912	17.81	12.05			
1300	6.86	7598	18.36	12.52			
1400	6.84	8284	18.86	12.95			
1500	6.82	8967	19.34	13.37			
1600	6.80	9648	19.78	13.75			
1700	4.97	45404	41.79	15.09			
1800	4.97	45901	42.08	16.58			
1900	4.98	46399	42.35	17.93			
2000	4.98	46897	42.60	19.16			
2100	4.99	47396	42.84	20.28			
2200	5.00	47896	43.08	21.31			
2300	5.01	48396	43.30	22.26			
2400	5.03	48898	43.51	23.14			
2500	5.05	49402	43.72	23.96			
2600	5.07	49908	43.92	24.73			
2700	5.10	50416	44.11	25.44			
2800	5.13	50928	44.30	26.12			
2900	5.17	51443	44.48	26.75			
3000	5.21	51962	44.65	27.33			

LITHIUM

LITHIUM Li
IDEAL MONATOMIC GAS

Reference State for Calculating ΔH_f°, ΔF_f°, and $\log_{10} K_p$: Solid from 298° to 453.7°, Liquid from 453.7° to 1604°, Ideal Monatomic Gas from 1604° to 3000°.

Gfw 6.940 GRAMS
$(H_{298.15}^\circ - H_0^\circ) = 1,481$ CAL./GFW.

M.P. °K
ΔH_m CAL./GFW.

B.P. °K
ΔH_v CAL./GFW.

S.P. °K
ΔH_s CAL./GFW.

T.P. °K
ΔH_f CAL./GFW.

T.P. °K
ΔH_t CAL./GFW.

$T_c =$ °K
$P_c =$ ATM.

T TEMPERATURE °K	C_p° HEAT CAPACITY CAL./DEG./GFW.	$H_T^\circ - H_{298.15}^\circ$ HEAT CONTENT CAL./GFW.	S_T° ENTROPY CAL./DEG./GFW.	$-(F_T^\circ - H_{298.15}^\circ)/T$ FREE ENERGY FUNCTION CAL./DEG./GFW.	FORMATION FROM REFERENCE STATE		
					HEAT ΔH_f° CAL./GFW.	FREE ENERGY ΔF_f° CAL./GFW.	$\log_{10} K_p$
298	4.97	0	33.14	33.14	38439	30570	-22.409
300	4.97	9	33.17	33.14	38437	30523	-22.237
400	4.97	506	34.60	33.34	38315	27903	-15.246
500	4.97	1003	35.71	33.71	37393	25393	-11.100
600	4.97	1500	36.62	34.12	37176	23010	-8.382
700	4.97	1996	37.38	34.53	36973	20670	-6.454
800	4.97	2493	38.05	34.94	36777	18345	-5.012
900	4.97	2990	38.63	35.31	36583	16063	-3.900
1000	4.97	3487	39.16	35.68	36390	13780	-3.011
1100	4.97	3984	39.63	36.01	36199	11537	-2.292
1200	4.97	4480	40.06	36.33	36007	9307	-1.695
1300	4.97	4977	40.46	36.64	35818	7088	-1.191
1400	4.97	5474	40.83	36.92	35629	4871	-.760
1500	4.97	5971	41.17	37.19	35443	2698	-.393
1600	4.97	6468	41.49	37.45	35259	523	-.071
1700	4.97	6965	41.79	37.70	0	0	0
1800	4.97	7462	42.08	37.94	0	0	0
1900	4.98	7959	42.35	38.17	0	0	0
2000	4.98	8458	42.60	38.38	0	0	0
2100	4.99	8956	42.84	38.58	0	0	0
2200	5.00	9456	43.08	38.79	0	0	0
2300	5.01	9957	43.30	38.98	0	0	0
2400	5.03	10459	43.51	39.16	0	0	0
2500	5.05	10963	43.72	39.34	0	0	0
2600	5.07	11469	43.92	39.51	0	0	0
2700	5.10	11981	44.11	39.68	0	0	0
2800	5.13	12492	44.30	39.84	0	0	0
2900	5.17	13006	44.48	40.00	0	0	0
3000	5.21	13525	44.65	40.15	0	0	0

THERMODYNAMIC PROPERTIES OF THE ELEMENTS

LITHIUM

LITHIUM	Li_2		Reference State for Calculating ΔH_f°, ΔF_f°, and $Log_{10}K_p$:
IDEAL DIATOMIC GAS			Solid from 298° to 453.7°, Liquid from 453.7° to 1604°, Ideal Monatomic Gas from 1604° to 3000°.

Gfw	13.880	GRAMS		
$(H^\circ_{298.15} - H^\circ_0)$ =	2,312	CAL./GFW.		

T TEMPERATURE °K	C_p° HEAT CAPACITY CAL./DEG./GFW.	$H^\circ_T - H^\circ_{298.15}$ HEAT CONTENT CAL./GFW.	S°_T ENTROPY CAL./DEG./GFW.	$-(F^\circ_T - H^\circ_{298.15})/T$ FREE ENERGY FUNCTION CAL./DEG./GFW.	FORMATION FROM REFERENCE STATE		
					HEAT ΔH°_f CAL./GFW.	FREE ENERGY ΔF°_f CAL./GFW.	$LOG_{10} K_p$
298	8.62	0	47.05	47.05	50467	40463	-29.661
300	8.63	16	47.12	47.07	50461	40399	-29.433
400	8.83	889	49.64	47.42	50096	37096	-20.269
500	8.94	1778	51.62	48.07	48147	34047	-14.882
600	9.02	2674	53.25	48.80	47615	31277	-11.393
700	9.08	3582	54.66	49.55	47125	28589	-8.926
800	9.13	4488	55.86	50.25	46645	25973	-7.096
900	9.18	5410	56.95	50.94	46185	23424	-5.688
1000	9.22	6328	57.92	51.60	45723	20903	-4.568
1100	9.26	7247	58.79	52.21	45266	18459	-3.667
1200	9.30	8176	59.60	52.79	44819	16043	-2.922
1300	9.34	9115	60.35	53.34	44386	13667	-2.297
1400	9.37	10050	61.05	53.88	43949	11287	-1.762
1500	9.41	10978	61.69	54.38	43511	8996	-1.310
1600	9.44	11921	62.30	54.85	43092	6708	-.916
1700	9.48	12867	62.87	55.31	-27474	7733	-.994
1800	9.51	13817	63.42	55.75	-27518	9814	-1.191
1900	9.55	14770	63.93	56.16	-27561	11902	-1.368
2000	9.58	15726	64.42	56.56	-27601	13959	-1.525

M.P.		°K
ΔH_m		CAL./GFW.
B.P.		°K
ΔH_v		CAL./GFW.
S.P.		°K
ΔH_s		CAL./GFW.
T.P.		°K
ΔH_t		CAL./GFW.
T.P.		°K
ΔH_t		CAL./GFW.
$T_c =$		°K
$P_c =$		ATM.

LUTETIUM

LUTETIUM Lu Solid from 298° to 2000°, Liquid from 2000° to 2200°, Ideal Monatomic Gas from 2200° to 3000°.

						FORMATION FROM REFERENCE STATE		
T TEMPERATURE °K	C_P° HEAT CAPACITY CAL./DEG./GFW.	$H_T^\circ - H_{298.15}^\circ$ HEAT CONTENT CAL./GFW.	S_T° ENTROPY CAL./DEG./GFW.	$-(F_T^\circ - H_{298.15}^\circ)/T$ FREE ENERGY FUNCTION CAL./DEG./GFW.		HEAT ΔH_f° CAL./GFW.	FREE ENERGY ΔF_f° CAL./GFW.	$\text{LOG}_{10} K_P$
298	6.45	0	11.75	11.75				
300	6.45	11	11.79	11.76				
400	6.60	665	13.66	12.00				
500	6.75	1330	15.15	12.49				
600	6.90	2015	16.40	13.05				
700	7.05	2710	17.47	13.60				
800	7.20	3425	18.42	14.14				
900	7.35	4150	19.28	14.67				
1000	7.50	4890	20.06	15.17				
1100	7.65	5650	20.78	15.65				
1200	7.80	6420	21.46	16.11				
1300	7.95	7210	22.09	16.55				
1400	8.10	8010	22.68	16.96				
1500	8.25	8830	23.25	17.37				
1600	8.40	9660	23.78	17.75				
1700	8.55	10510	24.30	18.12				
1800	8.70	11370	24.79	18.48				
1900	8.85	12250	25.26	18.82				
2000	8.00	17740	28.04	19.17				
2100	8.00	18540	28.41	19.59				
2200	8.00	19340	28.78	19.99				
2300	6.07	79250	55.98	21.53				
2400	6.05	79850	56.24	22.97				
2500	6.04	80460	56.49	24.31				
2600	6.02	81060	56.72	25.55				
2700	6.01	81660	56.95	26.71				
2800	6.01	82260	57.17	27.80				
2900	6.00	82860	57.38	28.81				
3000	6.00	83460	57.58	29.76				

Gfw 174.99 GRAMS
$(H^\circ_{298.15} - H^\circ_0) = $ CAL./GFW.
M.P. (2,000) °K
ΔH_m (4,600) CAL./GFW.
B.P. (2,200) °K
ΔH_v (59,000) CAL./GFW.
S.P. °K
ΔH_s CAL./GFW.
T.P. °K
ΔH_t CAL./GFW.
T.P. °K
ΔH_t CAL./GFW.
$T_c = $ °K
$P_c = $ ATM.

LUTETIUM

LUTETIUM Lu
IDEAL MONATOMIC GAS

Reference State for Calculating ΔH_f°, ΔF_f°, and $\log_{10} K_p$:
Solid from 298° to 2000°, Liquid from 2000° to 2200°, Ideal Monatomic Gas from 2200° to 3000°.

Gfw 174.99 GRAMS
$(H^\circ_{298.15} - H^\circ_0) = 1,482$ CAL./GFW.

T TEMPERATURE °K	C_p° HEAT CAPACITY CAL./DEG./GFW.	$H_T^\circ - H_{298.15}^\circ$ HEAT CONTENT CAL./GFW.	S_T° ENTROPY CAL./DEG./GFW.	$-(F_T^\circ - H_{298.15}^\circ)$ FREE ENERGY FUNCTION CAL./DEG./GFW.	FORMATION FROM REFERENCE STATE		
					HEAT ΔH_f° CAL./GFW.	FREE ENERGY ΔF_f° CAL./GFW.	$\log_{10} K_p$
298	4.99	0	44.14	44.14	67200	57543	-42.181
300	4.99	9	44.17	44.14	67198	57484	-41.880
400	5.09	512	45.62	44.34	67047	54263	-29.650
500	5.28	1030	46.77	44.71	66900	51090	-22.332
600	5.53	1570	47.76	45.15	66755	47939	-17.463
700	5.77	2136	48.63	45.58	66626	44814	-13.992
800	5.98	2723	49.41	46.01	66498	41706	-11.394
900	6.13	3329	50.13	46.44	66379	38614	-9.377
1000	6.23	3948	50.78	46.84	66258	35538	-7.767
1100	6.29	4574	51.38	47.23	66124	32464	-6.450
1200	6.32	5205	51.93	47.60	65985	29421	-5.358
1300	6.32	5837	52.43	47.94	65827	26385	-4.436
1400	6.31	6469	52.90	48.28	65659	23351	-3.645
1500	6.29	7100	53.33	48.60	65470	20350	-2.965
1600	6.27	7728	53.74	48.91	65268	17332	-2.367
1700	6.24	8353	54.12	49.22	65043	14349	-1.844
1800	6.21	8976	54.47	49.49	64806	11382	-1.381
1900	6.18	9595	54.81	49.76	64545	8400	-.966
2000	6.15	10212	55.13	50.03	59672	5492	-.600
2100	6.12	10825	55.43	50.28	59485	2743	-.285
2200	6.10	11436	55.71	50.52	59296	50	-.004
2300	6.07	12045	55.98	50.75	0	0	0
2400	6.05	12651	56.24	50.97	0	0	0
2500	6.04	13256	56.49	51.19	0	0	0
2600	6.02	13859	56.72	51.39	0	0	0
2700	6.01	14461	56.95	51.60	0	0	0
2800	6.01	15061	57.17	51.80	0	0	0
2900	6.00	15662	57.38	51.98	0	0	0
3000	6.00	16262	57.58	52.16	0	0	0

M.P. °K
ΔH_m CAL./GFW.

B.P. °K
ΔH_v CAL./GFW.

S.P. °K
ΔH_s CAL./GFW.

T.P. °K
ΔH_t CAL./GFW.

T.P. °K
ΔH_t CAL./GFW.

T_c = °K
P_c = ATM.

MAGNESIUM

MAGNESIUM	Mg		Solid from 298° to 923°, Liquid from 923° to 1390°, Ideal Monatomic Gas from 1390° to 3000°.						
Gfw	24.32	GRAMS						FORMATION FROM REFERENCE STATE	
REFERENCE STATE									
$(H°_{298.15} - H°_0) = 1,195$		CAL./GFW.							
			T °K TEMPERATURE	$C°_P$ HEAT CAPACITY CAL./DEG./GFW.	$H°_T - H°_{298.15}$ HEAT CONTENT CAL./GFW.	$S°_T$ ENTROPY CAL./DEG./GFW.	$-(F°_T - H°_{298.15})/T$ FREE ENERGY FUNCTION CAL./DEG./GFW.	HEAT $\Delta H°_f$ CAL./GFW.	FREE ENERGY $\Delta F°_f$ CAL./GFW.	$LOG_{10} K_P$
M.P.	923	°K	298	5.96	0	7.81	7.81			
ΔH_m	2,140	CAL./GFW.	300	5.97	10	7.85	7.82			
			400	6.24	620	9.60	8.05			
			500	6.48	1256	11.02	8.51			
			600	6.76	1920	12.23	9.03			
			700	7.08	2610	13.29	9.57			
B.P.	1,390	°K	800	7.42	3330	14.26	10.10			
ΔH_v	30,750	CAL./GFW.	900	7.81	4095	15.15	10.60			
			1000	7.88	7010	18.29	11.28			
			1100	8.14	7810	19.06	11.96			
			1200	8.40	8640	19.78	12.58			
S.P.		°K	1300	8.66	9490	20.46	13.16			
ΔH_s		CAL./GFW.	1400	4.97	41074	43.19	13.86			
			1500	4.97	41570	43.53	15.82			
			1600	4.97	42070	43.85	17.56			
			1700	4.97	42560	44.15	19.12			
T.P.		°K	1800	4.97	43060	44.44	20.52			
ΔH_t		CAL./GFW.	1900	4.97	43560	44.71	21.79			
			2000	4.97	44060	44.96	22.93			
			2100	4.97	44550	45.20	23.99			
T.P.		°K	2200	4.97	45050	45.43	24.96			
ΔH_t		CAL./GFW.	2300	4.97	45550	45.66	25.86			
			2400	4.97	46040	45.87	26.69			
			2500	4.98	46540	46.07	27.46			
			2600	4.98	47040	46.26	28.17			
$T_c =$		°K	2700	4.99	47540	46.45	28.85			
			2800	5.00	48040	46.63	29.48			
$P_c =$		ATM.	2900	5.01	48540	46.81	30.08			
			3000	5.02	49040	46.97	30.63			

124

MAGNESIUM

MAGNESIUM Mg

IDEAL MONATOMIC GAS

Gfw 24.32 GRAMS

$(H°_{298.15} - H°_0) = 1,481$ CAL./GFW

Reference State for Calculating $\Delta H°_f$, $\Delta F°_f$, and $\log_{10} K_p$:
Solid from 298° to 923°, Liquid from 923° to 1390°, Ideal Monatomic Gas from 1390° to 3000°.

T TEMPERATURE °K	$C°_p$ HEAT CAPACITY CAL./DEG./GFW	$H°_T - H°_{298.15}$ HEAT CONTENT CAL./GFW	$S°_T$ ENTROPY CAL./DEG./GFW	$-(F°_T - H°_{298.15})/T$ FREE ENERGY FUNCTION CAL./DEG./GFW	FORMATION FROM REFERENCE STATE		
					HEAT $\Delta H°_f$ CAL./GFW	FREE ENERGY $\Delta F°_f$ CAL./GFW	$\log_{10} K_p$
298	4.97	0	35.51	35.51	35600	27341	-20.042
300	4.97	9	35.54	35.51	35599	27292	-19.883
400	4.97	506	36.96	35.70	35486	24542	-13.410
500	4.97	1003	38.07	36.07	35347	21822	-9.539
600	4.97	1500	38.98	36.48	35180	19130	-6.968
700	4.97	1996	39.75	36.90	34986	16464	-5.140
800	4.97	2493	40.41	37.30	34763	13843	-3.782
900	4.97	2990	40.99	37.67	34495	11239	-2.729
1000	4.97	3487	41.52	38.04	32077	8847	-1.933
1100	4.97	3984	41.99	38.37	31774	6551	-1.301
1200	4.97	4480	42.42	38.69	31440	4272	-.778
1300	4.97	4977	42.82	39.00	31087	2019	-.339
1400	4.97	5474	43.19	39.28	0	0	0
1500	4.97	5971	43.53	39.55	0	0	0
1600	4.97	6468	43.85	39.81	0	0	0
1700	4.97	6964	44.15	40.06	0	0	0
1800	4.97	7461	44.44	40.30	0	0	0
1900	4.97	7958	44.71	40.53	0	0	0
2000	4.97	8455	44.96	40.74	0	0	0
2100	4.97	8952	45.20	40.94	0	0	0
2200	4.97	9448	45.43	41.14	0	0	0
2300	4.97	9945	45.66	41.34	0	0	0
2400	4.97	10443	45.87	41.52	0	0	0
2500	4.98	10940	46.07	41.70	0	0	0
2600	4.98	11438	46.26	41.87	0	0	0
2700	4.99	11937	46.45	42.03	0	0	0
2800	5.00	12436	46.63	42.19	0	0	0
2900	5.01	12937	46.81	42.35	0	0	0
3000	5.02	13438	46.97	42.48	0	0	0

M.P. °K
ΔH_m CAL./GFW

B.P. °K
ΔH_v CAL./GFW

S.P. °K
ΔH_s CAL./GFW

T.P. °K
ΔH_t CAL./GFW

T.P. °K
ΔH_t CAL./GFW

T_c = °K
P_c = ATM.

MANGANESE

MANGANESE Mn
Gfw 54.94 **GRAMS**
REFERENCE STATE
$(H^\circ_{298.15} - H^\circ_0) = 1,194$ CAL./GFW.

Solid I from 298° to 1000°, Solid II from 1000° to 1374°, Solid III from 1374° to 1410°, Solid IV from 1410° to 1517°, Liquid from 1517° to 2314°, Ideal Monatomic Gas from 2314° to 3000°.

					FORMATION FROM REFERENCE STATE		
T TEMPERATURE °K	C°_P HEAT CAPACITY CAL./DEG./GFW.	$H^\circ_T - H^\circ_{298.15}$ HEAT CONTENT CAL./GFW.	S°_T ENTROPY CAL./DEG./GFW.	$-(F^\circ_T - H^\circ_{298.15})$ FREE ENERGY FUNCTION CAL./DEG./GFW.	HEAT ΔH°_f CAL./GFW.	FREE ENERGY ΔF°_f CAL./GFW.	LOG$_{10}$ K$_P$
298	6.29	0	7.65	7.65			
300	6.29	11	7.69	7.66			
400	6.75	664	9.57	7.91			
500	7.18	1360	11.12	8.40			
600	7.54	2100	12.46	8.96			
700	7.87	2870	13.65	9.55			
800	8.21	3670	14.72	10.14			
900	8.60	4510	15.71	10.70			
1000	9.30	5920	17.18	11.26			
1100	9.30	6850	18.06	11.84			
1200	9.30	7780	18.87	12.39			
1300	9.30	8710	19.62	12.92			
1400	10.70	10220	20.73	13.43			
1500	11.30	11780	21.81	13.96			
1600	11.00	16380	24.94	14.71			
1700	11.00	17480	25.61	15.33			
1800	11.00	18580	26.24	15.92			
1900	11.00	19680	26.83	16.48			
2000	11.00	20780	27.40	17.01			
2100	11.00	21880	27.93	17.52			
2200	11.00	22980	28.45	18.01			
2300	11.00	24080	28.93	18.47			
2400	5.02	77180	51.86	19.71			
2500	5.04	77690	52.07	21.00			
2600	5.07	78190	52.26	22.19			
2700	5.10	78700	52.46	23.32			
2800	5.14	79210	52.64	24.36			
2900	5.19	79730	52.82	25.33			
3000	5.25	80250	53.00	26.25			

M.P. 1,517 °K
ΔH_m 3,500 CAL./GFW.

B.P. 2,314 °K
ΔH_v 52,520 CAL./GFW.

S.P. °K
ΔH_s CAL./GFW.

T.P. 1,000 °K
ΔH_t 535 CAL./GFW.

T.P. 1,374 °K
ΔH_t 545 CAL./GFW.

$T_c =$ 1,410 °K
$P_c =$ 430 ATM.

THERMODYNAMIC PROPERTIES OF THE ELEMENTS

MANGANESE

MANGANESE   Mn   Reference State for Calculating $\Delta H_f^°$, $\Delta F_f^°$, and $\text{Log}_{10}K_p$: Solid I 298° to 1000°, Solid II 1000° to 1374°, Solid III 1374° to 1410°, Solid IV 1410° to 1517°, Liquid 1517° to 2314°, Ideal Monatomic Gas 2314° to 3000°.

							FORMATION FROM REFERENCE STATE		
Gfw 54.94 GRAMS		T °K	$C_p^°$	$H_T^° - H_{298.15}^°$	$S_T^°$	$-(F_T^° - H_{298.15}^°)/T$	HEAT $\Delta H_f^°$	FREE ENERGY $\Delta F_f^°$	LOG$_{10}$ K$_p$
IDEAL MONATOMIC GAS			HEAT CAPACITY	HEAT CONTENT	ENTROPY	FREE ENERGY FUNCTION			
$(H_{298.15}^° - H_0^°) = 1,481$ CAL./GFW.		TEMPERATURE °K	CAL./DEG./GFW.	CAL./GFW.	CAL./DEG./GFW.	CAL./DEG./GFW.	CAL./GFW.	CAL./GFW.	
M.P.	°K	298	4.97	0	41.49	41.49	66730	56640	-41.519
		300	4.97	9	41.52	41.49	66728	56579	-41.221
ΔH_m	CAL./GFW.	400	4.97	506	42.95	41.69	66572	53220	-29.080
		500	4.97	1003	44.06	42.06	66373	49903	-21.814
		600	4.97	1500	44.97	42.47	66130	46624	-16.984
B.P.	°K	700	4.97	1996	45.73	42.88	65856	43400	-13.551
		800	4.97	2493	46.40	43.29	65553	40209	-10.985
ΔH_v	CAL./GFW.	900	4.97	2990	46.98	43.66	65210	37067	-9.001
		1000	4.97	3487	47.51	44.03	64297	33967	-7.424
		1100	4.97	3984	47.98	44.36	63864	30952	-6.150
S.P.	°K	1200	4.97	4480	48.41	44.68	63430	27982	-5.096
		1300	4.97	4977	48.81	44.99	62997	25050	-4.211
ΔH_s	CAL./GFW.	1400	4.97	5474	49.18	45.27	61984	22154	-3.458
		1500	4.97	5971	49.52	45.54	60921	19356	-2.820
		1600	4.97	6468	49.84	45.80	56818	16978	-2.319
T.P.	°K	1700	4.97	6965	50.14	46.05	56215	14514	-1.865
		1800	4.97	7462	50.43	46.29	55612	12070	-1.465
ΔH_t	CAL./GFW.	1900	4.97	7959	50.69	46.51	55009	9675	-1.112
		2000	4.98	8456	50.95	46.73	54406	7306	-.798
		2100	4.98	8953	51.19	46.93	53804	4958	-.515
T.P.	°K	2200	4.99	9453	51.42	47.13	53203	2669	-.265
		2300	5.00	9952	51.65	47.33	52602	346	-.032
ΔH_t	CAL./GFW.	2400	5.02	10453	51.86	47.51	0	0	0
		2500	5.04	10956	52.07	47.69	0	0	0
		2600	5.07	11461	52.26	47.86	0	0	0
		2700	5.10	11970	52.46	48.03	0	0	0
$T_c =$	°K	2800	5.14	12482	52.64	48.19	0	0	0
		2900	5.19	12998	52.82	48.34	0	0	0
$P_c =$	ATM.	3000	5.25	13520	53.00	48.50	0	0	0

MERCURY

MERCURY Hg
Liquid from 298° to 629.88°, Ideal Monatomic Gas from 629.88° to 3000°.

REFERENCE STATE

Gfw 200.61 GRAMS
$(H°_{298.15} - H°_0) = 2,232$ CAL./GFW

						FORMATION FROM REFERENCE STATE		
T TEMPERATURE °K	$C°_p$ HEAT CAPACITY CAL./DEG./GFW	$H°_T - H°_{298.15}$ HEAT CONTENT CAL./GFW	$S°_T$ ENTROPY CAL./DEG./GFW	$-(F°_T - H°_{298.15})/T$ FREE ENERGY FUNCTION CAL./DEG./GFW		HEAT $\Delta H°_f$ CAL./GFW	FREE ENERGY $\Delta F°_f$ CAL./GFW	LOG$_{10}$ K$_p$
298	6.69	0	18.19	18.19				
300	6.68	12	18.23	18.19				
400	6.54	672	20.13	18.45				
500	6.48	1323	21.58	18.94				
600	6.49	1970	22.77	19.49				
700	4.97	16650	46.03	22.25				
800	4.97	17140	46.70	25.28				
900	4.97	17640	47.28	27.68				
1000	4.97	18140	47.81	29.67				
1100	4.97	18630	48.28	31.35				
1200	4.97	19130	48.71	32.77				
1300	4.97	19630	49.11	34.01				
1400	4.97	20120	49.48	35.11				
1500	4.97	20620	49.82	36.08				
1600	4.97	21120	50.14	36.94				
1700	4.97	21610	50.44	37.73				
1800	4.97	22110	50.73	38.45				
1900	4.97	22610	50.99	39.09				
2000	4.97	23110	51.25	39.70				
2100	4.97	23600	51.49	40.26				
2200	4.97	24100	51.72	40.77				
2300	4.97	24600	51.94	41.25				
2400	4.97	25090	52.15	41.70				
2500	4.97	25590	52.36	42.13				
2600	4.97	26090	52.55	42.52				
2700	4.97	26580	52.74	42.90				
2800	4.97	27080	52.92	43.25				
2900	4.97	27580	53.09	43.58				
3000	4.97	28070	53.26	43.91				

M.P. 234.29 °K
ΔH_m 548.6 CAL./GFW

B.P. 629.88 °K
ΔH_v 14,137 CAL./GFW

S.P. °K
ΔH_s CAL./GFW

T.P. °K
ΔH_t CAL./GFW

T.P. °K
ΔH_t CAL./GFW

$T_c =$ °K
$P_c =$ ATM.

MERCURY

MERCURY　　Hg　　Reference State for Calculating ΔH_f°, ΔF_f°, and $Log_{10} K_p$: Liquid

IDEAL MONATOMIC GAS　　from 298° to 629.88°, Ideal Monatomic Gas from 629.88° to 3000°.

Gfw　200.61　GRAMS

$(H_{298.15}^\circ - H_0^\circ) = 1,481$ CAL./GFW.

						FORMATION FROM REFERENCE STATE		
T TEMPERATURE °K	C_p° HEAT CAPACITY CAL./DEG./GFW.	$H_T^\circ - H_{298.15}^\circ$ HEAT CONTENT CAL./GFW.	S_T° ENTROPY CAL./DEG./GFW.	$-(F_T^\circ - H_{298.15}^\circ)/T$ FREE ENERGY FUNCTION CAL./DEG./GFW.		HEAT ΔH_f° CAL./GFW.	FREE ENERGY ΔF_f° CAL./GFW.	$LOG_{10} K_p$
298	4.97	0	41.79	41.79		14650	7613	−5.580
300	4.97	9	41.82	41.79		14647	7570	−5.515
400	4.97	506	43.25	41.99		14484	5236	−2.861
500	4.97	1003	44.36	42.36		14330	2940	−1.285
600	4.97	1500	45.27	42.77		14180	680	−.247
700	4.97	1996	46.03	43.18		0	0	0
800	4.97	2493	46.70	43.59		0	0	0
900	4.97	2990	47.28	43.96		0	0	0
1000	4.97	3487	47.81	44.33		0	0	0
1100	4.97	3984	48.28	44.66		0	0	0
1200	4.97	4480	48.71	44.98		0	0	0
1300	4.97	4977	49.11	45.29		0	0	0
1400	4.97	5474	49.48	45.57		0	0	0
1500	4.97	5971	49.82	45.84		0	0	0
1600	4.97	6468	50.14	46.10		0	0	0
1700	4.97	6964	50.44	46.35		0	0	0
1800	4.97	7461	50.73	46.59		0	0	0
1900	4.97	7958	50.99	46.81		0	0	0
2000	4.97	8455	51.25	47.03		0	0	0
2100	4.97	8952	51.49	47.23		0	0	0
2200	4.97	9448	51.72	47.43		0	0	0
2300	4.97	9945	51.94	47.62		0	0	0
2400	4.97	10442	52.15	47.80		0	0	0
2500	4.97	10939	52.36	47.99		0	0	0
2600	4.97	11436	52.55	48.16		0	0	0
2700	4.97	11932	52.74	48.33		0	0	0
2800	4.97	12429	52.92	48.48		0	0	0
2900	4.97	12926	53.09	48.64		0	0	0
3000	4.97	13423	53.26	48.79		0	0	0

M.P. °K

ΔH_m CAL./GFW.

B.P. °K

ΔH_v CAL./GFW.

S.P. °K

ΔH_s CAL./GFW.

T.P. °K

ΔH_t CAL./GFW.

T.P. °K

ΔH_t CAL./GFW.

$T_c =$ °K

$P_c =$ ATM.

MOLYBDENUM

MOLYBDENUM	Mo		Solid from 298° to 2890°, Liquid from 2890° to 3000°.
GFW	95.95	GRAMS	
$(H°_{298.15} - H°_0)$ =	1,092	CAL./GFW.	
M.P.	2,890	°K	
ΔH_m	6,600	CAL./GFW.	
B.P.	5,100	°K	
ΔH_v	142,000	CAL./GFW.	
S.P.		°K	
ΔH_s		CAL./GFW.	
T.P.		°K	
ΔH_t		CAL./GFW.	
T.P.		°K	
ΔH_t		CAL./GFW.	
T_c =		°K	
P_c =		ATM.	

T TEMPERATURE °K	$C°_P$ HEAT CAPACITY CAL./DEG./GFW.	$H°_T - H°_{298.15}$ HEAT CONTENT CAL./GFW.	$S°_T$ ENTROPY CAL./DEG./GFW.	$-(F°_T - H°_{298.15})/T$ FREE ENERGY FUNCTION CAL./DEG./GFW.	FORMATION FROM REFERENCE STATE			
					HEAT $\Delta H°_f$ CAL./GFW.	FREE ENERGY $\Delta F°_f$ CAL./GFW.	LOG$_{10}$ K_P
298	5.68	0	6.83	6.83			
300	5.69	11	6.86	6.83			
400	5.97	595	8.54	7.06			
500	6.15	1203	9.89	7.49			
600	6.28	1825	11.03	7.99			
700	6.35	2460	12.00	8.49			
800	6.44	3100	12.85	8.98			
900	6.55	3750	13.62	9.46			
1000	6.70	4410	14.32	9.91			
1100	6.86	5090	14.96	10.34			
1200	7.05	5790	15.57	10.75			
1300	7.24	6510	16.14	11.14			
1400	7.45	7250	16.68	11.51			
1500	7.65	8000	17.21	11.88			
1600	7.83	8780	17.71	12.23			
1700	8.00	9570	18.19	12.57			
1800	8.18	10380	18.65	12.89			
1900	8.35	11200	19.10	13.21			
2000	8.52	12040	19.53	13.51			
2100	8.69	12900	19.95	13.81			
2200	8.85	13770	20.36	14.11			
2300	9.02	14680	20.75	14.37			
2400	9.19	15580	21.14	14.65			
2500	9.36	16510	21.52	14.92			
2600	9.53	17460	21.89	15.18			
2700	9.70	18420	22.25	15.43			
2800	9.87	19400	22.61	15.69			
2900	10.00	26990	26.46	17.16			
3000	10.00	27990	26.80	17.47			

MOLYBDENUM

MOLYBDENUM Mo Reference State for Calculating ΔH_f°, ΔF_f°, and $\log_{10} K_P$:
Gfw 95.95 GRAMS
IDEAL MONATOMIC GAS Solid from 298° to 2890°, Liquid from 2890° to 3000°.
$(H°_{298.15} - H°_0) = 1,481$ CAL./GFW.

T TEMPERATURE °K	C_P° HEAT CAPACITY CAL./DEG./GFW.	$H°_T - H°_{298.15}$ HEAT CONTENT CAL./GFW.	$S°_T$ ENTROPY CAL./DEG./GFW.	$-(F°_T - H°_{298.15})$ FREE ENERGY FUNCTION CAL./DEG./GFW.	FORMATION FROM REFERENCE STATE		
					HEAT Δ H°$_f$ CAL./GFW.	FREE ENERGY Δ F°$_f$ CAL./GFW.	LOG$_{10}$ K$_P$
298	4.97	0	43.46	43.46	157500	146578	-107.449
300	4.97	9	43.49	43.46	157498	146509	-106.740
400	4.97	506	44.92	43.66	157411	142859	-78.061
500	4.97	1003	46.03	44.03	157300	139230	-60.861
600	4.97	1500	46.94	44.44	157175	135629	-49.406
700	4.97	1996	47.70	44.85	157036	132046	-41.230
800	4.97	2493	48.37	45.26	156893	128477	-35.101
900	4.97	2990	48.95	45.63	156740	124943	-30.342
1000	4.97	3487	49.47	45.99	156577	121427	-26.540
1100	4.97	3984	49.95	46.33	156394	117905	-23.427
1200	4.97	4481	50.38	46.65	156191	114419	-20.840
1300	4.97	4978	50.78	46.96	155968	110936	-18.651
1400	4.98	5475	51.15	47.24	155725	107467	-16.777
1500	4.98	5973	51.49	47.51	155473	104053	-15.161
1600	5.00	6472	51.81	47.77	155192	100632	-13.745
1700	5.02	6973	52.12	48.02	154903	97222	-12.498
1800	5.04	7476	52.40	48.25	154596	93846	-11.393
1900	5.08	7982	52.68	48.48	154282	90480	-10.407
2000	5.13	8492	52.94	48.70	153952	87132	-9.520
2100	5.18	9007	53.19	48.91	153607	83803	-8.721
2200	5.25	9529	53.43	49.10	153259	80505	-7.997
2300	5.34	10059	53.67	49.30	152879	77163	-7.332
2400	5.44	10598	53.90	49.49	152518	73894	-6.728
2500	5.56	11147	54.12	49.67	152137	70637	-6.175
2600	5.69	11709	54.34	49.84	151749	67379	-5.663
2700	5.84	12286	54.56	50.01	151366	64129	-5.190
2800	6.00	12878	54.77	50.18	150978	60930	-4.755
2900	6.19	13487	54.99	50.34	143997	61260	-4.616
3000	6.39	14116	55.20	50.50	143626	58426	-4.256

M.P. °K
Δ H$_m$ CAL./GFW.

B.P. °K
Δ H$_v$ CAL./GFW.

S.P. °K
Δ H$_s$ CAL./GFW.

T.P. °K
Δ H$_t$ CAL./GFW.

T.P. °K
Δ H$_t$ CAL./GFW.

T$_c$ = °K
P$_c$ = ATM.

NEODYMIUM

NEODYMIUM Nd

Solid I from 298° to 1141°, Solid II from 1141° to 1297°, Liquid from 1297° to 3000°.

							FORMATION FROM REFERENCE STATE		
Gfw	144.27	GRAMS	T TEMPERATURE °K	$C_p°$ HEAT CAPACITY CAL./DEG./GFW.	$H_T° - H_{298.15}°$ HEAT CONTENT CAL./GFW.	$S_T°$ ENTROPY CAL./DEG./GFW.	$-(F_T°-H_0°)/T_{298.15}$ FREE ENERGY FUNCTION CAL./DEG./GFW.	HEAT $\Delta H_f°$ CAL./GFW.	FREE ENERGY $\Delta F_f°$ CAL./GFW.	$LOG_{10} K_p$
$(H°_{298.15} - H°_0) = 1,804$ CAL./GFW.			298	7.20	0	17.50	17.50			
			300	7.21	13	17.54	17.50			
M.P.	1,297	°K	400	7.75	761	19.69	17.79			
ΔH_m	(2,600)	CAL./GFW.	500	8.28	1563	21.48	18.36			
			600	8.81	2417	23.03	19.01			
			700	9.35	3325	24.43	19.68			
B.P.	3,360	°K	800	9.88	4287	25.72	20.37			
ΔH_v	67,800	CAL./GFW.	900	10.42	5302	26.91	21.02			
			1000	10.95	6370	28.04	21.67			
			1100	11.49	7492	29.10	22.29			
S.P.		°K	1200	8.00	8780	30.23	22.92			
			1300	8.00	12180	32.87	23.51			
ΔH_s		CAL./GFW.	1400	8.00	12980	33.47	24.20			
			1500	8.00	13780	34.02	24.84			
			1600	8.00	14580	34.53	25.42			
			1700	8.00	15380	35.02	25.98			
T.P.	1,141	°K	1800	8.00	16180	35.48	26.50			
ΔH_t	(340)	CAL./GFW.	1900	8.00	16980	35.91	26.98			
			2000	8.00	17780	36.32	27.43			
			2100	8.00	18580	36.71	27.87			
			2200	8.00	19380	37.08	28.28			
T.P.		°K	2300	8.00	20180	37.44	28.67			
ΔH_t		CAL./GFW.	2400	8.00	20980	37.78	29.04			
			2500	8.00	21780	38.10	29.39			
			2600	8.00	22580	38.42	29.74			
			2700	8.00	23380	38.72	30.07			
$T_c =$		°K	2800	8.00	24180	39.01	30.38			
$P_c =$		ATM.	2900	8.00	24980	39.29	30.68			
			3000	8.00	25780	39.56	30.97			

THERMODYNAMIC PROPERTIES OF THE ELEMENTS

NEODYMIUM

NEODYMIUM	Nd		Reference State for Calculating $\Delta H_f^°$, $\Delta F_f^°$, and
Gfw 144.27 GRAMS			$Log_{10} K_P$: Solid I from 298° to 1141°, Solid II
$(H°_{298.15} - H°_0) =$ 1,498 CAL./GFW.			from 1141° to 1297°, Liquid from 1297° to 3000°.

IDEAL MONATOMIC GAS

	T TEMPERATURE °K	$C_p°$ HEAT CAPACITY CAL./DEG./GFW.	$H°_T - H°_{298.15}$ HEAT CONTENT CAL./GFW.	$S°_T$ ENTROPY CAL./DEG./GFW.	$-(F°_T - H°_{298.15})/T$ FREE ENERGY FUNCTION CAL./DEG./GFW.	FORMATION FROM REFERENCE STATE		
						HEAT $\Delta H_f°$ CAL./GFW.	FREE ENERGY $\Delta F_f°$ CAL./GFW.	$LOG_{10} K_P$
	298	5.28	0	45.24	45.24	76800	68529	-50.235
	300	5.29	10	45.28	45.25	76797	68475	-49.888
	400	5.67	558	46.85	45.46	76597	65733	-35.917
M.P. °K	500	6.02	1143	48.15	45.87	76380	63045	-27.558
ΔH_m CAL./GFW.	600	6.28	1758	49.28	46.35	76141	60391	-21.999
	700	6.48	2397	50.26	46.84	75872	57791	-18.044
	800	6.63	3054	51.14	47.33	75567	55231	-15.089
B.P. °K	900	6.74	3723	51.93	47.80	75221	52703	-12.798
ΔH_v CAL./GFW.	1000	6.82	4401	52.64	48.24	74831	50231	-10.978
	1100	6.87	5086	53.29	48.67	74394	47785	-9.494
	1200	6.90	5774	53.89	49.08	73794	45402	-8.269
S.P. °K	1300	6.91	6465	54.44	49.47	71085	43044	-7.236
ΔH_s CAL./GFW.	1400	6.90	7155	54.96	49.85	70975	40889	-6.383
	1500	6.87	7843	55.43	50.21	70863	38748	-5.645
	1600	6.83	8528	55.87	50.54	70748	36604	-4.999
	1700	6.79	9210	56.29	50.88	70630	34471	-4.431
T.P. °K	1800	6.74	9886	56.67	51.18	70506	32364	-3.929
ΔH_t CAL./GFW.	1900	6.68	10557	57.03	51.48	70377	30249	-3.479
	2000	6.62	11222	57.38	51.77	70242	28122	-3.072
	2100	6.56	11881	57.70	52.05	70101	26022	-2.708
	2200	6.50	12533	58.00	52.31	69953	23929	-2.377
T.P. °K	2300	6.44	13180	58.29	52.56	69800	21845	-2.075
ΔH_t CAL./GFW.	2400	6.38	13821	58.56	52.81	69641	19769	-1.800
	2500	6.32	14456	58.82	53.04	69476	17676	-1.545
	2600	6.27	15086	59.07	53.27	69306	15616	-1.312
	2700	6.22	15711	59.30	53.49	69131	13565	-1.097
$T_c =$ °K	2800	6.18	16331	59.53	53.70	68951	11495	-.897
	2900	6.14	16947	59.75	53.91	68767	9433	-.710
$P_c =$ ATM.	3000	6.10	17559	59.95	54.10	68579	7409	-.539

NEON

Ideal Monatomic Gas from 298° to 3000°.

NEON	Ne							
GFW	20.183	GRAMS						
$(H°_{298.15} - H°_0)$ =	1,481	CAL./GFW.						

						FORMATION FROM REFERENCE STATE		
						HEAT Δ$H°_f$ CAL./GFW.	FREE ENERGY Δ$F°_f$ CAL./GFW.	$LOG_{10} K_p$

T TEMPERATURE °K	$C°_P$ HEAT CAPACITY CAL./DEG./GFW.	$H°_T - H°_{298.15}$ HEAT CONTENT CAL./GFW.	$S°_T$ ENTROPY CAL./DEG./GFW.	$-(F°_T - H°_{298.15})/T$ FREE ENERGY FUNCTION CAL./DEG./GFW.
298	4.97	0	34.95	34.95
300	4.97	9	34.98	34.95
400	4.97	506	36.41	35.15
500	4.97	1003	37.52	35.52
600	4.97	1500	38.42	35.92
700	4.97	1996	39.19	36.34
800	4.97	2493	39.85	36.74
900	4.97	2990	40.44	37.12
1000	4.97	3487	40.96	37.48
1100	4.97	3984	41.43	37.81
1200	4.97	4480	41.87	38.14
1300	4.97	4977	42.26	38.44
1400	4.97	5474	42.63	38.72
1500	4.97	5971	42.97	38.99
1600	4.97	6468	43.30	39.26
1700	4.97	6964	43.60	39.51
1800	4.97	7461	43.88	39.74
1900	4.97	7958	44.15	39.97
2000	4.97	8455	44.40	40.18
2100	4.97	8952	44.65	40.39
2200	4.97	9448	44.88	40.59
2300	4.97	9945	45.10	40.78
2400	4.97	10442	45.31	40.96
2500	4.97	10939	45.51	41.14
2600	4.97	11436	45.71	41.32
2700	4.97	11932	45.89	41.48
2800	4.97	12429	46.08	41.65
2900	4.97	12926	46.25	41.80
3000	4.97	13423	46.42	41.95

M.P.	24.55	°K
ΔH_m	80.1	CAL./GFW.
B.P.	27.07	°K
ΔH_v	422.	CAL./GFW.
S.P.		°K
ΔH_s		CAL./GFW.
T.P.		°K
ΔH_t		CAL./GFW.
T.P.		°K
ΔH_t		CAL./GFW.
T_c =	45.5	°K
P_c =	26.9	ATM.

NICKEL

Solid from 298° to 1728°, Liquid from 1728° to 3000°.

NICKEL	N1	GRAMS
Gfw	58.71	
$(H°_{298.15} - H°_0)$ =	1,144	CAL./GFW.

REFERENCE STATE

M.P.	1,728	°K
ΔH_m	4,210	CAL./GFW.
B.P.	3,110	°K
ΔH_v	88,870	CAL./GFW.
S.P.		°K
ΔH_s		CAL./GFW.
T.P.	630	°K
ΔH_t	0	CAL./GFW.
T.P.		°K
ΔH_t		CAL./GFW.
T_c =		°K
P_c =		ATM.

T TEMPERATURE °K	$C°_P$ HEAT CAPACITY CAL./DEG./GFW.	$H°_T - H°_{298.15}$ HEAT CONTENT CAL./GFW.	$S°_T$ ENTROPY CAL./DEG./GFW.	$-(F°_T - H°_{298.15})$ FREE ENERGY FUNCTION CAL./DEG./GFW.	FORMATION FROM REFERENCE STATE		
					HEAT $\Delta H°_f$ CAL./GFW.	FREE ENERGY $\Delta F°_f$ CAL./GFW.	$LOG_{10} K_P$
298	6.23	0	7.14	7.14			
300	6.24	12	7.18	7.14			
400	6.76	662	9.05	7.40			
500	7.47	1373	10.64	7.90			
600	8.37	2165	12.08	8.48			
700	7.35	2940	13.28	9.08			
800	7.44	3690	14.28	9.67			
900	7.62	4445	15.17	10.24			
1000	7.80	5210	15.98	10.77			
1100	7.98	5985	16.72	11.28			
1200	8.16	6780	17.41	11.76			
1300	8.34	7600	18.07	12.23			
1400	8.52	8450	18.70	12.67			
1500	8.70	9320	19.30	13.09			
1600	8.88	10210	19.87	13.49			
1700	9.06	11110	20.42	13.89			
1800	9.20	16230	23.37	14.36			
1900	9.20	17150	23.87	14.85			
2000	9.20	18070	24.34	15.31			
2100	9.20	18990	24.79	15.75			
2200	9.20	19910	25.22	16.17			
2300	9.20	20830	25.63	16.58			
2400	9.20	21750	26.02	16.96			
2500	9.20	22670	26.39	17.33			
2600	9.20	23590	26.76	17.69			
2700	9.20	24510	27.10	18.03			
2800	9.20	25430	27.44	18.36			
2900	9.20	26350	27.76	18.68			
3000	9.20	27270	28.07	18.98			

NICKEL

NICKEL N1 Reference State for Calculating ΔH_f°, ΔF_f°, and $\log_{10} K_p$:
IDEAL MONATOMIC GAS Solid from 298° to 1728°, Liquid from 1728° to 3000°.

Gfw 58.71 GRAMS
$(H_{298.15}^\circ - H_0^\circ)$ = 1,631 CAL./GFW.

T °K TEMPERATURE	C_p° HEAT CAPACITY CAL./DEG./GFW.	$H_T^\circ - H_{298.15}^\circ$ HEAT CONTENT CAL./GFW.	S_T° ENTROPY CAL./DEG./GFW.	$-(F_T^\circ - H_{298.15}^\circ)$ FREE ENERGY FUNCTION CAL./DEG./GFW.	FORMATION FROM REFERENCE STATE		
					HEAT ΔH_f° CAL./GFW.	FREE ENERGY ΔF_f° CAL./GFW.	$\log_{10} K_p$
298	5.58	0	43.52	43.52	101260	90413	-66.277
300	5.58	10	43.55	43.52	101258	90347	-65.823
400	5.70	574	45.18	43.74	101172	86720	-47.385
500	5.83	1151	46.46	44.16	101038	83128	-36.337
600	5.91	1738	47.53	44.64	100833	79563	-28.983
700	5.96	2332	48.45	45.12	100652	76033	-23.740
800	5.97	2929	49.24	45.58	100499	72531	-19.816
900	5.96	3525	49.95	46.04	100340	69038	-16.765
1000	5.94	4120	50.57	46.45	100170	65580	-14.333
1100	5.90	4712	51.14	46.86	99987	62125	-12.344
1200	5.86	5301	51.65	47.24	99781	58693	-10.690
1300	5.82	5885	52.12	47.60	99545	55280	-9.294
1400	5.78	6465	52.55	47.94	99275	51885	-8.100
1500	5.74	7041	52.95	48.25	98981	48506	-7.067
1600	5.70	7613	53.31	48.56	98663	45159	-6.168
1700	5.66	8181	53.66	48.85	98331	41823	-5.376
1800	5.62	8745	53.98	49.13	93775	38677	-4.695
1900	5.59	9306	54.28	49.39	93416	35637	-4.098
2000	5.56	9863	54.57	49.64	93053	32593	-3.561
2100	5.53	10417	54.84	49.88	92687	29582	-3.078
2200	5.50	10968	55.10	50.12	92318	26582	-2.640
2300	5.47	11517	55.34	50.34	91947	23614	-2.243
2400	5.45	12063	55.57	50.55	91573	20653	-1.880
2500	5.43	12607	55.80	50.75	91197	17672	-1.544
2600	5.41	13149	56.01	50.96	90819	14760	-1.241
2700	5.40	13690	56.21	51.14	90440	11843	-.958
2800	5.38	14229	56.41	51.33	90059	8943	-.698
2900	5.37	14767	56.60	51.51	89677	6041	-.455
3000	5.36	15303	56.78	51.68	89293	3163	-.230

M.P. °K
ΔH_m CAL./GFW.
B.P. °K
ΔH_v CAL./GFW.
S.P. °K
ΔH_s CAL./GFW.
T.P. °K
ΔH_t CAL./GFW.
T.P. °K
ΔH_t CAL./GFW.
T_c = °K
P_c = ATM.

NIOBIUM

THERMODYNAMIC PROPERTIES OF THE ELEMENTS

Solid from 298° to 2770°, Liquid from 2770° to 3000°.

NIOBIUM	Nb
G_{fw} 92.91 GRAMS	
$(H°_{298.15} - H°_0) = 1,264$ CAL./GFW.	
REFERENCE STATE	

M.P.	2,770	°K
ΔH_m	(6,400)	CAL./GFW.
B.P.	5,200	°K
ΔH_v	166,500	CAL./GFW.
S.P.		°K
ΔH_s		CAL./GFW.
T.P.		°K
ΔH_t		CAL./GFW.
T.P.		°K
ΔH_t		CAL./GFW.
T_c =		°K
P_c =		ATM.

T TEMPERATURE °K	$C_p°$ HEAT CAPACITY CAL./DEG./GFW.	$H°_T - H°_{298.15}$ HEAT CONTENT CAL./GFW.	$S°_T$ ENTROPY CAL./DEG./GFW.	$-(F°_T - H°_{298.15})/T$ FREE ENERGY FUNCTION CAL./DEG./GFW.	FORMATION FROM REFERENCE STATE			
					HEAT ΔH° CAL./GFW.	FREE ENERGY ΔF° CAL./GFW.	LOG₁₀ K_p
298	5.95	0	8.73	8.73			
300	5.95	11	8.77	8.74			
400	6.04	610	10.49	8.97			
500	6.14	1220	11.85	9.41			
600	6.24	1840	12.98	9.92			
700	6.33	2470	13.94	10.42			
800	6.43	3110	14.80	10.92			
900	6.52	3750	15.56	11.40			
1000	6.62	4410	16.25	11.84			
1100	6.72	5080	16.89	12.28			
1200	6.81	5750	17.48	12.69			
1300	6.91	6440	18.02	13.07			
1400	7.00	7130	18.54	13.45			
1500	7.10	7840	19.03	13.81			
1600	7.20	8550	19.49	14.15			
1700	7.29	9280	19.93	14.48			
1800	7.39	10010	20.35	14.79			
1900	7.48	10760	20.75	15.09			
2000	7.58	11510	21.13	15.38			
2100	7.68	12270	21.51	15.67			
2200	7.77	13050	21.87	15.94			
2300	7.87	13830	22.21	16.20			
2400	7.96	14620	22.55	16.46			
2500	8.06	15420	22.88	16.72			
2600	8.16	16230	23.20	16.96			
2700	8.25	17050	23.51	17.20			
2800	8.00	24270	26.12	17.46			
2900	8.00	25070	26.40	17.76			
3000	8.00	25870	26.67	18.05			

NIOBIUM

NIOBIUM   **Nb**

Reference State for Calculating ΔH_f°, ΔF_f°, and $\mathrm{Log}_{10} K_p$:
Solid from 298° to 2770°, Liquid from 2770° to 3000°.

Gfw	92.91 GRAMS
$(H^\circ_{298.15} - H^\circ_0) =$	1,997 CAL./GFW.
IDEAL MONATOMIC GAS	

						FORMATION FROM REFERENCE STATE		
	T TEMPERATURE °K	C_p° HEAT CAPACITY CAL./DEG./GFW.	$H^\circ_T - H^\circ_{298.15}$ HEAT CONTENT CAL./GFW.	S_T° ENTROPY CAL./DEG./GFW.	$-(F^\circ - H^\circ_{298.15})/T$ FREE ENERGY FUNCTION CAL./DEG./GFW.	HEAT ΔH_f° CAL./GFW.	FREE ENERGY ΔF_f° CAL./GFW.	$\mathrm{LOG}_{10} K_p$
	298	7.21	0	44.49	44.49	177500	166837	-122.299
	300	7.21	13	44.54	44.50	177502	166771	-121.502
M.P. °K	400	7.09	729	46.60	44.78	177619	163175	-89.162
ΔH_m CAL./GFW.	500	6.89	1428	48.16	45.31	177708	159553	-69.745
	600	6.70	2108	49.40	45.89	177768	155916	-56.797
	700	6.54	2770	50.42	46.47	177800	152264	-47.542
B.P. °K	800	6.40	3417	51.28	47.01	177807	148623	-40.605
ΔH_v CAL./GFW.	900	6.28	4051	52.03	47.53	177801	144978	-35.207
	1000	6.19	4674	52.69	48.02	177764	141324	-30.889
	1100	6.10	5288	53.27	48.47	177708	137690	-27.359
	1200	6.03	5895	53.80	48.89	177645	134061	-24.417
	1300	5.98	6496	54.28	49.29	177556	130418	-21.927
S.P. °K	1400	5.94	7092	54.72	49.66	177462	126810	-19.797
ΔH_s CAL./GFW.	1500	5.91	7684	55.13	50.01	177344	123194	-17.950
	1600	5.90	8275	55.51	50.34	177225	119593	-16.335
	1700	5.90	8865	55.87	50.66	177085	115987	-14.911
	1800	5.92	9456	56.21	50.96	176946	112398	-13.646
T.P. °K	1900	5.94	10049	56.53	51.25	176789	108807	-12.514
ΔH_t CAL./GFW.	2000	5.98	10646	56.83	51.51	176636	105236	-11.499
	2100	6.03	11246	57.13	51.78	176476	101674	-10.581
	2200	6.09	11853	57.41	52.03	176303	98115	-9.746
T.P. °K	2300	6.16	12465	57.68	52.27	176135	94554	-8.984
ΔH_t CAL./GFW.	2400	6.24	13086	57.94	52.49	175966	91030	-8.289
	2500	6.33	13714	58.20	52.72	175794	87494	-7.648
	2600	6.42	14351	58.45	52.94	175621	83971	-7.058
	2700	6.51	14998	58.70	53.15	175448	80435	-6.510
$T_c =$ °K	2800	6.61	15654	58.93	53.34	168884	77016	-6.011
$P_c =$ ATM.	2900	6.71	16321	59.17	53.55	168751	73718	-5.555
	3000	6.82	16997	59.40	53.74	168627	70437	-5.131

NITROGEN

NITROGEN N_2

REFERENCE STATE

Gfw	28.016	GRAMS
$(H°_{298.15} - H°_0)$ =	2,072	CAL./GFW.
M.P.	63.18	°K
ΔH_m	172.	CAL./GFW.
B.P.	77.36	°K
ΔH_v	1,335.	CAL./GFW.
S.P.		°K
ΔH_s		CAL./GFW.
T.P.	35.62	°K
ΔH_t	55.	CAL./GFW.
T.P.		°K
ΔH_t		CAL./GFW.
T_c =	126.26	°K
P_c =	33.54	ATM.

Ideal Diatomic Gas from 298° to 3000°.

T TEMPERATURE °K	$C°_P$ HEAT CAPACITY CAL./DEG./GFW.	$H°_T - H°_{298.15}$ HEAT CONTENT CAL./GFW.	$S°_T$ ENTROPY CAL./DEG./GFW.	$-(F°_T - H°_{298.15})/T$ FREE ENERGY FUNCTION CAL./DEG./GFW.	FORMATION FROM REFERENCE STATE HEAT $\Delta H°_f$ CAL./GFW.	FORMATION FROM REFERENCE STATE FREE ENERGY $\Delta F°_f$ CAL./GFW.	$LOG_{10} K_P$
298	6.96	0	45.77	45.77			
300	6.96	12	45.81	45.77			
400	6.99	709	47.82	46.05			
500	7.07	1412	49.39	46.57			
600	7.19	2125	50.69	47.67			
700	7.35	2852	51.81	47.72			
800	7.51	3596	52.80	48.31			
900	7.67	4355	53.69	48.86			
1000	7.81	5129	54.51	49.39			
1100	7.94	5918	55.26	49.88			
1200	8.06	6718	55.96	50.37			
1300	8.16	7530	56.61	50.82			
1400	8.25	8351	57.21	51.25			
1500	8.33	9180	57.79	51.67			
1600	8.39	10017	58.33	52.17			
1700	8.45	10860	58.84	52.46			
1800	8.51	11709	59.32	52.82			
1900	8.56	12563	59.78	53.17			
2000	8.60	13421	60.22	53.51			
2100	8.64	14283	60.64	53.84			
2200	8.67	15149	61.05	54.17			
2300	8.70	16018	61.43	54.47			
2400	8.73	16890	61.80	54.77			
2500	8.75	17764	62.16	55.06			
2600	8.78	18641	62.51	55.35			
2700	8.80	19521	62.84	55.61			
2800	8.82	20402	63.16	55.88			
2900	8.84	21286	63.47	56.13			
3000	8.86	22171	63.77	56.38			

NITROGEN

NITROGEN N Reference State for Calculating ΔH_f°, ΔF_f°, and $\text{Log}_{10} K_p$: Ideal Diatomic Gas from 298° to 3000°.

IDEAL MONATOMIC GAS

Gfw	14.008	GRAMS
$(H_{298.15}^\circ - H_0^\circ)$ =	1,481	CAL./GFW.

M.P.		°K
ΔH_m		CAL./GFW.
B.P.		°K
ΔH_v		CAL./GFW.
S.P.		°K
ΔH_s		CAL./GFW.
T.P.		°K
ΔH_t		CAL./GFW.
T.P.		°K
ΔH_t		CAL./GFW.
T_c =		°K
P_c =		ATM.

TEMPERATURE °K	C_p° HEAT CAPACITY CAL./DEG./GFW.	$H_T^\circ - H_{298.15}^\circ$ HEAT CONTENT CAL./GFW.	S_T° ENTROPY CAL./DEG./GFW.	$-(F^\circ - H_{298.15}^\circ)/T$ FREE ENERGY FUNCTION CAL./DEG./GFW.	FORMATION FROM REFERENCE STATE		
					HEAT ΔH_f° CAL./GFW.	FREE ENERGY ΔF_f° CAL./GFW.	$\text{LOG}_{10} K_p$
298	4.97	0	36.61	36.61	113000	108760	-79.726
300	4.97	9	36.65	36.62	113003	108881	-79.326
400	4.97	506	38.07	36.81	113151	107487	-58.733
500	4.97	1003	39.18	37.18	113297	106057	-46.360
600	4.97	1500	40.09	37.59	113437	104593	-38.101
700	4.97	1996	40.85	38.00	113570	103112	-32.195
800	4.97	2493	41.52	38.41	113695	101599	-27.757
900	4.97	2990	42.10	38.78	113812	100087	-24.306
1000	4.97	3487	42.63	39.15	113922	98552	-21.540
1100	4.97	3983	43.10	39.48	114024	97007	-19.275
1200	4.97	4480	43.53	39.80	114121	95461	-17.387
1300	4.97	4977	43.93	40.11	114212	93906	-15.788
1400	4.97	5474	44.30	40.39	114298	92332	-14.414
1500	4.97	5971	44.64	40.66	114381	90771	-13.226
1600	4.97	6468	44.96	40.92	114459	89195	-12.183
1700	4.97	6964	45.26	41.17	114534	87606	-11.262
1800	4.97	7461	45.55	41.41	114606	86004	-10.441
1900	4.97	7958	45.82	41.64	114676	84409	-9.708
2000	4.97	8455	46.07	41.85	114744	82824	-9.050
2100	4.97	8952	46.31	42.05	114810	81231	-8.453
2200	4.97	9449	46.54	42.25	114874	79652	-7.912
2300	4.97	9946	46.76	42.44	114937	78045	-7.415
2400	4.97	10443	46.98	42.63	114998	76406	-6.957
2500	4.98	10941	47.18	42.81	115059	74809	-6.539
2600	4.98	11439	47.37	42.98	115118	73232	-6.155
2700	4.99	11937	47.56	43.14	115176	71598	-5.795
2800	4.99	12436	47.74	43.30	115235	69987	-5.462
2900	5.00	12936	47.92	43.46	115293	68371	-5.152
3000	5.01	13437	48.09	43.62	115351	66751	-4.862

THERMODYNAMIC PROPERTIES OF THE ELEMENTS

OSMIUM

OSMIUM	Os		Solid from 298° to 3000°, Liquid at 3000°.

REFERENCE STATE

Gfw	190.2	GRAMS		
$(H°_{298.15} - H°_0) =$		CAL./GFW.		
M.P.	(3,000)	°K		
ΔH_m	(7,000)	CAL./GFW.		
B.P.	(4,500)	°K		
ΔH_v	(150,000)	CAL./GFW.		
S.P.		°K		
ΔH_s		CAL./GFW.		
T.P.		°K		
ΔH_t		CAL./GFW.		
T.P.		°K		
ΔH_t		CAL./GFW.		
$T_c =$		°K		
$P_c =$		ATM.		

T TEMPERATURE °K	$C°_p$ HEAT CAPACITY CAL./DEG./GFW.	$H°_T - H°_{298.15}$ HEAT CONTENT CAL./GFW.	$S°_T$ ENTROPY CAL./DEG./GFW.	$-(F°_T - H°_{298.15})/T$ FREE ENERGY FUNCTION CAL./DEG./GFW.	HEAT $\Delta H°_f$ CAL./GFW.	FREE ENERGY $\Delta F°_f$ CAL./GFW.	$LOG_{10} K_p$
298	5.95	0					
300	5.95	11	7.80	7.80			
400	6.04	610	7.84	7.81			
500	6.13	1210	9.56	8.04			
600	6.22	1830	10.90	8.48			
700	6.31	2460	12.03	8.98			
800	6.39	3100	13.00	9.49			
900	6.48	3740	13.86	9.99			
1000	6.57	4390	14.61	10.46			
1100	6.66	5060	15.29	10.90			
1200	6.75	5730	15.92	11.32			
1300	6.83	6410	16.50	11.73			
1400	6.92	7090	17.03	12.10			
1500	7.01	7790	17.54	12.48			
1600	7.10	8490	18.02	12.83			
1700	7.19	9210	18.49	13.19			
1800	7.27	9930	18.93	13.52			
1900	7.36	10660	19.34	13.83			
2000	7.45	11400	19.74	14.13			
2100	7.54	12150	20.12	14.42			
2200	7.63	12910	20.48	14.70			
2300	7.71	13680	20.83	14.97			
2400	7.80	14450	21.17	15.23			
2500	7.89	15240	21.50	15.48			
2600	7.98	16030	21.82	15.73			
2700	8.07	16840	22.14	15.98			
2800	8.15	17650	22.44	16.21			
2900	8.24	18470	22.73	16.43			
3000	8.30	26200	25.60	16.87			

OSMIUM

OSMIUM   Os

IDEAL MONATOMIC GAS

Reference State for Calculating ΔH_f°, ΔF_f°, and $\log_{10} K_p$:
Solid from 298° to 3000°, Liquid at 3000°.

Gw	190.2	GRAMS
$(H_{298.15}^\circ - H_0^\circ)$ =	1,481	CAL./GFW.

M.P.		°K
ΔH_m		CAL./GFW.
B.P.		°K
ΔH_v		CAL./GFW.
S.P.		°K
ΔH_s		CAL./GFW.
T.P.		°K
ΔH_t		CAL./GFW.
T.P.		°K
ΔH_t		CAL./GFW.
T_c =		°K
P_c =		ATM.

T TEMPERATURE °K	C_p° HEAT CAPACITY CAL./DEG./GFW.	$H_T^\circ - H_{298.15}^\circ$ HEAT CONTENT CAL./GFW.	S_T° ENTROPY CAL./DEG./GFW.	$-(F^\circ - H_{298.15}^\circ)/T$ FREE ENERGY FUNCTION CAL./DEG./GFW.	FORMATION FROM REFERENCE STATE		
					HEAT ΔH_f° CAL./GFW.	FREE ENERGY ΔF_f° CAL./GFW.	$\log_{10} K_p$
298	4.97	0	46.00	46.00	160000	148610	-108.938
300	4.97	9	46.03	46.00	159998	148541	-108.221
400	4.97	506	47.46	46.20	159896	144736	-79.086
500	5.00	1005	48.57	46.56	159795	140960	-61.617
600	5.04	1506	49.49	46.98	159676	137200	-49.979
700	5.12	2014	50.27	47.40	159554	133465	-41.673
800	5.23	2532	50.96	47.80	159432	129752	-35.449
900	5.37	3061	51.59	48.19	159321	126039	-30.608
1000	5.52	3606	52.16	48.56	159216	122346	-26.741
1100	5.69	4166	52.69	48.91	159106	118659	-23.577
1200	5.87	4744	53.20	49.25	159014	114974	-20.941
1300	6.04	5340	53.67	49.57	158930	111298	-18.712
1400	6.21	5952	54.13	49.88	158862	107636	-16.804
1500	6.37	6581	54.56	50.18	158791	103981	-15.151
1600	6.52	7226	54.98	50.47	158736	100352	-13.707
1700	6.66	7885	55.38	50.75	158675	96710	-12.433
1800	6.78	8557	55.76	51.01	158627	93071	-11.299
1900	6.89	9240	56.13	51.27	158580	89439	-10.287
2000	6.99	9934	56.48	51.52	158534	85814	-9.376
2100	7.07	10637	56.83	51.77	158487	82152	-8.549
2200	7.15	11348	57.16	52.01	158438	78512	-7.799
2300	7.22	12067	57.48	52.24	158387	74874	-7.114
2400	7.28	12792	57.79	52.46	158342	71246	-6.487
2500	7.34	13523	58.08	52.68	158283	67633	-5.912
2600	7.39	14260	58.37	52.89	158230	64032	-5.382
2700	7.44	15002	58.65	53.10	158162	60395	-4.888
2800	7.48	15748	58.93	53.31	158098	56738	-4.428
2900	7.52	16498	59.19	53.51	158028	53135	-4.004
3000	7.55	17251	59.44	53.69	151050	49530	-3.608

OXYGEN

Ideal Diatomic Gas from 298° to 3000°.

REFERENCE STATE

Gfw	32.000 GRAMS	
$(H^°_{298.15} - H^°_0) =$	2,075	CAL./GFW.
M.P.	54.36	°K
ΔH_m	106.3	CAL./GFW.
B.P.	90.19	°K
ΔH_v	1630.	CAL./GFW.
S.P.		°K
ΔH_s		CAL./GFW.
T.P.	23.89	°K
ΔH_t	22.4	CAL./GFW.
T.P.	43.80	°K
ΔH_t	177.6	CAL./GFW.
$T_c =$	154.78	°K
$P_c =$	50.14	ATM.

T TEMPERATURE °K	$C^°_P$ HEAT CAPACITY CAL./DEG./GFW.	$H^°_T - H^°_{298.15}$ HEAT CONTENT CAL./GFW.	$S^°_T$ ENTROPY CAL./DEG./GFW.	$-(F^°_T - H^°_{298.15})/T$ FREE ENERGY FUNCTION CAL./DEG./GFW.	FORMATION FROM REFERENCE STATE		
					HEAT $\Delta H^°_f$ CAL./GFW.	FREE ENERGY $\Delta F^°_f$ CAL./GFW.	$LOG_{10} K_P$
298	7.02	0	49.01	49.01			
300	7.02	13	49.06	49.02			
400	7.20	723	51.10	49.30			
500	7.43	1454	52.73	49.83			
600	7.67	2209	54.11	50.43			
700	7.88	2987	55.30	51.04			
800	8.06	3785	56.37	51.64			
900	8.21	4599	57.33	52.22			
1000	8.34	5427	58.20	52.78			
1100	8.44	6265	59.00	53.31			
1200	8.53	7114	59.74	53.82			
1300	8.60	7970	60.42	54.29			
1400	8.67	8834	61.06	54.75			
1500	8.74	9705	61.66	55.19			
1600	8.80	10582	62.23	55.62			
1700	8.86	11464	62.76	56.02			
1800	8.92	12353	63.27	56.41			
1900	8.97	13248	63.76	56.79			
2000	9.03	14148	64.22	57.15			
2100	9.08	15053	64.66	57.50			
2200	9.14	15965	65.08	57.83			
2300	9.19	16881	65.49	58.16			
2400	9.25	17803	65.88	58.47			
2500	9.30	18731	66.26	58.77			
2600	9.35	19663	66.63	59.17			
2700	9.41	20601	66.98	59.35			
2800	9.46	21544	67.32	59.63			
2900	9.50	22492	67.66	59.91			
3000	9.55	23445	67.98	60.17			

OXYGEN

OXYGEN   O

IDEAL MONATOMIC GAS

Gfw 16.000 GRAMS

$(H°_{298.15} - H°_0) = 1,607$ CAL./GFW.

Reference State for Calculating $\Delta H°_f$, $\Delta F°_f$, and $\log_{10} K_p$: Ideal Diatomic Gas from 298° to 3000°.

T TEMPERATURE °K	$C°_p$ HEAT CAPACITY CAL./DEG./GFW.	$H°_T - H°_{298.15}$ HEAT CONTENT CAL./GFW.	$S°_T$ ENTROPY CAL./DEG./GFW.	$-(F°_T - H°_{298.15})$ FREE ENERGY FUNCTION CAL./DEG./GFW.	FORMATION FROM REFERENCE STATE		
					HEAT $\Delta H°_f$ CAL./GFW.	FREE ENERGY $\Delta F°_f$ CAL./GFW.	LOG$_{10}$ K$_p$
298	5.24	0	38.47	38.47	59550	55387	-40.601
300	5.23	10	38.50	38.47	59553	55362	-40.334
400	5.13	528	39.99	38.67	59716	53940	-29.473
500	5.08	1038	41.13	39.06	59861	52476	-22.938
600	5.05	1544	42.05	39.48	59989	50989	-18.574
700	5.03	2048	42.83	39.91	60104	49478	-15.449
800	5.02	2550	43.50	40.32	60207	47951	-13.100
900	5.01	3051	44.09	40.70	60301	46414	-11.271
1000	5.00	3551	44.62	41.07	60387	44867	-9.806
1100	4.99	4051	45.09	41.41	60468	43319	-8.607
1200	4.99	4550	45.53	41.74	60543	41751	-7.604
1300	4.99	5049	45.93	42.05	60614	40178	-6.755
1400	4.98	5548	46.30	42.34	60681	38603	-6.026
1500	4.98	6046	46.64	42.61	60743	37028	-5.395
1600	4.98	6544	46.96	42.87	60803	35443	-4.841
1700	4.98	7042	47.27	43.13	60860	33860	-4.351
1800	4.98	7540	47.55	43.37	60913	32275	-3.918
1900	4.98	8038	47.82	43.59	60964	30678	-3.529
2000	4.98	8535	48.07	43.81	61011	29091	-3.178
2100	4.98	9033	48.32	44.02	61056	27496	-2.859
2200	4.98	9531	48.55	44.22	61098	25876	-2.570
2300	4.98	10029	48.77	44.41	61138	24292	-2.308
2400	4.98	10527	48.98	44.60	61175	22679	-2.065
2500	4.98	11025	49.19	44.78	61209	21060	-1.841
2600	4.99	11523	49.39	44.96	61241	19433	-1.633
2700	4.99	12022	49.59	45.12	61271	17855	-1.445
2800	4.99	12521	49.76	45.29	61299	16219	-1.265
2900	5.00	13021	49.93	45.44	61325	14635	-1.102
3000	5.00	13521	50.10	45.60	61348	13018	-.948

M.P. °K
ΔH_m CAL./GFW.

B.P. °K
ΔH_v CAL./GFW.

S.P. °K
ΔH_s CAL./GFW.

T.P. °K
ΔH_t CAL./GFW.

T.P. °K
ΔH_t CAL./GFW.

T_c = °K
P_c = ATM.

THERMODYNAMIC PROPERTIES OF THE ELEMENTS

PALLADIUM

PALLADIUM Pd Solid from 298° to 1823°, Liquid from 1823° to 3000°.

REFERENCE STATE

Gfw	106.4 GRAMS
$(H°_{298.15} - H°_0) =$	1,308 CAL./GFW.
M.P.	1,823 °K
ΔH_m	4,000 CAL./GFW.
B.P.	3,400 °K
ΔH_v	94,000 CAL./GFW.
S.P.	°K
ΔH_s	CAL./GFW.
T.P.	°K
ΔH_t	CAL./GFW.
T.P.	°K
ΔH_t	CAL./GFW.
$T_c =$	°K
$P_c =$	ATM.

T TEMPERATURE °K	$C°_P$ HEAT CAPACITY CAL./DEG./GFW.	$H°_T - H°_{298.15}$ HEAT CONTENT CAL./GFW.	$S°_T$ ENTROPY CAL./DEG./GFW.	$-(F°_T - H°_{298.15})/T$ FREE ENERGY FUNCTION CAL./DEG./GFW.	HEAT $\Delta H°_f$ CAL./GFW.	FREE ENERGY $\Delta F°_f$ CAL./GFW.	$LOG_{10} K_P$
298	6.26	0	9.05	9.05			
300	6.27	11	9.09	9.06			
400	6.35	640	10.89	9.29			
500	6.49	1280	12.32	9.76			
600	6.62	1940	13.52	10.29			
700	6.76	2610	14.55	10.83			
800	6.90	3290	15.46	11.35			
900	7.04	3980	16.28	11.86			
1000	7.18	4690	17.03	12.34			
1100	7.32	5420	17.72	12.80			
1200	7.46	6170	18.37	13.23			
1300	7.60	6930	18.97	13.64			
1400	7.74	7690	19.54	14.05			
1500	7.88	8460	20.08	14.44			
1600	8.02	9250	20.59	14.81			
1700	8.16	10060	21.08	15.17			
1800	8.30	10890	21.55	15.50			
1900	8.30	15720	24.19	15.92			
2000	8.30	16550	24.61	16.34			
2100	8.30	17380	25.02	16.75			
2200	8.30	18210	25.40	17.13			
2300	8.30	19040	25.77	17.50			
2400	8.30	19870	26.13	17.86			
2500	8.30	20700	26.47	18.19			
2600	8.30	21530	26.79	18.51			
2700	8.30	22360	27.10	18.82			
2800	8.30	23190	27.41	19.13			
2900	8.30	24020	27.70	19.42			
3000	8.30	24850	27.98	19.70			

PALLADIUM

PALLADIUM   Pd   Reference State for Calculating ΔH_f°, ΔF_f°, and $\log_{10} K_p$:

IDEAL MONATOMIC GAS   Solid from 298° to 1823°, Liquid from 1823° to 3000°.

Gfw 106.4 grams
$(H^\circ_{298.15} - H^\circ_0) = 1{,}481$ CAL./GFW.

	T TEMPERATURE °K	C_p° HEAT CAPACITY CAL./DEG./GFW.	$H_T^\circ - H_{298.15}^\circ$ HEAT CONTENT CAL./GFW.	S_T° ENTROPY CAL./DEG./GFW.	$-(F_T^\circ - H_{298.15}^\circ)/T$ FREE ENERGY FUNCTION CAL./DEG./GFW.	FORMATION FROM REFERENCE STATE		
						HEAT ΔH_f° CAL./GFW.	FREE ENERGY ΔF_f° CAL./GFW.	$\log_{10} K_p$
	298	4.97	0	39.90	39.90	94000	84802	−62.164
	300	4.97	9	39.93	39.90	93998	84746	−61.742
	400	4.97	506	41.36	40.10	93866	81678	−44.630
M.P.	500	4.97	1003	42.47	40.47	93723	78648	−34.379
ΔH_m	600	4.97	1500	43.38	40.88	93560	75644	−27.555
	700	4.97	1997	44.14	41.29	93387	72674	−22.691
	800	4.98	2494	44.81	41.70	93204	69724	−19.049
B.P.	900	5.02	2994	45.40	42.08	93014	66806	−16.223
ΔH_v	1000	5.08	3499	45.93	42.44	92809	63909	−13.968
	1100	5.20	4013	46.42	42.78	92593	61023	−12.125
	1200	5.38	4541	46.88	43.10	92371	58159	−10.593
S.P.	1300	5.63	5091	47.32	43.41	92161	55306	−9.298
ΔH_s	1400	5.94	5669	47.75	43.71	91979	52485	−8.193
	1500	6.32	6281	48.17	43.99	91821	49686	−7.239
	1600	6.74	6934	48.59	44.26	91684	46884	−6.403
	1700	7.21	7631	49.01	44.53	91571	44090	−5.668
	1800	7.69	8376	49.44	44.79	91486	41284	−5.012
T.P.	1900	8.17	9169	49.87	45.05	87449	38657	−4.446
ΔH_t	2000	8.63	10009	50.30	45.30	87459	36079	−3.942
	2100	9.06	10894	50.73	45.55	87514	33523	−3.488
	2200	9.44	11820	51.16	45.79	87610	30938	−3.073
T.P.	2300	9.77	12781	51.59	46.04	87741	28355	−2.694
ΔH_t	2400	10.03	13771	52.01	46.28	87901	25789	−2.348
	2500	10.24	14785	52.42	46.51	88085	23210	−2.029
	2600	10.38	15816	52.83	46.75	88286	20582	−1.730
	2700	10.46	16858	53.22	46.98	88498	17974	−1.454
T_c =	2800	10.49	17906	53.60	47.21	88716	15384	−1.200
P_c =	2900	10.47	18954	53.97	47.44	88934	12751	−.960
ATM.	3000	10.42	19999	54.32	47.66	89149	10129	−.737

PHOSPHORUS

PHOSPHORUS P
Red Triclinic Solid from 298° to 704°,
Ideal Diatomic Gas from 704° to 3000°.

REFERENCE STATE

Gfw	30.975	GRAMS
$(H°_{298.15} - H°_0)$ =		CAL./GFW
M.P.	870	°K
ΔH_m		CAL./GFW
B.P.		°K
ΔH_v		CAL./GFW
S.P.	704	°K
ΔH_s	7,200	CAL./GFW
T.P.		°K
ΔH_f		CAL./GFW
T.P.		°K
ΔH_f		CAL./GFW
T_c =		°K
P_c =		ATM

T TEMPERATURE °K	$C°_P$ HEAT CAPACITY CAL./DEG./GFW	$H°_T - H°_{298.15}$ HEAT CONTENT CAL./GFW	$S°_T$ ENTROPY CAL./DEG./GFW	$-(F°_T - H°_{298.15})$ FREE ENERGY FUNCTION CAL./DEG./GFW	FORMATION FROM REFERENCE STATE		
					HEAT $\Delta H°_f$ CAL./GFW	FREE ENERGY $\Delta F°_f$ CAL./GFW	LOG$_{10}$ K_P
298	4.98	0	5.46	5.46			
300	4.98	9	5.49	5.46			
400	5.37	527	6.98	5.67			
500	5.76	1083	8.22	6.06			
600	6.14	1678	9.30	6.51			
700	6.53	2312	10.28	6.98			
800	4.34	23450	30.13	.82			
900	4.37	23890	30.65	4.11			
1000	4.40	24330	31.10	6.77			
1100	4.42	24770	31.53	9.02			
1200	4.43	25210	31.92	10.92			
1300	4.44	25660	32.27	12.54			
1400	4.46	26100	32.60	13.96			
1500	4.47	26550	32.91	15.21			
1600	4.47	27000	33.19	16.32			
1700	4.48	27440	33.46	17.32			
1800	4.49	27890	33.72	18.23			
1900	4.50	28340	33.96	19.05			
2000	4.50	28790	34.19	19.80			
2100	4.51	29240	34.41	20.49			
2200	4.51	29690	34.62	21.13			
2300	4.52	30150	34.82	21.72			
2400	4.52	30600	35.02	22.27			
2500	4.53	31050	35.20	22.78			
2600	4.53	31500	35.38	23.27			
2700	4.53	31950	35.55	23.72			
2800	4.54	32410	35.71	24.14			
2900	4.54	32860	35.87	24.54			
3000	4.55	33320	36.03	24.93			

PHOSPHORUS

PHOSPHORUS P_2
IDEAL DIATOMIC GAS

Reference State for Calculating ΔH_f°, ΔF_f°, and $\log_{10} K_p$: Red Triclinic Solid from 298° to 704°, Ideal Diatomic Gas from 704° to 3000°.

G_{fw} 61.950 GRAMS
$(H^\circ_{298.15} - H^\circ_0) =$ 2,133 CAL./GFW.

T TEMPERATURE °K	C_p° HEAT CAPACITY CAL./DEG./GFW.	$H_T^\circ - H_{298.15}^\circ$ HEAT CONTENT CAL./GFW.	S_T° ENTROPY CAL./DEG./GFW.	$-(F_T^\circ - H_{298.15}^\circ)/T$ FREE ENERGY FUNCTION CAL./DEG./GFW.	FORMATION FROM REFERENCE STATE		
					HEAT ΔH_f° CAL./GFW.	FREE ENERGY ΔF_f° CAL./GFW.	$\log_{10} K_p$
298	7.65	0	52.11	52.11	42725	30442	-22.315
300	7.66	14	52.16	52.12	42721	30367	-22.124
400	8.04	799	54.42	52.43	42470	26286	-14.363
500	8.30	1618	56.25	53.02	42177	22272	-9.735
600	8.48	2459	57.78	53.69	41828	18320	-6.673
700	8.60	3314	59.11	54.38	41415	14430	-4.505
800	8.68	4179	60.26	55.04	0	0	0
900	8.74	5052	61.29	55.68	0	0	0
1000	8.79	5930	62.20	56.27	0	0	0
1100	8.83	6812	63.05	56.86	0	0	0
1200	8.86	7698	63.83	57.42	0	0	0
1300	8.88	8586	64.53	57.93	0	0	0
1400	8.91	9477	65.19	58.43	0	0	0
1500	8.93	10370	65.81	58.90	0	0	0
1600	8.94	11265	66.38	59.34	0	0	0
1700	8.96	12161	66.92	59.77	0	0	0
1800	8.97	13059	67.44	60.19	0	0	0
1900	8.99	13958	67.92	60.58	0	0	0
2000	9.00	14858	68.38	60.96	0	0	0
2100	9.01	15760	68.82	61.32	0	0	0
2200	9.02	16661	69.24	61.67	0	0	0
2300	9.03	17565	69.64	62.01	0	0	0
2400	9.04	18468	70.03	62.34	0	0	0
2500	9.05	19373	70.40	62.66	0	0	0
2600	9.05	20286	70.75	62.95	0	0	0
2700	9.06	21186	71.09	63.25	0	0	0
2800	9.07	22090	71.42	63.54	0	0	0
2900	9.08	23000	71.74	63.81	0	0	0
3000	9.09	23910	72.05	64.08	0	0	0

M.P. °K
ΔH_m CAL./GFW.

B.P. °K
ΔH_v CAL./GFW.

S.P. °K
ΔH_s CAL./GFW.

T.P. °K
ΔH_t CAL./GFW.

T.P. °K
ΔH_t CAL./GFW.

$T_c =$ °K
$P_c =$ ATM.

THERMODYNAMIC PROPERTIES OF THE ELEMENTS

PHOSPHORUS

PHOSPHORUS P_4

IDEAL TETRATOMIC GAS

Gfw 123.900 GRAMS
$(H°_{298.15} - H°_0) = 3,368$ CAL./GFW.

Reference State for Calculating $\Delta H_f°$, $\Delta F_f°$, and $\text{Log}_{10} K_p$: Red Triclinic Solid from 298° to 704°, Ideal Diatomic Gas from 704° to 2000°.

T TEMPERATURE °K	$C_p°$ HEAT CAPACITY CAL./DEG./GFW.	$H°_T - H°_{298.15}$ HEAT CONTENT CAL./GFW.	$S°_T$ ENTROPY CAL./DEG./GFW.	$-(F°_T - H°_{298.15})/T$ FREE ENERGY FUNCTION CAL./DEG./GFW.	FORMATION FROM REFERENCE STATE		
					HEAT $\Delta H_f°$ CAL./GFW.	FREE ENERGY $\Delta F_f°$ CAL./GFW.	LOG$_{10}$ K_p
298	16.05	0	66.85	66.85	30820	17397	−12.752
300	16.06	30	66.95	66.85	30814	17317	−12.616
400	17.51	1710	71.78	67.51	30422	12878	−7.036
500	18.28	3500	75.77	68.77	29988	8543	−3.734
600	18.73	5360	79.16	70.23	29468	4292	−1.563
700	19.02	7240	82.06	71.72	28812	154	−.048
800	19.21	9150	84.61	73.18	53830	−25102	6.858
900	19.35	11080	86.88	74.57	53660	−21512	5.224
1000	19.44	13020	88.92	75.90	53480	−18000	3.934
1100	19.52	14980	90.79	77.18	53280	−14417	2.864
1200	19.57	16940	92.50	78.39	53080	−10864	1.978
1300	19.62	18900	94.06	79.53	52920	−7394	1.243
1400	19.65	20860	95.52	80.62	52720	−3888	.606
1500	19.68	22830	96.88	81.66	52550	−410	.059
1600	19.70	24800	98.15	82.65	52380	2996	−.409
1700	19.72	26770	99.35	83.61	52170	6463	−.830
1800	19.74	28740	100.47	84.51	52000	9938	−1.206
1900	19.75	30720	101.54	85.38	51820	13350	−1.535
2000	19.76	32690	102.55	86.21	51650	16770	−1.832

M.P.		°K
ΔH_m		CAL./GFW.
B.P.		°K
ΔH_v		CAL./GFW.
S.P.		°K
ΔH_s		CAL./GFW.
T.P.		°K
ΔH_t		CAL./GFW.
T.P.		°K
ΔH_t		CAL./GFW.
$T_c =$		°K
$P_c =$		ATM.

PHOSPHORUS

PHOSPHORUS P
Gfw 30.975 GRAMS
IDEAL MONATOMIC GAS
$(H°_{298.15} - H°_0) = 1,481$ CAL./GFW.

Reference State for Calculating $\Delta H°_f$, $\Delta F°_f$, and $\text{Log}_{10} K_p$: Red Triclinic Solid from 298° to 704°, Ideal Diatomic Gas from 704° to 3000°.

T TEMPERATURE °K	$C°_p$ HEAT CAPACITY CAL./DEG./GFW.	$H°_T - H°_{298.15}$ HEAT CONTENT CAL./GFW.	$S°_T$ ENTROPY CAL./DEG./GFW.	$-(F°_T - H°_{298.15})/T$ FREE ENERGY FUNCTION CAL./DEG./GFW.	FORMATION FROM REFERENCE STATE		
					HEAT $\Delta H°_f$ CAL./GFW.	FREE ENERGY $\Delta F°_f$ CAL./GFW.	$\text{LOG}_{10} K_p$
298	4.97	0	38.98	38.98	79800	69805	-51.170
300	4.97	9	39.01	38.98	79800	69744	-50.812
400	4.97	506	40.44	39.18	79779	63395	-36.279
500	4.97	1003	41.55	39.55	79720	63055	-27.563
600	4.97	1500	42.45	39.95	79622	59732	-21.759
700	4.97	1996	43.22	40.37	79484	56426	-17.618
800	4.97	2493	43.88	40.77	58843	47843	-13.071
900	4.97	2990	44.47	41.15	58900	45462	-11.283
1000	4.97	3487	44.99	41.51	58957	45067	-9.850
1100	4.97	3984	45.47	41.85	59014	43680	-8.679
1200	4.97	4481	45.90	42.17	59071	42295	-7.703
1300	4.97	4978	46.30	42.48	59118	40879	-6.872
1400	4.97	5475	46.66	42.75	59175	39491	-6.165
1500	4.98	5972	47.01	43.03	59222	38072	-5.547
1600	4.99	6471	47.33	43.29	59271	36647	-5.005
1700	5.01	6970	47.63	43.53	59330	35241	-4.530
1800	5.01	7470	47.92	43.77	59380	33820	-4.106
1900	5.04	7973	48.19	44.00	59433	32396	-3.726
2000	5.06	8478	48.45	44.22	59488	30968	-3.383
2100	5.09	8985	48.70	44.43	59545	29536	-3.073
2200	5.13	9497	48.93	44.62	59607	28125	-2.793
2300	5.18	10012	49.16	44.81	59662	26680	-2.535
2400	5.22	10532	49.38	45.00	59732	25268	-2.300
2500	5.28	11057	49.60	45.18	59807	23807	-2.081
2600	5.34	11588	49.81	45.36	59888	22370	-1.880
2700	5.40	12125	50.01	45.52	59975	20933	-1.694
2800	5.47	12669	50.21	45.69	60059	19459	-1.518
2900	5.54	13219	50.40	45.85	60159	18022	-1.358
3000	5.62	13777	50.59	46.00	60257	16577	-1.207

M.P. °K
ΔH_m CAL./GFW.
B.P. °K
ΔH_v CAL./GFW.
S.P. °K
ΔH_s CAL./GFW.
T.P. °K
ΔH_t CAL./GFW.
T.P. °K
ΔH_t CAL./GFW.
$T_c =$ °K
$P_c =$ ATM.

THERMODYNAMIC PROPERTIES OF THE ELEMENTS

PHOSPHORUS

PHOSPHORUS P
Reference State for Calculating ΔH_f°, ΔF_f°, and $\log_{10} K_p$: Red Triclinic Solid from 298° to 704°, Ideal Diatomic Gas from 704° to 3000°.

CUBIC WHITE (α) SOLID

Gfw 30.975 GRAMS
$(H^\circ_{298.15} - H^\circ_0) =$ CAL./GFW.

					FORMATION FROM REFERENCE STATE		
T TEMPERATURE °K	C_p° HEAT CAPACITY CAL./DEG./GFW.	$H_T^\circ - H_{298.15}^\circ$ HEAT CONTENT CAL./GFW.	S_T° ENTROPY CAL./DEG./GFW.	$-(F_T^\circ - H_{298.15}^\circ)/T$ FREE ENERGY FUNCTION CAL./DEG./GFW.	HEAT ΔH_f° CAL./GFW.	FREE ENERGY ΔF_f° CAL./GFW.	$\log_{10} K_p$
298	5.63	0	9.80	9.80	4180	2886	− 2.115
300	5.63	10	9.83	9.80	4181	2879	− 2.097
400	6.00	750	12.01	10.14	4403	2391	− 1.306
500	6.15	1360	13.37	10.65	4457	1882	− .822
600	6.30	1980	14.50	11.20	4482	1362	− .496
700	6.45	2620	15.48	11.74	4488	848	− .264
800	6.60	3270	16.35	12.27	−16000	−4976	1.359
900	6.75	3940	17.14	12.77	−15770	−3611	.876

M.P. 317.4 °K
ΔH_m 150. CAL./GFW.

B.P. 554. °K
ΔH_v 2,960. CAL./GFW.

S.P. °K
ΔH_s CAL./GFW.

T.P. °K
ΔH_t CAL./GFW.

T.P. °K
ΔH_t CAL./GFW.

$T_c =$ °K
$P_c =$ ATM.

PLATINUM

PLATINUM Pt Solid from 298° to 2043°, Liquid from 2043° to 3000°.

Gfw 195.09 GRAMS
$(H°_{298.15} - H°_0) = 1,384$ CAL./GFW.

M.P. 2,043 °K
ΔH_m 4,700 CAL./GFW.

B.P. 4,100 °K
ΔH_v 122,000 CAL./GFW.

S.P. °K
ΔH_s CAL./GFW.

T.P. °K
ΔH_t CAL./GFW.

T.P. °K
ΔH_t CAL./GFW.

$T_c =$ °K
$P_c =$ ATM.

T TEMPERATURE °K	$C°_P$ HEAT CAPACITY CAL./DEG./GFW.	$H°_T - H°_{298.15}$ HEAT CONTENT CAL./GFW.	$S°_T$ ENTROPY CAL./DEG./GFW.	$-(F°_T - H°_{298.15})$ FREE ENERGY FUNCTION CAL./DEG./GFW.	FORMATION FROM REFERENCE STATE		
					HEAT $\Delta H°_f$ CAL./GFW.	FREE ENERGY $\Delta F°_f$ CAL./GFW.	LOG$_{10}$ K$_P$
298	6.19	0	10.00	10.00			
300	6.20	11	10.04	10.01			
400	6.34	645	11.86	10.25			
500	6.45	1280	13.28	10.72			
600	6.57	1920	14.44	11.24			
700	6.70	2580	15.46	11.78			
800	6.83	3260	16.37	12.30			
900	6.96	3950	17.18	12.80			
1000	7.09	4660	17.93	13.27			
1100	7.22	5380	18.61	13.72			
1200	7.36	6110	19.25	14.16			
1300	7.49	6850	19.84	14.58			
1400	7.62	7600	20.39	14.97			
1500	7.75	8370	20.93	15.35			
1600	7.85	9150	21.43	15.72			
1700	7.95	9940	21.91	16.07			
1800	8.05	10740	22.37	16.41			
1900	8.15	11550	22.81	16.74			
2000	8.25	12370	25.52	19.34			
2100	8.30	17900	25.92	17.40			
2200	8.30	18730	26.31	17.80			
2300	8.30	19560	26.68	18.18			
2400	8.30	20390	27.03	18.54			
2500	8.30	21220	27.37	18.89			
2600	8.30	22050	27.69	19.21			
2700	8.30	22880	28.01	19.54			
2800	8.30	23710	28.31	19.85			
2900	8.30	24540	28.60	20.14			
3000	8.30	25370	28.88	20.43			

THERMODYNAMIC PROPERTIES OF THE ELEMENTS

PLATINUM

PLATINUM Pt Reference State for Calculating ΔH_f°, ΔF_f°, and $Log_{10} K_p$:
IDEAL MONATOMIC GAS Solid from 298° to 2043°, Liquid from 2043° to 3000°.

Gfw 195.09 GRAMS
$(H°_{298.15} - H°_0) = 1,572$ CAL./GFW.

M.P. °K
ΔH_m CAL./GFW.

B.P. °K
ΔH_v CAL./GFW.

S.P. °K
ΔH_s CAL./GFW.

T.P. °K
ΔH_t CAL./GFW.

T.P. °K
ΔH_t CAL./GFW.

$T_c =$ °K
$P_c =$ ATM.

T TEMPERATURE °K	C_p° HEAT CAPACITY CAL./DEG./GFW.	$H°_T - H°_{298.15}$ HEAT CONTENT CAL./GFW.	$S°_T$ ENTROPY CAL./DEG./GFW.	$-(F°_T - H°_{298.15})$ FREE ENERGY FUNCTION CAL./DEG./GFW.	FORMATION FROM REFERENCE STATE		
					HEAT ΔH_f° CAL./GFW.	FREE ENERGY ΔF_f° CAL./GFW.	LOG$_{10}$ K$_p$
298	6.10	0	45.96	45.96	134800	124078	-90.955
300	6.11	11	46.00	45.97	134800	124012	-90.342
400	6.46	655	47.80	46.19	134810	120434	-65.807
500	6.43	1291	49.26	46.68	134811	116821	-51.065
600	6.26	1926	50.42	47.21	134806	113218	-41.243
700	6.06	2542	51.37	47.74	134762	109625	-34.229
800	5.88	3138	52.16	48.25	134678	106046	-28.972
900	5.73	3718	52.85	48.72	134568	102465	-24.883
1000	5.61	4285	53.44	49.17	134425	98915	-21.619
1100	5.52	4841	53.97	49.58	134261	95365	-18.949
1200	5.45	5389	54.45	49.96	134079	91839	-16.727
1300	5.36	5931	54.88	50.67	133881	88329	-14.850
1400	5.33	6469	55.28	51.00	133669	84823	-13.242
1500	5.32	7003	55.65	51.29	133433	81353	-11.853
1600	5.31	7536	55.99	51.58	133186	77890	-10.638
1700	5.31	8067	56.32	51.85	132927	74430	-9.568
1800	5.32	8598	56.62	52.11	132658	71008	-8.621
1900	5.32	9129	56.91	52.35	132379	67589	-7.774
2000	5.33	9661	57.18	52.59	132091	68771	-7.514
2100	5.34	10194	57.44	52.82	127094	60902	-6.338
2200	5.36	10729	57.69	53.04	126799	57763	-5.738
2300	5.38	11266	57.93	53.25	126506	54631	-5.191
2400	5.40	11804	58.16	53.45	126214	51502	-4.689
2500	5.42	12345	58.38	53.65	125925	48400	-4.231
2600	5.44	12888	58.59	53.83	125638	45298	-3.807
2700	5.47	13434	58.80	54.01	125354	42221	-3.417
2800	5.50	13982	59.00	54.18	125072	39140	-3.054
2900	5.52	14533	59.19	54.36	124793	36082	-2.719
3000	5.55	15087	59.38	54.36	124517	33017	-2.405

POLONIUM

POLONIUM Po
REFERENCE STATE Gfw 210.
$(H°_T - H°_{298.15}) =$ CAL./GFW.

Solid from 298° to 527°, Liquid from 527° to 1235°, Ideal Monatomic Gas from 1235° to 3000°.

T TEMPERATURE °K	$C°_p$ HEAT CAPACITY CAL./DEG./GFW.	$H°_T - H°_{298.15}$ HEAT CONTENT CAL./GFW.	$S°_T$ ENTROPY CAL./DEG./GFW.	$-(F°_T - H°_{298.15})/T$ FREE ENERGY FUNCTION CAL./DEG./GFW.	HEAT $\Delta H°_f$ CAL./GFW.	FREE ENERGY $\Delta F°_f$ CAL./GFW.	$LOG_{10} K_p$
298	6.30	0	15.00	15.00			
300	6.30	11	15.04	15.01			
400	6.80	670	16.92	15.25			
500	7.30	1370	18.49	15.75			
600	7.50	2120	25.55	17.02			
700	7.50	5870	26.71	18.33			
800	7.50	6620	27.71	19.44			
900	7.50	7370	28.60	20.42			
1000	7.50	8120	29.39	21.27			
1100	7.50	8870	30.10	22.04			
1200	7.50	9620	30.75	22.74			
1300	4.98	39430	52.64	22.31			
1400	4.98	39930	53.00	24.48			
1500	4.99	40420	53.35	26.41			
1600	5.01	40920	53.67	28.10			
1700	5.02	41430	53.97	29.60			
1800	5.04	41930	54.26	30.97			
1900	5.07	42430	54.54	32.21			
2000	5.10	42940	54.80	33.33			
2100	5.14	43460	55.05	34.36			
2200	5.18	43970	55.29	35.31			
2300	5.23	44490	55.52	36.18			
2400	5.28	45020	55.74	36.99			
2500	5.34	45550	55.96	37.74			
2600	5.39	46080	56.17	38.45			
2700	5.45	46630	56.37	39.10			
2800	5.51	47170	56.57	39.73			
2900	5.57	47730	56.77	40.32			
3000	5.62	48290	56.96	40.87			

M.P. 527 °K
ΔH_m (3,000) CAL./GFW.

B.P. 1,235 °K
ΔH_v 14,400 CAL./GFW.

S.P. °K
ΔH_s CAL./GFW.

T.P. 370 °K
ΔH_t CAL./GFW.

T.P. °K
ΔH_t CAL./GFW.

$T_c =$ °K
$P_c =$ ATM.

THERMODYNAMIC PROPERTIES OF THE ELEMENTS

POLONIUM

POLONIUM	Po		Reference State for Calculating ΔH_f°, ΔF_f°, and $\log_{10} K_p$: Solid from 298° to 527°, Liquid from 527° to 1235°, Ideal Monatomic Gas from 1235° to 3000°.
IDEAL MONATOMIC GAS			
Gfw	210.	GRAMS	
$(H_{298.15}^\circ - H_0^\circ)$ =	1,481	CAL./GFW.	

T TEMPERATURE °K	C_p° HEAT CAPACITY CAL./DEG./GFW.	$H_T^\circ - H_{298.15}^\circ$ HEAT CONTENT CAL./GFW.	S_T° ENTROPY CAL./DEG./GFW.	$-(F^\circ - H_{298.15}^\circ)/T$ FREE ENERGY FUNCTION CAL./DEG./GFW.	FORMATION FROM REFERENCE STATE		
					HEAT ΔH_f° CAL./GFW.	FREE ENERGY ΔF_f° CAL./GFW.	LOG$_{10}$ K$_p$
298	4.97	0	45.13	45.13	34450	25467	-18.668
300	4.97	9	45.16	45.13	34448	25412	-18.514
400	4.97	506	46.59	45.33	34286	22418	-12.249
500	4.97	1003	47.70	45.70	34083	19478	-8.514
600	4.97	1500	48.79	46.29	30830	16886	-6.151
700	4.97	1997	49.56	46.71	30577	14582	-4.553
800	4.97	2493	50.22	47.11	30323	12315	-3.364
900	4.97	2990	50.81	47.49	30070	10081	-2.448
1000	4.97	3487	51.33	47.85	29817	7877	-1.721
1100	4.97	3984	51.80	48.18	29564	5694	-1.131
1200	4.97	4480	52.24	48.51	29310	3522	-.641
1300	4.98	4978	52.64	48.82		0	0
1400	4.98	5476	53.00	49.09		0	0
1500	4.99	5974	53.35	49.37		0	0
1600	5.01	6474	53.67	49.63		0	0
1700	5.02	6976	53.97	49.87		0	0
1800	5.04	7479	54.26	50.11		0	0
1900	5.07	7984	54.54	50.34		0	0
2000	5.10	8493	54.80	50.56		0	0
2100	5.14	9005	55.05	50.77		0	0
2200	5.18	9521	55.29	50.97		0	0
2300	5.23	10041	55.52	51.16		0	0
2400	5.28	10567	55.74	51.34		0	0
2500	5.34	11098	55.96	51.53		0	0
2600	5.39	11634	56.17	51.70		0	0
2700	5.45	12176	56.37	51.87		0	0
2800	5.51	12724	56.57	52.03		0	0
2900	5.57	13278	56.77	52.20		0	0
3000	5.62	13838	56.96	52.35		0	0

M.P.		°K
ΔH_m		CAL./GFW.
B.P.		°K
ΔH_v		CAL./GFW.
S.P.		°K
ΔH_s		CAL./GFW.
T.P.		°K
ΔH_t		CAL./GFW.
T.P.		°K
ΔH_t		CAL./GFW.
T_c =		°K
P_c =		ATM.

POLONIUM

POLONIUM Po_2
IDEAL DIATOMIC GAS

G_{fw} 420. GRAMS
$(H°_{298.15} - H°_0) =$ CAL./GFW.

M.P. °K
ΔH_m CAL./GFW.

B.P. °K
ΔH_v CAL./GFW.

S.P. °K
ΔH_s CAL./GFW.

T.P. °K
ΔH_t CAL./GFW.

T.P. °K
ΔH_t CAL./GFW.

$T_c =$ °K
$P_c =$ ATM.

Reference State for Calculating $\Delta H°_f$, $\Delta F°_f$, and $Log_{10} K_p$: Solid from 298° to 527°, Liquid from 527° to 1235°, Ideal Monatomic Gas from 1235° to 3000°.

T TEMPERATURE °K	$C°_P$ HEAT CAPACITY CAL./DEG./GFW.	$H°_T - H°_{298.15}$ HEAT CONTENT CAL./GFW.	$S°_T$ ENTROPY CAL./DEG./GFW.	$-(F°_T - H°_{298.15})/T$ FREE ENERGY FUNCTION CAL./DEG./GFW.	FORMATION FROM REFERENCE STATE		
					HEAT $\Delta H°_f$ CAL./GFW.	FREE ENERGY $\Delta F°_f$ CAL./GFW.	$LOG_{10} K_p$
298	8.84	0	65.80	65.80	32900	22226	-16.292
300	8.84	16	65.86	65.81	32894	22160	-16.144
400	8.89	900	68.41	66.16	32460	18632	-10.180
500	8.91	1790	70.40	66.82	31950	15240	-6.661
600	8.92	2680	72.02	67.56	25340	12788	-4.658
700	8.92	3580	73.40	68.29	24740	10754	-3.357
800	8.93	4470	74.59	69.01	24130	8794	-2.402
900	8.93	5360	75.64	69.69	23520	6924	-1.681
1000	8.93	6250	76.58	70.33	22910	5110	-1.116
1100	8.93	7150	77.43	70.93	22310	3357	-.667
1200	8.94	8040	78.21	71.51	21700	1648	-.300
1300	8.94	8940	78.92	72.05	-37020	-2752	.462
1400	8.94	9830	79.59	72.57	-37130	-156	.024
1500	8.94	10720	80.20	73.06	-37220	-2530	.368

THERMODYNAMIC PROPERTIES OF THE ELEMENTS

POTASSIUM

POTASSIUM	K		
REFERENCE STATE			
Gfw	39.100	GRAMS	
$(H°_{298.15} - H°_0) =$	1,695	CAL./GFW.	
M.P.	336.4	°K	
ΔH_m	554.	CAL./GFW.	
B.P.	1039.	°K	
ΔH_v	18,530.	CAL./GFW.	
S.P.		°K	
ΔH_s		CAL./GFW.	
T.P.		°K	
ΔH_t		CAL./GFW.	
T.P.		°K	
ΔH_t		CAL./GFW.	
$T_c =$		°K	
$P_c =$		ATM.	

Solid from 298° to 336.4°, Liquid from 336.4° to 1039°, Ideal Monatomic Gas from 1039° to 3000°.

T TEMPERATURE °K	$C_P°$ HEAT CAPACITY CAL./DEG./GFW.	$H°_T - H°_{298.15}$ HEAT CONTENT CAL./GFW.	$S°_T$ ENTROPY CAL./DEG./GFW.	$-(F°_T - H°_{298.15})/T$ FREE ENERGY FUNCTION CAL./DEG./GFW.	HEAT $\Delta H°_f$ CAL./GFW.	FREE ENERGY $\Delta F°_f$ CAL./GFW.	$LOG_{10} K_P$
298	7.16	0	15.39	15.39			
300	7.18	13	15.43	15.39			
400	7.53	1324	19.26	15.95			
500	7.34	2067	20.92	16.79			
600	7.20	2793	22.24	17.59			
700	7.13	3509	23.35	18.34			
800	7.11	4220	24.30	19.03			
900	7.16	4934	25.14	19.66			
1000	7.26	5654	25.90	20.25			
1100	4.97	25404	44.78	21.69			
1200	4.97	25900	45.21	23.63			
1300	4.97	26397	45.61	25.31			
1400	4.97	26894	45.98	26.77			
1500	4.97	27391	46.32	28.06			
1600	4.97	27889	46.65	29.22			
1700	4.98	28386	46.95	30.26			
1800	4.99	28885	47.23	31.19			
1900	5.00	29384	47.50	32.04			
2000	5.01	29884	47.76	32.82			
2100	5.03	30387	48.00	33.53			
2200	5.06	30891	48.24	34.20			
2300	5.09	31398	48.46	34.81			
2400	5.12	31909	48.68	35.39			
2500	5.16	32423	48.89	35.93			
2600	5.20	32941	49.10	36.44			
2700	5.25	33463	49.30	36.91			
2800	5.31	33991	49.49	37.36			
2900	5.38	34526	49.68	37.78			
3000	5.47	35068	49.86	38.18			

POTASSIUM

POTASSIUM   K
IDEAL MONATOMIC GAS
Gfw 39.100 grams
$(H°_{298.15} - H°_0) = 1{,}481$ cal./gfw.

Reference State for Calculating $\Delta H°_f$, $\Delta F°_f$, and $\log_{10} K_p$:
Solid from 298° to 336.4°, Liquid from 336.4° to 1039°, Ideal Monatomic Gas from 1039° to 3000°.

T TEMPERATURE °K	$C°_p$ HEAT CAPACITY CAL./DEG./GFW.	$H°_T - H°_{298.15}$ HEAT CONTENT CAL./GFW.	$S°_T$ ENTROPY CAL./DEG./GFW.	$-(F°_T - H°_{298.15})/T$ FREE ENERGY FUNCTION CAL./DEG./GFW.	FORMATION FROM REFERENCE STATE		
					HEAT $\Delta H°_f$ CAL./GFW.	FREE ENERGY $\Delta F°_f$ CAL./GFW.	$\log_{10} K_p$
298	4.97	0	38.30	38.30	21420	14589	−10.694
300	4.97	9	38.33	38.30	21416	14546	−10.597
400	4.97	506	39.76	38.50	20602	12402	−6.776
500	4.97	1003	40.87	38.87	20356	10381	−4.537
600	4.97	1500	41.77	39.27	20127	8409	−3.063
700	4.97	1996	42.54	39.69	19907	6474	−2.021
800	4.97	2493	43.20	40.09	19693	4573	−1.249
900	4.97	2990	43.79	40.47	19476	2691	−.653
1000	4.97	3487	44.31	40.83	19253	843	−.184
1100	4.97	3984	44.78	41.16	0	0	0
1200	4.97	4480	45.21	41.48	0	0	0
1300	4.97	4977	45.61	41.79	0	0	0
1400	4.97	5474	45.98	42.07	0	0	0
1500	4.97	5971	46.32	42.34	0	0	0
1600	4.97	6469	46.65	42.61	0	0	0
1700	4.98	6966	46.95	42.86	0	0	0
1800	4.99	7465	47.23	43.09	0	0	0
1900	5.00	7964	47.50	43.31	0	0	0
2000	5.01	8464	47.76	43.53	0	0	0
2100	5.03	8967	48.00	43.73	0	0	0
2200	5.06	9471	48.24	43.94	0	0	0
2300	5.09	9978	48.46	44.13	0	0	0
2400	5.12	10489	48.68	44.31	0	0	0
2500	5.16	11003	48.89	44.49	0	0	0
2600	5.20	11521	49.10	44.67	0	0	0
2700	5.25	12043	49.30	44.84	0	0	0
2800	5.31	12571	49.49	45.01	0	0	0
2900	5.38	13106	49.68	45.17	0	0	0
3000	5.47	13648	49.86	45.32	0	0	0

M.P. °K
ΔH_m CAL./GFW.
B.P. °K
ΔH_v CAL./GFW.
S.P. °K
ΔH_s CAL./GFW.
T.P. °K
ΔH_t CAL./GFW.
T.P. °K
ΔH_t CAL./GFW.
$T_c =$ °K
$P_c =$ ATM.

THERMODYNAMIC PROPERTIES OF THE ELEMENTS

POTASSIUM

POTASSIUM K_2

IDEAL DIATOMIC GAS

Gfw 78.200 GRAMS
$(H°_{298.15} − H°_0) = 2,566$ CAL./GFW.

Reference State for Calculating $\Delta H°_f$, $\Delta F°_f$, and $\text{Log}_{10} K_p$:
Solid from 298° to 336.4°, Liquid from 336.4° to 1039°, Ideal Monatomic Gas from 1039° to 3000°.

		T TEMPERATURE °K	$C°_p$ HEAT CAPACITY CAL./DEG./GFW.	$H°_T − H°_{298.15}$ HEAT CONTENT CAL./GFW.	$S°_T$ ENTROPY CAL./DEG./GFW.	$-(F°_T − H°_{298.15})/T$ FREE ENERGY FUNCTION CAL./DEG./GFW.	FORMATION FROM REFERENCE STATE		
							HEAT $\Delta H°_f$ CAL./GFW.	FREE ENERGY $\Delta F°_f$ CAL./GFW.	$\text{LOG}_{10} K_p$
		298	9.06	0	59.67	59.67	30580	21965	−16.101
		300	9.06	17	59.72	59.67	30571	21913	−15.964
M.P.	°K	400	9.12	926	62.34	60.03	28858	19330	−10.562
ΔH_m	CAL./GFW.	500	9.18	1841	64.38	60.70	28287	17017	−7.438
B.P.	°K	600	9.23	2762	66.06	61.46	27756	14808	−5.394
ΔH_v	CAL./GFW.	700	9.28	3685	67.48	62.22	27247	12701	−3.965
S.P.	°K	800	9.33	4618	68.72	62.95	26758	10662	−2.912
ΔH_s	CAL./GFW.	900	9.38	5552	69.83	63.67	26264	8669	−2.105
T.P.	°K	1000	9.43	6494	70.82	64.33	25766	6746	−1.474
ΔH_t	CAL./GFW.	1100	9.48	7444	71.72	64.96	−12784	6840	−1.359
T.P.	°K	1200	9.53	8390	72.55	65.56	−12830	8614	−1.568
ΔH_t	CAL./GFW.	1300	9.58	9342	73.31	66.13	−12872	10411	−1.750
		1400	9.63	10300	74.02	66.67	−12908	12208	−1.905
$T_c =$	°K	1500	9.68	11279	74.69	67.18	−12923	14002	−2.040
$P_c =$	ATM.								

PRASEODYMIUM

PRASEODYMIUM Pr Solid I from 298° to 1071°, Solid II from 1071° to 1208°, Liquid from 1208° to 3000°.

REFERENCE STATE

G_{fw} 140.92 GRAMS
$(H°_T - H°_{298.15}) = 1,697$ CAL./GFW.

M.P.	1,208	°K
ΔH_m	(2,400)	CAL./GFW.
B.P.	3,290	°K
ΔH_v	79,500	CAL./GFW.
S.P.		°K
ΔH_s		CAL./GFW.
T.P.	1,071	°K
ΔH_t	(320)	CAL./GFW.
T.P.		°K
ΔH_t		CAL./GFW.
$T_c =$		°K
$P_c =$		ATM.

T TEMPERATURE °K	$C°_P$ HEAT CAPACITY CAL./DEG./GFW.	$H°_T - H°_{298.15}$ HEAT CONTENT CAL./GFW.	$S°_T$ ENTROPY CAL./DEG./GFW.	$-(F°_T - H°_{298.15})/T$ FREE ENERGY FUNCTION CAL./DEG./GFW.	FORMATION FROM REFERENCE STATE		
					HEAT $\Delta H°_f$ CAL./GFW.	FREE ENERGY $\Delta F°_f$ CAL./GFW.	LOG$_{10}$ K_P
298	6.45	0	17.45	17.45			
300	6.46	11	17.49	17.46			
400	6.78	670	19.39	17.72			
500	7.10	1370	20.94	18.20			
600	7.42	2090	22.26	18.78			
700	7.74	2850	23.43	19.36			
800	8.06	3640	24.48	19.93			
900	8.38	4460	25.45	20.50			
1000	8.70	5320	26.35	21.03			
1100	8.00	6500	27.47	21.57			
1200	8.00	7300	28.16	22.08			
1300	8.00	8100	30.79	24.56			
1400	8.00	8900	31.39	25.04			
1500	8.00	9700	31.94	25.48			
1600	8.00	10500	32.46	25.90			
1700	8.00	11300	32.94	26.30			
1800	8.00	12100	33.40	26.68			
1900	8.00	12900	33.83	27.05			
2000	8.00	13700	34.24	27.39			
2100	8.00	14500	34.63	27.73			
2200	8.00	15300	35.00	28.05			
2300	8.00	16100	35.36	28.36			
2400	8.00	16900	35.70	28.66			
2500	8.00	17700	36.03	28.95			
2600	8.00	18500	36.34	29.23			
2700	8.00	19300	36.64	29.50			
2800	8.00	20100	36.93	29.76			
2900	8.00	20900	37.21	30.01			
3000	8.00	21700	37.48	30.25			

THERMODYNAMIC PROPERTIES OF THE ELEMENTS

PROMETHIUM

PROMETHIUM Pm Solid from 298° to 1300°, Liquid from 1300° to 3000°.

REFERENCE STATE

Gfw	145*	GRAMS
$(H°_{298.15} - H°_0)$ =		CAL./GFW.

M.P.	(1,300)	°K
ΔH_m	(3,000)	CAL./GFW.
B.P.	(3,000)	°K
ΔH_v	(70,000)	CAL./GFW.
S.P.		°K
ΔH_s		CAL./GFW.
T.P.		°K
ΔH_t		CAL./GFW.
T.P.		°K
ΔH_t		CAL./GFW.
T_c =		°K
P_c =		ATM.

*Isotope of Longest Known Half Life

T TEMPERATURE °K	$C_P°$ HEAT CAPACITY CAL./DEG./GFW.	$H°_T - H°_{298.15}$ HEAT CONTENT CAL./GFW.	$S°_T$ ENTROPY CAL./DEG./GFW.	$-(F°_T - H°_{298.15})/T$ FREE ENERGY FUNCTION CAL./DEG./GFW.	HEAT $\Delta H°_f$ CAL./GFW.	FREE ENERGY $\Delta F°_f$ CAL./GFW.	LOG$_{10}$ K$_P$
298	6.50	0	17.21	17.21			
300	6.50	12	17.25	17.21			
400	6.75	670	19.15	17.48			
500	7.00	1360	20.69	17.97			
600	7.25	2070	21.98	18.53			
700	7.50	2810	23.12	19.11			
800	7.75	3570	24.14	19.68			
900	8.00	4360	25.07	20.23			
1000	8.25	5170	25.92	20.75			
1100	8.50	6010	26.72	21.26			
1200	8.75	6870	27.47	21.75			
1300	8.00	10760	30.49	22.22			
1400	8.00	11560	31.08	22.83			
1500	8.00	12360	31.64	23.40			
1600	8.00	13160	32.15	23.93			
1700	8.00	13960	32.64	24.43			
1800	8.00	14760	33.09	24.89			
1900	8.00	15560	33.53	25.35			
2000	8.00	16360	33.94	25.76			
2100	8.00	17160	34.33	26.16			
2200	8.00	17960	34.70	26.54			
2300	8.00	18760	35.05	26.90			
2400	8.00	19560	35.40	27.25			
2500	8.00	20360	35.72	27.58			
2600	8.00	21160	36.04	27.91			
2700	8.00	21960	36.34	28.21			
2800	8.00	22760	36.63	28.51			
2900	8.00	23560	36.71	28.59			
3000	8.00	24360	36.98	28.86			

PROTACTINIUM

PROTACTINIUM Pa

Solid from 298° to 1500°, Liquid from 1500° to 3000°.

Gfw	231	GRAMS
$(H°_{298.15} - H°_0)$ =		CAL./GFW.
M.P.	(1,500)	°K
ΔH_m	(3,500)	CAL./GFW.
B.P.	(4,300)	°K
ΔH_v	(110,000)	CAL./GFW.
S.P.		°K
ΔH_s		CAL./GFW.
T.P.		°K
ΔH_f		CAL./GFW.
T.P.		°K
ΔH_f		CAL./GFW.
T_c =		°K
P_c =		ATM.

T TEMPERATURE °K	$C°_p$ HEAT CAPACITY CAL./DEG./GFW.	$H°_T - H°_{298.15}$ HEAT CONTENT CAL./GFW.	$S°_T$ ENTROPY CAL./DEG./GFW.	$-(F°_T - H°_{298.15})$ FREE ENERGY FUNCTION CAL./DEG./GFW.	FORMATION FROM REFERENCE STATE		
					HEAT $\Delta H°_f$ CAL./GFW.	FREE ENERGY $\Delta F°_f$ CAL./GFW.	$\log_{10} K_p$
298	6.79	0	12.40	12.40			
300	6.80	12	12.44	12.40			
400	7.10	710	14.44	12.67			
500	7.40	1430	16.05	13.19			
600	7.70	2190	17.43	13.78			
700	8.00	2970	18.64	14.40			
800	8.30	3790	19.73	15.00			
900	8.60	4630	20.72	15.58			
1000	8.90	5510	21.64	16.13			
1100	9.20	6410	22.51	16.69			
1200	9.50	7350	23.32	17.20			
1300	9.80	8310	24.09	17.70			
1400	10.10	9310	24.83	18.18			
1500	10.00	13830	27.87	18.65			
1600	10.00	14830	28.51	19.25			
1700	10.00	15830	29.12	19.81			
1800	10.00	16830	29.69	20.34			
1900	10.00	17830	30.23	20.85			
2000	10.00	18830	30.74	21.33			
2100	10.00	19830	31.23	21.79			
2200	10.00	20830	31.70	22.24			
2300	10.00	21830	32.14	22.65			
2400	10.00	22830	32.57	23.06			
2500	10.00	23830	32.97	23.44			
2600	10.00	24830	33.37	23.82			
2700	10.00	25830	33.74	24.18			
2800	10.00	26830	34.11	24.53			
2900	10.00	27830	34.46	24.87			
3000	10.00	28830	34.80	25.19			

THERMODYNAMIC PROPERTIES OF THE ELEMENTS

RADIUM

RADIUM	Ra		Solid from 298° to 973°, Liquid from 973° to 1800°, Ideal Monatomic Gas from 1800° to 3000°.

								FORMATION FROM REFERENCE STATE		
REFERENCE STATE								HEAT Δ H°_f	FREE ENERGY Δ F°_f	LOG₁₀ K_p
Gfw	226.05	GRAMS	T TEMPERATURE °K	C°_p HEAT CAPACITY CAL./DEG./GFW.	H°_T − H°_298.15 HEAT CONTENT CAL./GFW.	S°_T ENTROPY CAL./DEG./GFW.	−(F°_T−H°_298.15)/T FREE ENERGY FUNCTION CAL./DEG./GFW.	CAL./GFW.	CAL./GFW.	
(H°_298.15 − H°_0) =		CAL./GFW.								
M.P.	973	°K	298	6.49	0	17.00	17.00			
ΔH_m	(2,000)	CAL./GFW.	300	6.50	12	17.04	17.00			
			400	7.00	690	18.98	17.26			
			500	7.50	1410	20.59	17.77			
			600	8.00	2190	22.01	18.36			
			700	8.50	3010	23.28	18.98			
B.P.	(1,800)	°K	800	9.00	3890	24.44	19.58			
ΔH_v	(32,700)	CAL./GFW.	900	9.50	4810	25.53	20.19			
			1000	7.50	7520	28.55	21.03			
			1100	7.50	8270	29.26	21.75			
			1200	7.50	9020	29.92	22.41			
			1300	7.50	9770	30.52	23.01			
S.P.		°K	1400	7.50	10520	31.07	23.56			
ΔH_s		CAL./GFW.	1500	7.50	11270	31.59	24.08			
			1600	7.50	12020	32.07	24.56			
			1700	7.50	12770	32.53	25.02			
			1800	7.50	13520	32.96	25.45			
T.P.		°K	1900	5.09	46690	51.37	26.80			
ΔH_t		CAL./GFW.	2000	5.15	47200	51.64	28.04			
			2100	5.24	47720	51.89	29.17			
			2200	5.35	48250	52.14	30.21			
T.P.		°K	2300	5.49	48790	52.38	31.17			
ΔH_t		CAL./GFW.	2400	5.65	49350	52.61	32.05			
			2500	5.84	49920	52.85	32.89			
			2600	6.04	50520	53.08	33.65			
			2700	6.26	51130	53.31	34.38			
T_c =		°K	2800	6.53	51770	53.55	35.07			
P_c =		ATM.	2900	6.85	52440	53.78	35.70			
			3000	7.25	53150	54.02	36.31			

RADIUM

RADIUM   Ra   Reference State for Calculating ΔH_f°, ΔF_f°, and $\log_{10} K_p$: Solid from 298° to 973°, Liquid from 973° to 1800°, Ideal Monatomic Gas from 1800° to 3000°.

IDEAL MONATOMIC GAS

Gfw 226.05 GRAMS
$(H^\circ_{298.15} - H^\circ_0)$ = 1,481 CAL./GFW.

T TEMPERATURE °K	C_p° HEAT CAPACITY CAL./DEG./GFW.	$H^\circ_T - H^\circ_{298.15}$ HEAT CONTENT CAL./GFW.	S°_T ENTROPY CAL./DEG./GFW.	$-(F^\circ_T - H^\circ_{298.15})$ FREE ENERGY FUNCTION CAL./DEG./GFW.	FORMATION FROM REFERENCE STATE		
					HEAT ΔH_f° CAL./GFW.	FREE ENERGY ΔF_f° CAL./GFW.	$\log_{10} K_p$
298	4.97	0	42.15	42.15	38700	31201	−22.871
300	4.97	9	42.18	42.15	38697	31155	−22.696
400	4.97	506	43.60	42.34	38516	28668	−15.564
500	4.97	1003	44.71	42.71	38293	26233	−11.457
600	4.97	1500	45.62	43.12	38010	23844	−8.685
700	4.97	1996	46.39	43.54	37685	21509	−6.715
800	4.97	2493	47.05	43.94	37303	19215	−5.249
900	4.97	2990	47.63	44.31	36880	16990	−4.126
1000	4.97	3487	48.16	44.68	34667	15057	−3.291
1100	4.97	3984	48.63	45.01	34414	13107	−2.604
1200	4.97	4481	49.07	45.34	34161	11181	−2.036
1300	4.98	4978	49.47	45.65	33908	9273	−1.559
1400	4.99	5477	49.84	45.93	33657	7379	−1.152
1500	5.00	5976	50.18	46.20	33406	5521	−.804
1600	5.01	6476	50.51	46.47	33156	3652	−.498
1700	5.03	6978	50.81	46.71	32908	1832	−.235
1800	5.05	7482	51.10	46.95	32662	10	−.001
1900	5.09	7989	51.37	47.17	0	0	0
2000	5.15	8501	51.64	47.39	0	0	0
2100	5.24	9020	51.89	47.60	0	0	0
2200	5.35	9550	52.14	47.80	0	0	0
2300	5.49	10092	52.38	48.00	0	0	0
2400	5.65	10649	52.61	48.18	0	0	0
2500	5.84	11224	52.85	48.37	0	0	0
2600	6.04	11818	53.08	48.54	0	0	0
2700	6.26	12433	53.31	48.71	0	0	0
2800	6.53	13072	53.55	48.89	0	0	0
2900	6.85	13741	53.78	49.05	0	0	0
3000	7.25	14446	54.02	49.21	0	0	0

M.P.   °K ΔH_m CAL./GFW.
B.P.   °K ΔH_v CAL./GFW.
S.P.   °K ΔH_s CAL./GFW.
T.P.   °K ΔH_t CAL./GFW.
T.P.   °K ΔH_t CAL./GFW.
T_c =   °K
P_c =   ATM.

THERMODYNAMIC PROPERTIES OF THE ELEMENTS

RADON

RADON Rn

REFERENCE STATE

Gfw	222	GRAMS
$(H°_{298.15} - H°_0) =$	1,481	CAL./GFW.
M.P.	(202)	°K
ΔH_m	(693)	CAL./GFW.
B.P.	(211)	°K
ΔH_v	(3,920)	CAL./GFW.
S.P.		°K
ΔH_s		CAL./GFW.
T.P.		°K
ΔH_t		CAL./GFW.
T.P.		°K
ΔH_t		CAL./GFW.
$T_c =$		°K
$P_c =$		ATM.

Ideal Monatomic Gas from 298° to 3000°.

T TEMPERATURE °K	$C°_P$ HEAT CAPACITY CAL./DEG./GFW.	$H°_T - H°_{298.15}$ HEAT CONTENT CAL./GFW.	$S°_T$ ENTROPY CAL./DEG./GFW.	$-(F°_T - H°_{298.15})/T$ FREE ENERGY FUNCTION CAL./DEG./GFW.	HEAT $\Delta H°_f$ CAL./GFW.	FREE ENERGY $\Delta F°_f$ CAL./GFW.	$LOG_{10} K_P$
298	4.97	0	42.10	42.10			
300	4.97	9	42.13	42.10			
400	4.97	506	43.56	42.30			
500	4.97	1003	44.66	42.66			
600	4.97	1500	45.57	43.07			
700	4.97	1996	46.34	43.49			
800	4.97	2493	47.00	43.89			
900	4.97	2990	47.58	44.26			
1000	4.97	3487	48.11	44.63			
1100	4.97	3984	48.58	44.96			
1200	4.97	4480	49.01	45.28			
1300	4.97	4977	49.41	45.59			
1400	4.97	5474	49.78	45.87			
1500	4.97	5971	50.12	46.14			
1600	4.97	6468	50.44	46.40			
1700	4.97	6964	50.74	46.65			
1800	4.97	7461	51.03	46.89			
1900	4.97	7958	51.30	47.12			
2000	4.97	8455	51.55	47.33			
2100	4.97	8952	51.79	47.53			
2200	4.97	9448	52.03	47.74			
2300	4.97	9945	52.25	47.93			
2400	4.97	10442	52.46	48.11			
2500	4.97	10939	52.66	48.29			
2600	4.97	11436	52.85	48.46			
2700	4.97	11932	53.04	48.63			
2800	4.97	12429	53.22	48.79			
2900	4.97	12926	53.40	48.95			
3000	4.97	13423	53.57	49.10			

RHENIUM

RHENIUM Re
Gfw 186.22 **GRAMS**
$(H°_{298.15} - H°_0) = 1,307$ CAL./GFW.

REFERENCE STATE Solid from 298° to 3000°.

					FORMATION FROM REFERENCE STATE		
T TEMPERATURE °K	$C°_p$ HEAT CAPACITY CAL./DEG./GFW.	$H°_T - H°_{298.15}$ HEAT CONTENT CAL./GFW.	$S°_T$ ENTROPY CAL./DEG./GFW.	$-(F°_T - H°_{298.15})/T$ FREE ENERGY FUNCTION CAL./DEG./GFW.	HEAT $\Delta H°_f$ CAL./GFW.	FREE ENERGY $\Delta F°_f$ CAL./GFW.	$LOG_{10} K_p$
298	6.14	0	8.89	8.89			
300	6.14	11	8.93	8.90			
400	6.18	620	10.68	9.13			
500	6.31	1240	12.06	9.58			
600	6.44	1890	13.25	10.10			
700	6.57	2550	14.26	10.62			
800	6.70	3210	15.14	11.13			
900	6.83	3880	15.93	11.62			
1000	6.96	4570	16.66	12.09			
1100	7.09	5270	17.33	12.54			
1200	7.22	5980	17.94	12.96			
1300	7.35	6710	18.53	13.37			
1400	7.48	7460	19.08	13.76			
1500	7.61	8220	19.61	14.13			
1600	7.74	8990	20.09	14.48			
1700	7.87	9770	20.56	14.82			
1800	8.00	10560	21.02	15.16			
1900	8.13	11370	21.45	15.47			
2000	8.26	12180	21.87	15.78			
2100	8.39	13010	22.28	16.09			
2200	8.52	13850	22.67	16.38			
2300	8.65	14710	23.05	16.66			
2400	8.78	15580	23.43	16.94			
2500	8.91	16470	23.79	17.21			
2600	9.04	17360	24.14	17.47			
2700	9.17	18280	24.48	17.71			
2800	9.30	19200	24.82	17.97			
2900	9.43	20140	25.15	18.21			
3000	9.56	21080	25.47	18.45			

M.P. 3,453 °K
ΔH_m (7,900) CAL./GFW.

B.P. 5,900 °K
ΔH_v 169,000 CAL./GFW.

S.P. °K
ΔH_s CAL./GFW.

T.P. °K
ΔH_t CAL./GFW.

T.P. °K
ΔH_t CAL./GFW.

$T_c =$ °K
$P_c =$ ATM.

THERMODYNAMIC PROPERTIES OF THE ELEMENTS

RHENIUM

RHENIUM   Re

IDEAL MONATOMIC GAS

Gfw 186.22 GRAMS

$(H°_{298.15} - H°_0) = 1,481$ CAL./GFW.

Reference State for Calculating $\Delta H_f°$, $\Delta F_f°$, and $\text{Log}_{10} K_p$: Solid from 298° to 3000°.

T TEMPERATURE °K	$C_p°$ HEAT CAPACITY CAL./DEG./GFW.	$H°_T - H°_{298.15}$ HEAT CONTENT CAL./GFW.	$S°_T$ ENTROPY CAL./DEG./GFW.	$-(F°_T - H°_{298.15})/T$ FREE ENERGY FUNCTION CAL./DEG./GFW.	FORMATION FROM REFERENCE STATE		
					HEAT $\Delta H°$ CAL./GFW.	FREE ENERGY $\Delta F_f°$ CAL./GFW.	$\text{LOG}_{10} K_p$
298	4.97	0	45.13	45.13	185650	174845	-128.170
300	4.97	9	45.16	45.13	185648	174779	-127.336
400	4.97	506	46.59	45.33	185536	171172	-93.531
500	4.97	1003	47.70	45.70	185413	167593	-73.259
600	4.97	1500	48.61	46.11	185260	164044	-59.757
700	4.97	1996	49.37	46.52	185096	160519	-50.120
800	4.97	2493	50.04	46.93	184933	157013	-42.897
900	4.97	2990	50.62	47.30	184760	153539	-37.286
1000	4.97	3487	51.14	47.67	184587	150087	-32.804
1100	4.97	3984	51.62	48.00	184364	146645	-29.138
1200	4.97	4481	52.05	48.32	184151	143219	-26.085
1300	4.97	4977	52.45	48.63	183917	139821	-23.508
1400	4.97	5475	52.82	48.91	183665	136429	-21.299
1500	4.98	5972	53.16	49.18	183402	133077	-19.390
1600	4.99	6471	53.48	49.44	183131	129707	-17.716
1700	5.00	6970	53.78	49.69	182850	126376	-16.246
1800	5.02	7472	54.07	49.92	182562	123072	-14.942
1900	5.06	7975	54.34	50.15	182255	119764	-13.775
2000	5.10	8483	54.60	50.36	181953	116493	-12.729
2100	5.15	8995	54.85	50.57	181635	113238	-11.784
2200	5.22	9513	55.09	50.78	181313	109989	-10.926
2300	5.30	10039	55.33	50.97	180979	106735	-10.141
2400	5.40	10573	55.56	51.16	180643	103531	-9.427
2500	5.51	11118	55.78	51.34	180298	100323	-8.770
2600	5.51	11676	56.00	51.51	179966	97130	-8.164
2700	5.79	12247	56.21	51.68	179617	93946	-7.603
2800	5.96	12835	56.43	51.85	179285	90777	-7.085
2900	6.15	13440	56.64	52.01	178950	87629	-6.603
3000	6.35	14065	56.85	52.17	178635	84495	-6.155

M.P. °K
ΔH_m CAL./GFW.

B.P. °K
ΔH_v CAL./GFW.

S.P. °K
ΔH_s CAL./GFW.

T.P. °K
ΔH_t CAL./GFW.

T.P. °K
ΔH_t CAL./GFW.

$T_c =$ °K
$P_c =$ ATM.

RHODIUM

RHODIUM Rh

REFERENCE STATE

Gfw 102.91 GRAMS CAL./GFW.

$(H°_{298.15} - H°_0) =$

M.P. 2,239 °K
ΔH_m (5,200) CAL./GFW.

B.P. (4,000) °K
ΔH_v (118,400) CAL./GFW.

S.P. °K
ΔH_s CAL./GFW.

T.P. °K
ΔH_t CAL./GFW.

T.P. °K
ΔH_t CAL./GFW.

$T_c =$ °K
$P_c =$ ATM.

Solid from 298° to 2239°, Liquid from 2239° to 3000°.

T TEMPERATURE °K	$C°_p$ HEAT CAPACITY CAL./DEG./GFW.	$H°_T - H°_{298.15}$ HEAT CONTENT CAL./GFW.	$S°_T$ ENTROPY CAL./DEG./GFW.	$-(F°_T - H°_{298.15})/T$ FREE ENERGY FUNCTION CAL./DEG./GFW.	FORMATION FROM REFERENCE STATE		
					HEAT $\Delta H°_f$ CAL./GFW.	FREE ENERGY $\Delta F°_f$ CAL./GFW.	LOG$_{10}$ K_p
298	6.11	0	7.60	7.60			
300	6.11	11	7.64	7.61			
400	6.31	630	9.41	7.84			
500	6.52	1260	10.87	8.30			
600	6.73	1920	12.02	8.82			
700	6.93	2600	13.07	9.36			
800	7.14	3300	14.00	9.88			
900	7.35	4030	14.86	10.39			
1000	7.55	4790	15.66	10.87			
1100	7.76	5570	16.40	11.34			
1200	7.96	6370	17.10	11.80			
1300	8.17	7180	17.75	12.23			
1400	8.38	8010	18.36	12.64			
1500	8.58	8860	18.95	13.05			
1600	8.80	9720	19.50	13.43			
1700	9.00	10600	20.04	13.81			
1800	9.20	11490	20.55	14.17			
1900	9.40	12420	21.07	14.54			
2000	9.60	13370	21.56	14.88			
2100	9.80	14350	22.03	15.20			
2200	10.00	15340	22.49	15.52			
2300	10.00	21540	25.26	15.90			
2400	10.00	22540	25.68	16.29			
2500	10.00	23540	26.09	16.68			
2600	10.00	24540	26.48	17.05			
2700	10.00	25540	26.86	17.41			
2800	10.00	26540	27.22	17.75			
2900	10.00	27540	27.57	18.08			
3000	10.00	28540	27.91	18.40			

THERMODYNAMIC PROPERTIES OF THE ELEMENTS

RHODIUM

RHODIUM	Rh	Reference State for Calculating ΔH_f°, ΔF_f°, and $\log_{10} K_p$:
IDEAL MONATOMIC GAS		Solid from 298° to 2239°, Liquid from 2239° to 3000°.

G/w 102.91 GRAMS
($H^\circ_{298.15} - H^\circ_0$) = 1,483 CAL./GFW

T TEMPERATURE °K	C_p° HEAT CAPACITY CAL./DEG./GFW	$H^\circ_T - H^\circ_{298.15}$ HEAT CONTENT CAL./GFW	S°_T ENTROPY CAL./DEG./GFW	$-(F^\circ - H^\circ_{298.15})/T$ FREE ENERGY FUNCTION CAL./DEG./GFW	FORMATION FROM REFERENCE STATE			
					HEAT ΔH_f° CAL./GFW	FREE ENERGY ΔF_f° CAL./GFW	$\log_{10} K_p$
298	5.02	0	44.39	44.39	133000	122031	−89.454
300	5.02	9	44.41	44.38	132998	121967	−88.860
400	5.17	519	45.88	44.59	132889	118301	−64.642
500	5.39	1046	47.06	44.97	132786	114666	−50.123
600	5.62	1596	48.06	45.40	132676	111052	−40.454
700	5.84	2169	48.94	45.85	132569	107460	−33.553
800	6.03	2763	49.74	46.29	132463	103871	−28.378
900	6.20	3375	50.46	46.71	132345	100305	−24.359
1000	6.33	4002	51.12	47.12	132212	96752	−21.147
1100	6.43	4640	51.73	47.52	132070	93207	−18.520
1200	6.50	5287	52.29	47.89	131917	89689	−16.335
1300	6.56	5940	52.81	48.25	131760	86182	−14.489
1400	6.59	6598	53.30	48.59	131588	82672	−12.906
1500	6.62	7259	53.75	48.92	131399	79199	−11.540
1600	6.63	7921	54.18	49.23	131201	75713	−10.341
1700	6.63	8584	54.58	49.54	130984	72266	−9.290
1800	6.64	9248	54.96	49.83	130758	68820	−8.355
1900	6.63	9911	55.32	50.11	130491	65416	−7.524
2000	6.63	10574	55.66	50.38	130204	62004	−6.775
2100	6.63	11237	55.99	50.64	129887	58571	−6.095
2200	6.62	11900	56.29	50.89	129560	55200	−5.483
2300	6.62	12562	56.59	51.13	124022	51963	−4.937
2400	6.62	13224	56.87	51.36	123684	48828	−4.446
2500	6.62	13885	57.14	51.59	123345	45720	−3.996
2600	6.62	14547	57.40	51.81	123007	42615	−3.582
2700	6.62	15209	57.65	52.02	122669	39536	−3.200
2800	6.63	15871	57.89	52.23	122331	36455	−2.845
2900	6.63	16534	58.12	52.42	121994	33399	−2.516
3000	6.64	17198	58.35	52.62	121658	30338	−2.210

M.P. °K
ΔH_m CAL./GFW

B.P. °K
ΔH_v CAL./GFW

S.P. °K
ΔH_s CAL./GFW

T.P. °K
ΔH_t CAL./GFW

T.P. °K
ΔH_t CAL./GFW

T_c = °K
P_c = ATM.

RUBIDIUM

Solid from 298° to 312.0°, Liquid from 312.0° to 974°, Ideal Monatomic Gas from 974° to 3000°.

					FORMATION FROM REFERENCE STATE		
T TEMPERATURE °K	C_p° HEAT CAPACITY CAL./DEG./GFW.	$H_T^\circ - H_{298.15}^\circ$ HEAT CONTENT CAL./GFW.	S_T° ENTROPY CAL./DEG./GFW.	$-(F_T^\circ - H_{298.15}^\circ)/T$ FREE ENERGY FUNCTION CAL./DEG./GFW.	HEAT ΔH_f° CAL./GFW.	FREE ENERGY ΔF_f° CAL./GFW.	$LOG_{10} K_p$
298	7.50	0	18.22	18.22			
300	7.55	14	18.27	18.23			
400	7.50	1324	22.22	18.91			
500	7.50	2074	23.90	19.76			
600	7.50	2824	25.26	20.56			
700	7.50	3574	26.42	21.32			
800	7.50	4324	27.42	22.02			
900	7.50	5074	28.30	22.67			
1000	4.97	23087	46.64	23.56			
1100	4.97	23584	47.11	25.67			
1200	4.97	24080	47.55	27.49			
1300	4.97	24577	47.94	29.04			
1400	4.97	25074	48.31	30.40			
1500	4.97	25571	48.66	31.62			
1600	4.97	26069	48.98	32.69			
1700	4.98	26567	49.28	33.66			
1800	4.99	27065	49.56	34.53			
1900	5.01	27565	49.83	35.33			
2000	5.02	28067	50.09	36.06			
2100	5.05	28570	50.34	36.74			
2200	5.07	29076	50.57	37.36			
2300	5.11	29585	50.80	37.94			
2400	5.15	30098	51.02	38.48			
2500	5.20	30616	51.23	38.99			
2600	5.26	31139	51.44	39.47			
2700	5.32	31668	51.63	39.91			
2800	5.39	32203	51.83	40.33			
2900	5.47	32746	52.02	40.73			
3000	5.56	33298	52.21	41.12			

RUBIDIUM Rb

Gfw 85.48 GRAMS

REFERENCE STATE

$(H_{298.15}^\circ - H_0^\circ) = 1,790$ CAL./GFW.

M.P. 312.0 °K
ΔH_m 560. CAL./GFW.

B.P. 974. °K
ΔH_v 16,540. CAL./GFW.

S.P. °K
ΔH_s CAL./GFW.

T.P. °K
ΔH_t CAL./GFW.

T.P. °K
ΔH_t CAL./GFW.

$T_c =$ °K
$P_c =$ ATM.

THERMODYNAMIC PROPERTIES OF THE ELEMENTS

RUBIDIUM

RUBIDIUM	Rb		Reference State for Calculating ΔH_f°, ΔF_f°, and $Log_{10}K_p$:
IDEAL MONATOMIC GAS			Solid from 298° to 312.0°, Liquid from 312.0° to 974°, Ideal Monatomic Gas from 974° to 3000°.
Gfw	85.48	GRAMS	
$(H_{298.15}^\circ - H_0^\circ)$ =	1,481	CAL./GFW.	

							FORMATION FROM REFERENCE STATE		
T TEMPERATURE °K	C_p° HEAT CAPACITY CAL./DEG./GFW.	$H_T^\circ - H_{298.15}^\circ$ HEAT CONTENT CAL./GFW.	S_T° ENTROPY CAL./DEG./GFW.	$-(F^\circ-H^\circ_{298.15})$ FREE ENERGY FUNCTION CAL./DEG./GFW.		HEAT ΔH_f° CAL./GFW.	FREE ENERGY ΔF_f° CAL./GFW.		$LOG_{10} K_p$
298	4.97	0	40.63	40.63		19600	12918		9.469
300	4.97	9	40.66	40.63		19595	12878		9.382
400	4.97	506	42.09	40.83		18782	10834		5.919
500	4.97	1003	43.20	41.20		18529	8879		3.881
600	4.97	1500	44.10	41.60		18276	6972		2.539
700	4.97	1996	44.87	42.02		18022	5107		1.594
800	4.97	2493	45.53	42.42		17769	3281		.896
900	4.97	2990	46.12	42.80		17516	1478		.358
1000	4.97	3487	46.64	43.16		0	0		0
1100	4.97	3984	47.11	43.49		0	0		0
1200	4.97	4480	47.55	43.82		0	0		0
1300	4.97	4977	47.94	44.12		0	0		0
1400	4.97	5474	48.31	44.40		0	0		0
1500	4.97	5971	48.66	44.68		0	0		0
1600	4.98	6469	48.98	44.94		0	0		0
1700	4.98	6967	49.28	45.19		0	0		0
1800	4.99	7465	49.56	45.42		0	0		0
1900	5.01	7965	49.83	45.64		0	0		0
2000	5.02	8467	50.09	45.86		0	0		0
2100	5.05	8970	50.34	46.07		0	0		0
2200	5.07	9476	50.57	46.27		0	0		0
2300	5.11	9985	50.80	46.46		0	0		0
2400	5.15	10498	51.02	46.65		0	0		0
2500	5.20	11016	51.23	46.83		0	0		0
2600	5.26	11539	51.44	47.01		0	0		0
2700	5.32	12068	51.63	47.17		0	0		0
2800	5.39	12603	51.83	47.33		0	0		0
2900	5.47	13146	52.02	47.49		0	0		0
3000	5.56	13698	52.21	47.65		0	0		0

M.P.		°K
ΔH_m		CAL./GFW.
B.P.		°K
ΔH_v		CAL./GFW.
S.P.		°K
ΔH_s		CAL./GFW.
T.P.		°K
ΔH_t		CAL./GFW.
T.P.		°K
ΔH_t		CAL./GFW.
T_c =		°K
P_c =		ATM.

RUBIDIUM

RUBIDIUM Rb_2
IDEAL DIATOMIC GAS

Reference State for Calculating ΔH_f°, ΔF_f°, and $Log_{10} K_P$: Solid from 298° to 312.0°, Liquid from 312.0° to 974°, Ideal Monatomic Gas from 974° to 3000°.

Gfw 85.48 GRAMS
$(H_{298.15}^\circ - H_0^\circ) = 2,608$ CAL./GFW.

						FORMATION FROM REFERENCE STATE		
T TEMPERATURE °K	C_P° HEAT CAPACITY CAL./DEG./GFW.	$H_T^\circ - H_{298.15}^\circ$ HEAT CONTENT CAL./GFW.	S_T° ENTROPY CAL./DEG./GFW.	$-(F^\circ - H_{298.15}^\circ)/T$ FREE ENERGY FUNCTION CAL./DEG./GFW.	HEAT ΔH_f° CAL./GFW.	FREE ENERGY ΔF_f° CAL./GFW.	$LOG_{10} K_P$
298	9.06	0	64.69	64.69	27550	19127	-14.021
300	9.06	17	64.75	64.70	27539	19076	-13.898
400	9.11	926	67.36	65.05	25828	16660	-9.103
500	9.16	1839	69.40	65.73	25241	14441	-6.312
600	9.20	2756	71.07	66.48	24658	12328	-4.490
700	9.25	3678	72.49	67.24	24080	10325	-3.223
800	9.29	4608	73.73	67.97	23510	8398	-2.294
900	9.34	5537	74.83	68.68	22939	6532	-1.586
1000	9.38	6472	75.81	69.34	-12152	5318	-1.162
1100	9.43	7413	76.71	69.98	-12205	7056	-1.402
1200	9.47	8360	77.53	70.57	-12250	8834	-1.609
1300	9.51	9313	78.30	71.14	-12291	10563	-1.775
1400	9.56	10258	79.00	71.68	-12340	12328	-1.924
1500	9.60	11222	79.66	72.18	-12370	14120	-2.057

M.P. °K
ΔH_m CAL./GFW.

B.P. °K
ΔH_v CAL./GFW.

S.P. °K
ΔH_s CAL./GFW.

T.P. °K
ΔH_t CAL./GFW.

T.P. °K
ΔH_t CAL./GFW.

T_c = °K
P_c = ATM.

RUTHENIUM

RUTHENIUM Ru
REFERENCE STATE

Solid I from 298° to 1308°, Solid II from 1308° to 1473°, Solid III from 1473° to 1773°, Solid IV from 1773° to 2700°, Liquid from 2700° to 3000°.

Gfw	101.1	GRAMS
$(H°_{298.15} - H°_0) =$		CAL./GFW.

M.P.	(2,700)	°K
ΔH_m	(6,100)	CAL./GFW.
B.P.	(4,000)	°K
ΔH_v	(135,700)	CAL./GFW.
S.P.		°K
ΔH_s		CAL./GFW.
T.P.	1,308	°K
ΔH_t	60	CAL./GFW.
T.P.	1,473	°K
ΔH_t	0	CAL./GFW.
$T_c =$	1,773	°K
$P_c =$	320	ATM.

T TEMPERATURE °K	$C_p^°$ HEAT CAPACITY CAL./DEG./GFW.	$H° - H°_{298.15}$ HEAT CONTENT CAL./GFW.	$S_T°$ ENTROPY CAL./DEG./GFW.	$-(F° - H°_{298.15})/T$ FREE ENERGY FUNCTION CAL./DEG./GFW.	FORMATION FROM REFERENCE STATE	
					HEAT $\Delta H°_f$ CAL./GFW.	FREE ENERGY $\Delta F°_f$ CAL./GFW.	LOG$_{10}$ K$_p$
298	5.70	0	6.90	6.90			
300	5.70	10	6.93	6.90			
400	5.85	590	8.60	7.13			
500	6.00	1180	9.92	7.56			
600	6.15	1780	11.01	8.05			
700	6.30	2400	11.96	8.54			
800	6.45	3040	12.82	9.02			
900	6.60	3690	13.58	9.48			
1000	6.75	4360	14.29	9.93			
1100	6.90	5050	14.95	10.36			
1200	7.05	5750	15.56	10.77			
1300	7.20	6460	16.13	11.17			
1400	7.20	7240	16.71	11.54			
1500	7.20	7960	17.21	11.91			
1600	7.20	8680	17.67	12.25			
1700	7.20	9400	18.11	12.59			
1800	7.50	10450	18.70	12.90			
1900	7.50	11200	19.11	13.22			
2000	7.50	11950	19.50	13.53			
2100	7.50	12700	19.87	13.83			
2200	7.50	13450	20.21	14.10			
2300	7.50	14200	20.55	14.38			
2400	7.50	14950	20.87	14.65			
2500	7.50	15700	21.17	14.89			
2600	7.50	16450	21.47	15.15			
2700	7.50	17200	21.75	15.38			
2800	7.50	24150	24.32	15.70			
2900	7.50	24900	24.59	16.01			
3000	7.50	25650	24.84	16.29			

RUTHENIUM

RUTHENIUM Ru
IDEAL MONATOMIC GAS

Gfw 101.1 GRAMS
$(H°_{298.15} - H°_0) = 1,490$ CAL./GFW.

Reference State for Calculating $\Delta H_f°$, $\Delta F_f°$, and $\text{Log}_{10} K_p$: Solid I from 298° to 1308°, Solid II from 1308° to 1473°, Solid III from 1473° to 1773°, Solid IV from 1773° to 2700°, Liquid from 2700° to 3000°.

T TEMPERATURE °K	$C_p°$ HEAT CAPACITY CAL./DEG./GFW.	$H°_T - H°_{298.15}$ HEAT CONTENT CAL./GFW.	$S°_T$ ENTROPY CAL./DEG./GFW.	$-(F°_T - H°_{298.15})/T$ FREE ENERGY FUNCTION CAL./DEG./GFW.	HEAT $\Delta H_f°$ CAL./GFW.	FREE ENERGY $\Delta F_f°$ CAL./GFW.	$\text{LOG}_{10} K_p$
298	5.14	0	44.55	44.55	144000	132774	-97.329
300	5.15	10	44.58	44.55	144000	132705	-96.683
400	5.41	537	46.10	44.76	143947	128947	-70.459
500	5.69	1093	47.34	45.16	143913	125203	-54.729
600	5.92	1673	48.40	45.62	143893	121459	-44.245
700	6.09	2274	49.32	46.08	143874	117722	-36.757
800	6.20	2889	50.14	46.53	143849	113993	-31.144
900	6.27	3513	50.88	46.98	143823	110253	-26.774
1000	6.30	4142	51.54	47.40	143782	106532	-23.284
1100	6.30	4772	52.14	47.81	143722	102813	-20.428
1200	6.30	5402	52.69	48.19	143652	99096	-18.049
1300	6.29	6032	53.19	48.55	143572	95394	-16.038
1400	6.29	6661	53.66	48.91	143421	91691	-14.314
1500	6.30	7291	54.09	49.23	143331	88011	-12.824
1600	6.32	7922	54.50	49.55	143242	84314	-11.516
1700	6.35	8555	54.88	49.85	143155	80646	-10.367
1800	6.39	9192	55.25	50.15	142742	76952	-9.342
1900	6.45	9834	55.60	50.43	142634	73303	-8.431
2000	6.51	10481	55.93	50.69	142531	69671	-7.612
2100	6.58	11136	56.25	50.95	142436	66038	-6.872
2200	6.66	11798	56.56	51.20	142348	62378	-6.196
2300	6.74	12468	56.85	51.43	142268	58778	-5.585
2400	6.83	13146	57.14	51.67	142196	55148	-5.021
2500	6.91	13833	57.42	51.89	142133	51508	-4.502
2600	7.00	14529	57.69	52.11	142079	47907	-4.027
2700	7.08	15233	57.96	52.32	142033	44266	-3.582
2800	7.17	15945	58.22	52.53	135795	40875	-3.190
2900	7.24	16666	58.47	52.73	135766	37514	-2.827
3000	7.32	17394	58.72	52.93	135744	34104	-2.484

M.P. °K
ΔH_m CAL./GFW.

B.P. °K
ΔH_v CAL./GFW.

S.P. °K
ΔH_s CAL./GFW.

T.P. °K
ΔH_t CAL./GFW.

T.P. °K
ΔH_t CAL./GFW.

$T_c =$ °K
$P_c =$ ATM.

SAMARIUM

THERMODYNAMIC PROPERTIES OF THE ELEMENTS

SAMARIUM Sm
G|w 150.35 GRAMS
($H^°_{298.15} - H^°_0$) =   CAL./GFW.

Solid I from 298° to 1190°, Solid II from 1190° to 1325°, Liquid from 1325° to 1860°, Ideal Monatomic Gas from 1860° to 3000°.

T °K	$C^°_P$ CAL./DEG./GFW.	$H^°_T - H^°_{298.15}$ HEAT CONTENT CAL./GFW.	$S^°_T$ ENTROPY CAL./DEG./GFW.	$-(F^°_T - H^°_{298.15})/T$ FREE ENERGY FUNCTION CAL./DEG./GFW.	HEAT Δ$H^°_f$ CAL./GFW.	FREE ENERGY Δ$F^°_f$ CAL./GFW.	LOG$_{10}$ K_P
298	6.49	0	16.28	16.28			
300	6.50	11	16.32	16.29			
400	6.78	675	18.23	16.55			
500	7.06	1370	19.77	17.03			
600	7.34	2090	21.08	17.60			
700	7.62	2835	22.23	18.18			
800	7.90	3610	23.27	18.76			
900	8.18	4415	24.22	19.32			
1000	8.46	5250	25.09	19.84			
1100	8.74	6110	25.91	20.36			
1200	8.00	7350	26.98	20.86			
1300	8.00	8150	27.62	21.36			
1400	8.00	11600	30.21	21.93			
1500	8.00	12400	30.76	22.50			
1600	8.00	13200	31.28	23.03			
1700	8.00	14000	31.76	23.53			
1800	8.00	14800	32.22	24.00			
1900	6.39	61340	57.04	24.76			
2000	6.31	61970	57.36	26.38			
2100	6.24	62600	57.67	27.87			
2200	6.18	63220	57.96	29.23			
2300	6.13	63830	58.23	30.48			
2400	6.09	64440	58.49	31.64			
2500	6.06	65050	58.74	32.72			
2600	6.04	65660	58.98	33.73			
2700	6.04	66260	59.20	34.66			
2800	6.04	66860	59.42	35.55			
2900	6.05	67470	59.64	36.38			
3000	6.08	68080	59.84	37.15			

M.P. 1,325 °K
ΔH_m (2,650) CAL./GFW.
B.P. (1,860) °K
ΔH_v (45,800) CAL./GFW.
S.P.   °K
ΔH_s CAL./GFW.
T.P. 1,190 °K
ΔH_t (360) CAL./GFW.
T.P.   °K
ΔH_t CAL./GFW.
T_c =   °K
P_c =   ATM.

SAMARIUM

Reference State for Calculating ΔH_f°, ΔF_f°, and $\log_{10} K_p$:
Solid I from 298° to 1190°, Solid II from 1190° to 1325°, Liquid from 1325° to 1860°, Ideal Monatomic Gas from 1860° to 3000°.

SAMARIUM	Sm
IDEAL MONATOMIC GAS	
Gfw 150.35 GRAMS	CAL./GFW
$(H°_{298.15} - H°_0)$ =	

T °K	C_p° HEAT CAPACITY CAL./DEG./GFW	$H_T^\circ - H_{298.15}^\circ$ HEAT CONTENT CAL./GFW	S_T° ENTROPY CAL./DEG./GFW	$-(F_T^\circ - H_{298.15}^\circ)/T$ FREE ENERGY FUNCTION CAL./DEG./GFW	HEAT ΔH_f° CAL./GFW	FREE ENERGY ΔF_f° CAL./GFW	$\log_{10} K_p$
298	7.25	0	43.72	43.72	50000	41819	−30.655
300	7.25	14	43.77	43.73	50003	41768	−30.430
400	7.28	740	45.86	44.01	50065	39013	−21.317
500	7.33	1471	47.49	44.55	50101	36241	−15.842
600	7.37	2205	48.83	45.16	50115	33465	−12.190
700	7.40	2944	49.97	45.77	50109	30691	−9.582
800	7.40	3684	50.95	46.35	50074	27930	−7.630
900	7.37	4423	51.82	46.91	50008	25168	−6.112
1000	7.32	5157	52.60	47.45	49907	22397	−4.895
1100	7.24	5885	53.29	47.94	49775	19657	−3.905
1200	7.14	6605	53.92	48.42	49255	16927	−3.083
1300	7.04	7314	54.49	48.87	49164	14233	−2.392
1400	6.92	8012	55.00	49.28	46412	11706	−1.827
1500	6.81	8698	55.48	49.69	46298	9218	−1.343
1600	6.69	9373	55.91	50.06	46173	6765	−.924
1700	6.59	10037	56.31	50.41	46037	4302	−.553
1800	6.49	10691	56.69	50.76	45891	1845	−.224
1900	6.39	11335	57.04	51.08	0	0	0
2000	6.31	11970	57.36	51.38	0	0	0
2100	6.24	12597	57.67	51.68	0	0	0
2200	6.18	13218	57.96	51.96	0	0	0
2300	6.13	13834	58.23	52.22	0	0	0
2400	6.09	14444	58.49	52.48	0	0	0
2500	6.06	15052	58.74	52.72	0	0	0
2600	6.04	15657	58.98	52.96	0	0	0
2700	6.04	16260	59.20	53.18	0	0	0
2800	6.04	16864	59.42	53.40	0	0	0
2900	6.05	17469	59.64	53.62	0	0	0
3000	6.08	18075	59.84	53.82	0	0	0

M.P.	°K	
ΔH_m	CAL./GFW	
B.P.	°K	
ΔH_v	CAL./GFW	
S.P.	°K	
ΔH_s	CAL./GFW	
T.P.	°K	
ΔH_t	CAL./GFW	
T.P.	°K	
ΔH_t	CAL./GFW	
T_c =	°K	
P_c =	ATM.	

SCANDIUM

THERMODYNAMIC PROPERTIES OF THE ELEMENTS

SCANDIUM	Sc		Solid from 298° to 1673°, Liquid from 1673° to 2750°, Ideal Monatomic Gas from 2750° to 3000°.
Gfw	44.96	GRAMS	
REFERENCE STATE			
$(H^°_{298.15} - H^°_0)$ =		CAL./GFW.	

						FORMATION FROM REFERENCE STATE		
T TEMPERATURE °K	$C^°_P$ HEAT CAPACITY CAL./DEG./GFW.	$H^°_T - H^°_{298.15}$ HEAT CONTENT CAL./GFW.	$S^°_T$ ENTROPY CAL./DEG./GFW.	$-(F^°_T - H^°_{298.15})$ FREE ENERGY FUNCTION CAL./DEG./GFW.		HEAT $\Delta H^°_f$ CAL./GFW.	FREE ENERGY $\Delta F^°_f$ CAL./GFW.	$LOG_{10} K_P$
298	6.01	0	9.00	9.00				
300	6.01	11	9.04	9.01				
400	6.12	617	10.78	9.24				
500	6.23	1235	12.16	9.69				
600	6.33	1860	13.30	10.20				
700	6.44	2500	14.29	10.72				
800	6.55	3150	15.15	11.22				
900	6.66	3810	15.93	11.70				
1000	6.77	4480	16.64	12.16				
1100	6.88	5170	17.29	12.59				
1200	6.99	5860	17.89	13.01				
1300	7.10	6560	18.46	13.42				
1400	7.21	7280	18.99	13.79				
1500	7.32	8010	19.49	14.15				
1600	7.42	8740	19.96	14.50				
1700	8.00	13350	22.72	14.87				
1800	8.00	14150	23.18	15.32				
1900	8.00	14950	23.61	15.75				
2000	8.00	15750	24.02	16.15				
2100	8.00	16550	24.41	16.53				
2200	8.00	17350	24.79	16.91				
2300	8.00	18150	25.14	17.25				
2400	8.00	18950	25.48	17.59				
2500	8.00	19750	25.81	17.91				
2600	8.00	20550	26.12	18.22				
2700	8.00	21350	26.42	18.52				
2800	5.87	94890	53.19	19.31				
2900	6.03	95490	53.40	20.48				
3000	6.21	96100	53.61	21.58				

M.P.	(1,673)	°K
ΔH_m	(3,850)	CAL./GFW.
B.P.	(2,750)	°K
ΔH_v	(72,850)	CAL./GFW.
S.P.		°K
ΔH_s		CAL./GFW.
T.P.		°K
ΔH_t		CAL./GFW.
T.P.		°K
ΔH_t		CAL./GFW.
T_c =		°K
P_c =		ATM.

SCANDIUM

SCANDIUM   Sc

IDEAL MONATOMIC GAS

Gfw	44.96	GRAMS
$(H°_{298.15} - H°_0)$ =	1,674	CAL./GFW.

Reference State for Calculating $\Delta H_f^°$, $\Delta F_f^°$, and $\log_{10} K_p$: Solid from 298° to 1673°, Liquid from 1673° to 2750°, Ideal Monatomic Gas from 2750° to 3000°.

T TEMPERATURE °K	$C_p^°$ HEAT CAPACITY CAL./DEG./GFW.	$H°_T - H°_{298.15}$ HEAT CONTENT CAL./GFW.	$S°_T$ ENTROPY CAL./DEG./GFW.	$-(F°_T - H°_{298.15})/T$ FREE ENERGY FUNCTION CAL./DEG./GFW.	FORMATION FROM REFERENCE STATE		
					HEAT $\Delta H_f^°$ CAL./GFW.	FREE ENERGY $\Delta F_f^°$ CAL./GFW.	$\log_{10} K_p$
298	5.28	0	41.75	41.75	82000	72235	−52.951
300	5.28	10	41.78	41.75	81999	72177	−52.585
400	5.15	530	43.28	41.96	81913	68913	−37.655
500	5.08	1042	44.42	42.34	81807	65677	−28.709
600	5.05	1548	45.35	42.77	81688	62458	−22.752
700	5.03	2052	46.12	43.19	81552	59271	−18.506
800	5.01	2554	46.79	43.60	81404	56092	−15.324
900	5.00	3055	47.38	43.99	81245	52940	−12.856
1000	5.00	3555	47.91	44.36	81075	49805	−10.885
1100	4.99	4054	48.38	44.70	80884	46685	−9.276
1200	4.99	4553	48.82	45.03	80693	43577	−7.937
1300	4.99	5051	49.22	45.34	80492	40504	−6.809
1400	4.99	5551	49.59	45.63	80271	37431	−5.843
1500	4.99	6050	49.93	45.90	80040	34380	−5.009
1600	5.00	6550	50.25	46.16	79810	31346	−4.281
1700	5.01	7050	50.56	46.42	75700	28372	−3.647
1800	5.03	7553	50.85	46.66	75403	25597	−3.107
1900	5.06	8057	51.12	46.88	75107	22838	−2.626
2000	5.10	8565	51.38	47.10	74815	20095	−2.195
2100	5.15	9078	51.63	47.31	74528	17366	−1.807
2200	5.21	9595	51.87	47.51	74245	14669	−1.457
2300	5.28	10120	52.10	47.70	73970	11962	−1.136
2400	5.37	10652	52.33	47.90	73702	9262	−.843
2500	5.47	11194	52.55	48.08	73444	6594	−.576
2600	5.59	11747	52.77	48.26	73197	3907	−.328
2700	5.72	12312	52.98	48.42	72962	1250	−.101
2800	5.87	12892	53.19	48.59	0	0	0
2900	6.03	13487	53.40	48.75	0	0	0
3000	6.21	14098	53.61	48.92	0	0	0

M.P. °K
ΔH_m CAL./GFW.

B.P. °K
ΔH_v CAL./GFW.

S.P. °K
ΔH_s CAL./GFW.

T.P. °K
ΔH_t CAL./GFW.

T.P. °K
ΔH_t CAL./GFW.

T_c = °K
P_c = ATM.

SELENIUM

SELENIUM Se

REFERENCE STATE

Solid from 298° to 490°, Liquid from 490° to 958°, Ideal Diatomic Gas from 958° to 3000°.

Gfw	78.96 GRAMS
$(H°_{298.15} - H°_0)$ =	1,319 CAL./GFW.
M.P.	490 °K
ΔH_m	1,300 CAL./GFW.
B.P.	958 °K
ΔH_v	6,290 CAL./GFW.
S.P.	°K
ΔH_s	CAL./GFW.
T.P.	°K
ΔH_t	CAL./GFW.
T.P.	°K
ΔH_t	CAL./GFW.
T_c =	°K
P_c =	ATM.

T TEMPERATURE °K	$C°_p$ HEAT CAPACITY CAL./DEG./GFW.	$H°_T - H°_{298.15}$ HEAT CONTENT CAL./GFW.	$S°_T$ ENTROPY CAL./DEG./GFW.	$-(F°_T - H°_{298.15})/T$ FREE ENERGY FUNCTION CAL./DEG./GFW.	HEAT $\Delta H°_f$ CAL./GFW.	FREE ENERGY $\Delta F°_f$ CAL./GFW.	$LOG_{10} K_p$
298	6.09	0	10.15	10.15			
300	6.10	11	10.19	10.16			
400	6.65	650	12.02	10.40			
500	8.40	2650	16.21	10.01			
600	8.40	3490	17.74	11.93			
700	8.40	4330	19.04	12.86			
800	8.40	5170	20.16	13.70			
900	8.40	6010	21.15	14.48			
1000	4.49	20160	35.43	15.27			
1100	4.50	20610	35.86	17.13			
1200	4.52	21060	36.25	18.70			
1300	4.53	21510	36.62	20.08			
1400	4.55	21970	36.95	21.26			
1500	4.58	22420	37.27	22.33			
1600	4.60	22880	37.56	23.26			
1700	4.62	23340	37.84	24.12			
1800	4.65	23810	38.11	24.89			
1900	4.67	24270	38.36	25.59			
2000	4.69	24740	38.60	26.23			
2100	4.71	25210	38.83	26.83			
2200	4.73	25680	39.05	27.38			
2300	4.75	26150	39.26	27.90			
2400	4.77	26630	39.46	28.37			
2500	4.78	27110	39.66	28.82			
2600	4.80	27590	39.84	29.23			
2700	4.82	28070	40.02	29.63			
2800	4.83	28550	40.20	30.01			
2900	4.84	29030	40.37	30.36			
3000	4.86	29520	40.53	30.69			

SELENIUM

SELENIUM Se_2
IDEAL DIATOMIC GAS

Reference State for Calculating ΔH_f°, ΔF_f°, and $\log_{10} K_p$: Solid from 298° to 490°, Liquid from 490° to 958°, Ideal Diatomic Gas from 958° to 3000°.

Gfw 157.92 GRAMS
$(H^\circ_{298.15} - H^\circ_0) = 2{,}570$ CAL./GFW.

		T TEMPERATURE °K	C_p° HEAT CAPACITY CAL./DEG./GFW.	$H^\circ_T - H^\circ_{298.15}$ HEAT CONTENT CAL./GFW.	S°_T ENTROPY CAL./DEG./GFW.	$-(F^\circ_T - H^\circ_{298.15})/T$ FREE ENERGY FUNCTION CAL./DEG./GFW.	FORMATION FROM REFERENCE STATE		
							HEAT ΔH_f° CAL./GFW.	FREE ENERGY ΔF_f° CAL./GFW.	$\log_{10} K_p$
		298	8.47	0	60.23	60.23	34120	22214	−16.283
		300	8.47	16	60.28	60.23	34114	22144	−16.133
M.P.	°K	400	8.67	874	62.75	60.57	33694	18210	−9.950
ΔH_m	CAL./GFW.	500	8.78	1748	64.70	61.21	30568	14428	−6.306
		600	8.85	2630	66.31	61.93	29770	11272	−4.106
		700	8.89	3516	67.68	62.66	28976	8256	−2.577
B.P.	°K	800	8.92	4407	68.87	63.37	28187	5347	−1.460
ΔH_v	CAL./GFW.	900	8.95	5301	69.92	64.03	27401	2543	−.617
		1000	8.98	6197	70.86	64.67	0	0	0
		1100	9.00	7096	71.72	65.27	0	0	0
		1200	9.03	7997	72.50	65.84	0	0	0
S.P.	°K	1300	9.06	8902	73.23	66.39	0	0	0
ΔH_s	CAL./GFW.	1400	9.10	9810	73.90	66.90	0	0	0
		1500	9.15	10720	74.53	67.39	0	0	0
		1600	9.20	11640	75.12	67.85	0	0	0
		1700	9.24	12560	75.68	68.30	0	0	0
T.P.	°K	1800	9.29	13490	76.21	68.72	0	0	0
ΔH_t	CAL./GFW.	1900	9.33	14420	76.71	69.13	0	0	0
		2000	9.37	15350	77.19	69.52	0	0	0
		2100	9.41	16290	77.65	69.90	0	0	0
		2200	9.45	17240	78.09	70.26	0	0	0
T.P.	°K	2300	9.49	18180	78.51	70.61	0	0	0
ΔH_t	CAL./GFW.	2400	9.53	19130	78.92	70.95	0	0	0
		2500	9.56	20090	79.31	71.28	0	0	0
		2600	9.60	21050	79.68	71.59	0	0	0
		2700	9.63	22010	80.04	71.89	0	0	0
$T_c =$	°K	2800	9.66	22970	80.39	72.19	0	0	0
$P_c =$	ATM.	2900	9.68	23940	80.73	72.48	0	0	0
		3000	9.71	24910	81.06	72.76	0	0	0

SELENIUM

SELENIUM Se$_6$
IDEAL HEXATOMIC GAS

Gfw 473.76 GRAMS
($H°_{298.15} - H°_0$) = CAL./GFW.

Reference State for Calculating $\Delta H°_f$, $\Delta F°_f$, and Log$_{10}$Kp: Solid from 298° to 490°, Liquid from 490° to 958°, Ideal Diatomic Gas from 958° to 1500°.

T TEMPERATURE °K	C°$_p$ HEAT CAPACITY CAL./DEG./GFW.	H°$_T$ - H°$_{298.15}$ HEAT CONTENT CAL./GFW.	S°$_T$ ENTROPY CAL./DEG./GFW.	$-(F°_T-H°_{298.15})/T$ FREE ENERGY FUNCTION CAL./DEG./GFW.	FORMATION FROM REFERENCE STATE		
					HEAT $\Delta H°_f$ CAL./GFW.	FREE ENERGY $\Delta F°_f$ CAL./GFW.	LOG$_{10}$ K$_p$
298	29.00	0	110.00	110.00	35380	20739	-15.202
300	29.02	54	110.19	110.01	35368	20653	-15.046
400	29.90	3000	118.67	111.17	34480	15860	-8.666
500	30.50	6030	125.40	113.34	25510	11440	-5.000
600	30.90	9100	131.00	115.84	23540	8804	-3.207
700	31.20	12200	135.80	118.38	21600	6508	-2.032
800	31.40	15330	140.00	120.84	19690	4458	-1.217
900	31.55	18480	143.70	123.17	17800	2680	-.650
1000	31.65	21640	147.00	125.36	-63940	1640	-.358
1100	31.75	24810	150.00	127.45	-63470	8206	-1.630
1200	31.82	27980	152.80	129.49	-63000	14640	-2.666
1300	31.88	31170	155.40	131.43	-62510	21106	-3.548
1400	31.95	34360	157.70	133.16	-62080	27520	-4.296
1500	32.00	37560	159.90	134.86	-61580	34000	-4.954

M.P. °K
ΔH_m CAL./GFW.

B.P. °K
ΔH_v CAL./GFW.

S.P. °K
ΔH_s CAL./GFW.

T.P. °K
ΔH_t CAL./GFW.

T.P. °K
ΔH_t CAL./GFW.

T_c = °K
P_c = ATM.

SELENIUM

SELENIUM Se

IDEAL MONATOMIC GAS

Gfw 78.96 GRAMS
$(H^°_{298.15} - H^°_0) = 1,481$ CAL./GFW.

Reference State for Calculating $\Delta H^°_f$, $\Delta F^°_f$, and $\text{Log}_{10} K_p$: Solid from 298° to 490°, Liquid from 490° to 958°, Ideal Diatomic Gas from 958° to 3000°.

T TEMPERATURE °K	$C^°_P$ HEAT CAPACITY CAL./DEG./GFW.	$H^°_T - H^°_{298.15}$ HEAT CONTENT CAL./GFW.	$S^°_T$ ENTROPY CAL./DEG./GFW.	$-(F^°_T - H^°_{298.15})/T$ FREE ENERGY FUNCTION CAL./DEG./GFW.	FORMATION FROM REFERENCE STATE ||| FREE ENERGY $\Delta F^°_f$ CAL./GFW.	$\text{LOG}_{10} K_p$
					HEAT $\Delta H^°_f$ CAL./GFW.	FREE ENERGY $\Delta F^°_f$ CAL./GFW.	
298	4.98	0	42.21	42.21	49400	39841	-29.205
300	4.98	9	42.24	42.21	49398	39783	-28.984
400	5.02	509	43.68	42.41	49259	36595	-19.996
500	5.11	1015	44.81	42.78	47765	33465	-14.628
600	5.23	1532	45.75	43.20	47442	30636	-11.160
700	5.35	2061	46.57	43.63	47131	27860	-8.699
800	5.46	2602	47.29	44.04	46832	25128	-6.865
900	5.55	3152	47.94	44.44	46542	22431	-5.447
1000	5.61	3711	48.53	44.82	32951	19851	-4.338
1100	5.65	4274	49.06	45.18	33064	18544	-3.684
1200	5.68	4841	49.56	45.53	33181	17209	-3.134
1300	5.69	5409	50.01	45.85	33299	15892	-2.671
1400	5.69	5978	50.43	46.16	33408	14536	-2.269
1500	5.68	6546	50.82	46.46	33526	13201	-1.923
1600	5.67	7113	51.19	46.75	33633	11825	-1.615
1700	5.66	7679	51.53	47.02	33739	10466	-1.345
1800	5.64	8244	51.86	47.28	33834	9084	-1.102
1900	5.63	8808	52.16	47.53	33938	7718	-.887
2000	5.62	9371	52.45	47.77	34031	6331	-.691
2100	5.61	9932	52.72	48.00	34122	4953	-.515
2200	5.61	10493	52.98	48.22	34213	3567	-.354
2300	5.60	11054	53.23	48.43	34304	2173	-.206
2400	5.60	11614	53.47	48.64	34384	760	-.069
2500	5.60	12174	53.70	48.84	34464	-636	.055
2600	5.60	12734	53.92	49.03	34544	-2064	.173
2700	5.60	13295	54.13	49.21	34625	-3472	.281
2800	5.61	13855	54.34	49.40	34705	-4887	.381
2900	5.61	14416	54.53	49.56	34786	-6278	.473
3000	5.62	14977	54.72	49.73	34857	-7713	.561

M.P. °K
ΔH_m CAL./GFW.

B.P. °K
ΔH_v CAL./GFW.

S.P. °K
ΔH_s CAL./GFW.

T.P. °K
ΔH_t CAL./GFW.

T.P. °K
ΔH_t CAL./GFW.

$T_c =$ °K
$P_c =$ ATM.

THERMODYNAMIC PROPERTIES OF THE ELEMENTS

SILICON

SILICON Si

REFERENCE STATE

Solid from 298° to 1683°, Liquid from 1683° to 3000°.

Gfw	28.09	GRAMS
$(H°_{298.15} - H°_0) =$	769	CAL./GFW.
M.P.	1,683	°K
ΔH_m	11,100	CAL./GFW.
B.P.	(2,950)	°K
ΔH_v		CAL./GFW.
S.P.		°K
ΔH_s		CAL./GFW.
T.P.		°K
ΔH_t		CAL./GFW.
T.P.		°K
ΔH_t		CAL./GFW.
$T_c =$		°K
$P_c =$		ATM.

T TEMPERATURE °K	$C°_P$ HEAT CAPACITY CAL./DEG./GFW.	$H°_T - H°_{298.15}$ HEAT CONTENT CAL./GFW.	$S°_T$ ENTROPY CAL./DEG./GFW.	$-(F°_T - H°_{298.15})/T$ FREE ENERGY FUNCTION CAL./DEG./GFW.	HEAT $\Delta H°_f$ CAL./GFW.	FREE ENERGY $\Delta F°_f$ CAL./GFW.	LOG$_{10}$ K_P
298	4.80	0	4.53	4.53			
300	4.81	8	4.56	4.54			
400	5.34	518	6.02	4.73			
500	5.63	1070	7.25	5.11			
600	5.83	1640	8.29	5.56			
700	5.98	2230	9.20	6.02			
800	6.11	2840	10.01	6.46			
900	6.23	3455	10.73	6.90			
1000	6.34	4080	11.40	7.32			
1100	6.44	4720	12.01	7.72			
1200	6.54	5370	12.57	8.10			
1300	6.64	6030	13.10	8.47			
1400	6.74	6700	13.59	8.81			
1500	6.83	7380	14.06	9.14			
1600	6.92	8070	14.51	9.47			
1700	7.00	19860	21.53	9.85			
1800	7.00	20560	21.93	10.51			
1900	7.00	21260	22.31	11.13			
2000	7.00	21960	22.66	11.68			
2100	7.00	22660	23.01	12.22			
2200	7.00	23360	23.33	12.72			
2300	7.00	24060	23.64	13.18			
2400	7.00	24760	23.94	13.63			
2500	7.00	25460	24.23	14.05			
2600	7.00	26160	24.50	14.44			
2700	7.00	26860	24.77	14.83			
2800	7.00	27560	25.02	15.18			
2900	7.00	28260	25.27	15.53			
3000	7.00	28960	25.51	15.86			

SILICON

SILICON   S1

IDEAL MONATOMIC GAS

Reference State for Calculating ΔH_f°, ΔF_f°, and $\log_{10} K_p$:
Solid from 298° to 1683°, Liquid from 1683° to 3000°.

					FORMATION FROM REFERENCE STATE		
T TEMPERATURE °K	C_p° HEAT CAPACITY CAL./DEG./GFW.	$H_T^\circ - H_{298.15}^\circ$ HEAT CONTENT CAL./GFW.	S_T° ENTROPY CAL./DEG./GFW.	$-(F^\circ - H_{298.15}^\circ)/T$ FREE ENERGY FUNCTION CAL./DEG./GFW.	HEAT ΔH_f° CAL./GFW.	FREE ENERGY ΔF_f° CAL./GFW.	$\log_{10} K_p$
298	5.32	0	40.12	40.12	105000	94388	-69.191
300	5.31	10	40.16	40.13	105002	94322	-68.719
400	5.17	533	41.66	40.33	105015	90759	-49.592
500	5.09	1046	42.80	40.71	104976	87201	-38.118
600	5.06	1553	43.73	41.15	104913	83649	-30.471
700	5.03	2057	44.51	41.58	104827	80110	-25.013
800	5.02	2560	45.18	41.98	104720	76584	-20.923
900	5.01	3061	45.77	42.37	104606	73070	-17.745
1000	5.01	3562	46.30	42.74	104482	69582	-15.208
1100	5.02	4064	46.78	43.09	104344	66097	-13.133
1200	5.03	4566	47.21	43.41	104196	62628	-11.407
1300	5.04	5069	47.62	43.73	104039	59163	-9.947
1400	5.06	5575	47.99	44.01	103875	55715	-8.698
1500	5.09	6082	48.34	44.29	103702	52282	-7.618
1600	5.11	6592	48.67	44.55	103522	48866	-6.674
1700	5.14	7105	48.98	44.81	92245	45580	-5.859
1800	5.17	7621	49.27	45.04	92061	42849	-5.202
1900	5.20	8139	49.56	45.28	91879	40104	-4.612
2000	5.23	8661	49.82	45.49	91701	37381	-4.084
2100	5.26	9186	50.08	45.71	91526	34679	-3.609
2200	5.29	9713	50.32	45.91	91353	31975	-3.176
2300	5.32	10243	50.56	46.11	91183	29267	-2.780
2400	5.34	10776	50.79	46.30	91016	26576	-2.420
2500	5.36	11311	51.01	46.49	90851	23901	-2.089
2600	5.39	11849	51.22	46.67	90689	21217	-1.783
2700	5.41	12389	51.42	46.84	90529	18574	-1.503
2800	5.42	12930	51.62	47.01	90370	15890	-1.240
2900	5.44	13473	51.81	47.17	90213	13247	-.998
3000	5.45	14018	51.99	47.32	90058	10618	-.773

G.W. 28.09 GRAMS
$(H_{298.15}^\circ - H_0^\circ) = 1,805$ CAL./GFW.

M.P. °K
ΔH_m CAL./GFW.

B.P. °K
ΔH_v CAL./GFW.

S.P. °K
ΔH_s CAL./GFW.

T.P. °K
ΔH_t CAL./GFW.

T.P. °K
ΔH_t CAL./GFW.

$T_c =$ °K
$P_c =$ ATM.

THERMODYNAMIC PROPERTIES OF THE ELEMENTS 185

SILVER

SILVER	Ag		Solid from 298° to 1234°, Liquid from 1234° to 2450°, Ideal Monatomic Gas from 2450° to 3000°.
Gfw	107.880	GRAMS	
REFERENCE STATE			
$(H°_{298.15} - H°_0)$ =	1,373	CAL./GFW.	
M.P.	1,234.0	°K	
ΔH_m	2,700.	CAL./GFW.	
B.P.	2,450.	°K	
ΔH_v	60,960.	CAL./GFW.	
S.P.		°K	
ΔH_s		CAL./GFW.	
T.P.		°K	
ΔH_t		CAL./GFW.	
T.P.		°K	
ΔH_t		CAL./GFW.	
T_c =		°K	
P_c =		ATM.	

T TEMPERATURE °K	$C°_P$ HEAT CAPACITY CAL./DEG./GFW.	$H°_T - H°_{298.15}$ HEAT CONTENT CAL./GFW.	$S°_T$ ENTROPY CAL./DEG./GFW.	$-(F°_T-H°_{298.15})/T$ FREE ENERGY FUNCTION CAL./DEG./GFW.	HEAT $\Delta H°_f$ CAL./GFW.	FREE ENERGY $\Delta F°_f$ CAL./GFW.	LOG$_{10}$ K_P
298	6.09	0	10.20	10.20			
300	6.10	11	10.24	10.21			
400	6.19	625	12.01	10.45			
500	6.32	1250	13.40	10.90			
600	6.48	1890	14.57	11.42			
700	6.64	2550	15.58	11.94			
800	6.80	3220	16.48	12.46			
900	6.76	3910	17.29	12.95			
1000	7.12	4610	18.03	13.42			
1100	7.28	5330	18.72	13.88			
1200	7.44	6060	19.35	14.30			
1300	7.50	9510	22.14	14.83			
1400	7.50	10260	22.70	15.38			
1500	7.50	11010	23.21	15.87			
1600	7.50	11760	23.70	16.35			
1700	7.50	12510	24.15	16.80			
1800	7.50	13260	24.58	17.22			
1900	7.50	14010	24.99	17.62			
2000	7.50	14760	25.37	17.99			
2100	7.50	15510	25.74	18.36			
2200	7.50	16260	26.09	18.70			
2300	7.50	17010	26.42	19.03			
2400	7.50	17760	26.74	19.34			
2500	4.97	79340	51.89	20.15			
2600	4.97	79840	52.08	21.38			
2700	4.97	80330	52.27	22.52			
2800	4.97	80830	52.45	23.59			
2900	4.97	81330	52.62	24.58			
3000	4.97	81820	52.79	25.52			

SILVER

SILVER   Ag   Reference State for Calculating ΔH_f°, ΔF_f°, and $\text{Log}_{10} K_p$: Solid from 298° to 1234°, Liquid from 1234° to 2450°, Ideal Monatomic Gas from 2450° to 3000°.

Gfw 107.880 GRAMS
$(H^\circ_{298.15} - H^\circ_0) = 1,481$ CAL./GFW.

IDEAL MONATOMIC GAS

T TEMPERATURE °K	C_p° HEAT CAPACITY CAL./DEG./GFW.	$H_T^\circ - H_{298.15}^\circ$ HEAT CONTENT CAL./GFW.	S_T° ENTROPY CAL./DEG./GFW.	$-(F_T^\circ - H_{298.15}^\circ)$ FREE ENERGY FUNCTION CAL./DEG./GFW.	HEAT ΔH_f° CAL./GFW.	FREE ENERGY ΔF_f° CAL./GFW.	$\text{LOG}_{10} K_p$
298	4.97	0	41.32	41.32	68400	59122	-43.339
300	4.97	9	41.35	41.32	68398	59065	-43.032
400	4.97	506	42.78	41.52	68281	55973	-30.584
500	4.97	1003	43.89	41.89	68153	52908	-23.127
600	4.97	1500	44.80	42.30	68010	49872	-18.167
700	4.97	1996	45.56	42.71	67846	46860	-14.631
800	4.97	2493	46.23	43.12	67673	43873	-11.986
900	4.97	2990	46.81	43.49	67480	40912	-9.935
1000	4.97	3487	47.33	43.85	67277	37977	-8.300
1100	4.97	3984	47.81	44.19	67054	35055	-6.965
1200	4.97	4480	48.24	44.51	66820	32152	-5.856
1300	4.97	4977	48.64	44.82	63867	29417	-4.945
1400	4.97	5474	49.01	45.10	63614	26780	-4.180
1500	4.97	5971	49.35	45.37	63361	24151	-3.519
1600	4.97	6468	49.67	45.63	63108	21556	-2.944
1700	4.97	6964	49.97	45.88	62854	18960	-2.437
1800	4.97	7461	50.25	46.11	62601	16395	-1.990
1900	4.97	7958	50.52	46.34	62348	13841	-1.591
2000	4.97	8455	50.78	46.56	62095	11275	-1.232
2100	4.97	8952	51.02	46.76	61842	8754	-.911
2200	4.97	9448	51.25	46.96	61588	6236	-.619
2300	4.97	9945	51.47	47.15	61335	3720	-.353
2400	4.97	10442	51.68	47.33	61082	1226	-.111
2500	4.97	10939	51.89	47.52	0	0	0
2600	4.97	11436	52.08	47.69	0	0	0
2700	4.97	11932	52.27	47.86	0	0	0
2800	4.97	12429	52.45	48.02	0	0	0
2900	4.97	12926	52.62	48.17	0	0	0
3000	4.97	13423	52.79	48.32	0	0	0

M.P. °K
ΔH_m CAL./GFW.

B.P. °K
ΔH_v CAL./GFW.

S.P. °K
ΔH_s CAL./GFW.

T.P. °K
ΔH_t CAL./GFW.

T.P. °K
ΔH_t CAL./GFW.

$T_c =$ °K
$P_c =$ ATM.

THERMODYNAMIC PROPERTIES OF THE ELEMENTS

SODIUM

SODIUM	Na		Solid from 298° to 371°, Liquid from 371° to 1163°, Ideal Monatomic Gas from 1163° to 3000°.						
								FORMATION FROM REFERENCE STATE	
Gfw	22.991	GRAMS							
REFERENCE STATE									
$(H°_{298.15} - H°_0) =$	1,532	CAL./GFW.							
			T TEMPERATURE °K	$C°_P$ HEAT CAPACITY CAL./DEG./GFW.	$H°_T - H°_{298.15}$ HEAT CONTENT CAL./GFW.	$S°_T$ ENTROPY CAL./DEG./GFW.	$-(F°_T - H°_{298.15})$ FREE ENERGY FUNCTION CAL./DEG./GFW.	HEAT $\Delta H°_f$ CAL./GFW.	FREE ENERGY $\Delta F°_f$ CAL./GFW.	$LOG_{10} K_P$
M.P.	370.97	°K	298	6.74	0	12.21	12.21			
ΔH_m	621.8	CAL./GFW.	300	6.75	12	12.25	12.21			
			400	7.52	1355	16.00	12.62			
			500	7.32	2096	17.65	13.46			
			600	7.10	2819	18.97	14.28			
			700	6.96	3521	20.05	15.02			
B.P.	1,163.	°K	800	6.90	4218	20.98	15.71			
ΔH_v	21,280.	CAL./GFW.	900	6.89	4909	21.80	16.35			
			1000	6.93	5597	22.52	16.93			
			1100	7.01	6293	23.19	17.47			
			1200	4.97	30380	43.63	18.32			
S.P.		°K	1300	4.97	30877	44.03	20.28			
ΔH_s		CAL./GFW.	1400	4.97	31374	44.40	21.99			
			1500	4.97	31871	44.74	23.50			
			1600	4.97	32368	45.06	24.83			
			1700	4.97	32864	45.36	26.03			
T.P.		°K	1800	4.97	33361	45.65	27.12			
ΔH_t		CAL./GFW.	1900	4.97	33858	45.92	28.10			
			2000	4.97	34355	46.17	29.00			
			2100	4.98	34853	46.41	29.82			
			2200	4.98	35351	46.64	30.58			
T.P.		°K	2300	4.98	35849	46.87	31.29			
ΔH_t		CAL./GFW.	2400	4.99	36347	47.08	31.94			
			2500	5.00	36847	47.28	32.55			
			2600	5.01	37348	47.48	33.12			
			2700	5.04	37850	47.67	33.66			
$T_c =$		°K	2800	5.04	38353	47.85	34.16			
$P_c =$		ATM.	2900	5.06	38858	48.03	34.64			
			3000	5.08	39365	48.20	35.08			

SODIUM

SODIUM   **Na**   Reference State for Calculating ΔH_f°, ΔF_f°, and $\text{Log}_{10} K_p$: Solid from 298° to 371°, Liquid from 371° to 1163°, Ideal Monatomic Gas from 1163° to 3000°.

IDEAL MONATOMIC GAS

Gfw 22.991 GRAMS
$(H^\circ_{298.15} - H^\circ_0) = 1{,}481$ CAL./GFW.

						FORMATION FROM REFERENCE STATE		
T TEMPERATURE °K	C_p° HEAT CAPACITY CAL./DEG./GFW.	$H^\circ_T - H^\circ_{298.15}$ HEAT CONTENT CAL./GFW.	S°_T ENTROPY CAL./DEG./GFW.	$-(F^\circ_T - H^\circ_{298.15})/T$ FREE ENERGY FUNCTION CAL./DEG./GFW.		HEAT ΔH_f° CAL./GFW.	FREE ENERGY ΔF_f° CAL./GFW.	$\text{LOG}_{10} K_p$
298	4.97	0	36.71	36.71		25900	18595	-13.631
300	4.97	9	36.74	36.71		25897	18550	-13.514
400	4.97	506	38.17	36.91		25051	16183	-8.842
500	4.97	1003	39.28	37.28		24807	13992	-6.116
600	4.97	1500	40.19	37.69		24581	11849	-4.316
700	4.97	1996	40.95	38.10		24375	9745	-3.042
800	4.97	2493	41.62	38.51		24175	7663	-2.093
900	4.97	2990	42.20	38.88		23981	5621	-1.365
1000	4.97	3487	42.73	39.25		23790	3580	-.782
1100	4.97	3984	43.20	39.58		23591	1580	-.313
1200	4.97	4480	43.63	39.90		0	0	0
1300	4.97	4977	44.03	40.21		0	0	0
1400	4.97	5474	44.40	40.49		0	0	0
1500	4.97	5971	44.74	40.76		0	0	0
1600	4.97	6468	45.06	41.02		0	0	0
1700	4.97	6964	45.36	41.27		0	0	0
1800	4.97	7461	45.65	41.51		0	0	0
1900	4.97	7958	45.92	41.74		0	0	0
2000	4.97	8455	46.17	41.95		0	0	0
2100	4.98	8953	46.41	42.15		0	0	0
2200	4.98	9451	46.64	42.35		0	0	0
2300	4.98	9949	46.87	42.55		0	0	0
2400	4.99	10447	47.08	42.73		0	0	0
2500	5.00	10947	47.28	42.91		0	0	0
2600	5.01	11448	47.48	43.08		0	0	0
2700	5.03	11950	47.67	43.25		0	0	0
2800	5.04	12453	47.85	43.41		0	0	0
2900	5.06	12958	48.03	43.57		0	0	0
3000	5.08	13465	48.20	43.72		0	0	0

M.P. °K
ΔH_m CAL./GFW.

B.P. °K
ΔH_v CAL./GFW.

S.P. °K
ΔH_s CAL./GFW.

T.P. °K
ΔH_t CAL./GFW.

T.P. °K
ΔH_t CAL./GFW.

$T_c =$ °K
$P_c =$ ATM.

THERMODYNAMIC PROPERTIES OF THE ELEMENTS

SODIUM

SODIUM Na₂
IDEAL DIATOMIC GAS

Gfw 45.982 GRAMS
$(H°_{298.15} - H°_0) = 2,484$ CAL./GFW.

Reference State for Calculating $\Delta H°_f$, $\Delta F°_f$, and $\log_{10} K_p$: Solid from 298° to 371°, Liquid from 371° to 1163°, Ideal Monatomic Gas from 1163° to 1500°.

T TEMPERATURE °K	$C°_P$ HEAT CAPACITY CAL./DEG./GFW.	$H°_T - H°_{298.15}$ HEAT CONTENT CAL./GFW.	$S°_T$ ENTROPY CAL./DEG./GFW.	$-(F°_T - H°_{298.15})$ FREE ENERGY FUNCTION CAL./DEG./GFW.	FORMATION FROM REFERENCE STATE		
					HEAT $\Delta H°_f$ CAL./GFW.	FREE ENERGY $\Delta F°_f$ CAL./GFW.	$\log_{10} K_P$
298	8.96	0	54.99	54.99	33800	24685	-18.095
300	8.96	17	55.05	55.00	33793	24628	-17.942
400	9.04	918	57.64	55.35	32008	21752	-11.885
500	9.10	1825	59.66	56.01	31433	19253	-8.416
600	9.15	2736	61.32	56.76	30898	16870	-6.145
700	9.19	3655	62.74	57.52	30413	14565	-4.547
800	9.24	4580	63.98	58.26	29944	12328	-3.368
900	9.28	5499	65.06	58.95	29481	10167	-2.469
1000	9.32	6436	66.05	59.62	29042	8032	-1.755
1100	9.36	7361	66.93	60.24	28575	5970	-1.186
1200	9.40	8304	67.75	60.83	-18656	4756	-.866
1300	9.44	9242	68.50	61.40	-18712	6716	-1.129
1400	9.47	10186	69.20	61.93	-18762	8678	-1.354
1500	9.51	11136	69.85	62.43	-18806	10639	-1.550

M.P. °K
ΔH_m CAL./GFW.

B.P. °K
ΔH_v CAL./GFW.

S.P. °K
ΔH_s CAL./GFW.

T.P. °K
ΔH_t CAL./GFW.

T.P. °K
ΔH_t CAL./GFW.

$T_c =$ °K
$P_c =$ ATM.

STRONTIUM

STRONTIUM Sr Solid I from 298° to 862°, Solid II from 862° to 1043°, Liquid
REFERENCE STATE from 1043° to 1640°, Ideal Monatomic Gas from 1640° to 3000°.

Gfw 87.63 GRAMS	
$(H°_{298.15} - H°_0) =$ CAL./GFW.	
M.P. 1,043 °K	
ΔH_m (2,200) CAL./GFW.	
B.P. 1,640 °K	
ΔH_v 33,200 CAL./GFW.	
S.P. °K	
ΔH_s CAL./GFW.	
T.P. 862 °K	
ΔH_t (200) CAL./GFW.	
T.P. °K	
ΔH_t CAL./GFW.	
$T_c =$ °K	
$P_c =$ ATM.	

T TEMPERATURE °K	$C°_p$ HEAT CAPACITY CAL./DEG./GFW.	$H°_T - H°_{298.15}$ HEAT CONTENT CAL./GFW.	$S°_T$ ENTROPY CAL./DEG./GFW.	$-(F°_T - H°_{298.15})/T$ FREE ENERGY FUNCTION CAL./DEG./GFW.	FORMATION FROM REFERENCE STATE		
					HEAT $\Delta H°_f$ CAL./GFW.	FREE ENERGY $\Delta F°_f$ CAL./GFW.	$\text{LOG}_{10} K_p$
298	6.30	0	12.50	12.50			
300	6.31	11	12.54	12.51			
400	6.65	660	14.40	12.75			
500	7.00	1340	15.92	13.24			
600	7.30	2050	17.22	13.81			
700	7.65	2800	18.37	14.37			
800	8.00	3580	19.41	14.94			
900	8.80	4610	20.62	15.50			
1000	9.50	5520	21.58	16.06			
1100	7.40	8550	24.56	16.79			
1200	7.40	9290	25.21	17.47			
1300	7.40	10040	25.80	18.08			
1400	7.40	10780	26.35	18.65			
1500	7.40	11520	26.86	19.18			
1600	7.40	12260	27.34	19.68			
1700	4.98	46070	47.97	20.87			
1800	4.99	46570	48.26	22.39			
1900	5.01	47070	48.53	23.76			
2000	5.03	47570	48.79	25.01			
2100	5.06	48071	49.03	26.14			
2200	5.11	48580	49.27	27.19			
2300	5.17	49090	49.50	28.16			
2400	5.25	49620	49.72	29.05			
2500	5.34	50140	49.94	29.89			
2600	5.46	50680	50.15	30.66			
2700	5.60	51300	50.36	31.36			
2800	5.76	51800	50.56	32.06			
2900	5.95	52390	50.77	32.71			
3000	6.16	53000	50.97	33.31			

STRONTIUM

STRONTIUM Sr
IDEAL MONATOMIC GAS

Gfw 87.63 GRAMS
$(H°_{298.15} - H°_0) = 1,481$ CAL./GFW.

Reference State for Calculating $\Delta H°_f$, $\Delta F°_f$, and $\text{Log}_{10} K_p$:
Solid I from 298° to 862°, Solid II from 862° to 1043°, Liquid from 1043° to 1640°, Ideal Monatomic Gas from 1640° to 3000°.

T TEMPERATURE °K	$C°_p$ HEAT CAPACITY CAL./DEG./GFW.	$H°_T - H°_{298.15}$ HEAT CONTENT CAL./GFW.	$S°_T$ ENTROPY CAL./DEG./GFW.	$-(F°_T-H°_{298.15})$ FREE ENERGY FUNCTION CAL./DEG./GFW.	FORMATION FROM REFERENCE STATE HEAT $\Delta H°_f$ CAL./GFW.	FORMATION FROM REFERENCE STATE FREE ENERGY $\Delta F°_f$ CAL./GFW.	LOG$_{10}$ K_p
298	4.97	0	39.32	39.32	39100	31104	-22.800
300	4.97	9	39.36	39.35	39098	31052	-22.623
400	4.97	506	40.78	39.52	38946	28394	-15.515
500	4.97	1003	41.89	39.89	38763	25778	-11.268
600	4.97	1500	42.80	40.30	38550	23202	-8.452
700	4.97	1996	43.56	40.71	38296	20663	-6.451
800	4.97	2493	44.23	41.12	38013	18157	-4.960
900	4.97	2990	44.81	41.49	37480	15709	-3.814
1000	4.97	3487	45.34	41.86	37067	13307	-2.908
1100	4.97	3984	45.81	42.19	34534	11159	-2.217
1200	4.97	4480	46.24	42.51	34290	9054	-1.649
1300	4.97	4977	46.64	42.82	34037	6945	-1.167
1400	4.97	5474	47.01	43.10	33794	4816	-.760
1500	4.97	5971	47.35	43.37	33551	2816	-.410
1600	4.97	6468	47.67	43.63	33308	780	-.106
1700	4.98	6966	47.97	43.88	0	0	0
1800	4.99	7465	48.26	44.12	0	0	0
1900	5.01	7965	48.53	44.34	0	0	0
2000	5.03	8466	48.79	44.56	0	0	0
2100	5.06	8971	49.03	44.76	0	0	0
2200	5.11	9480	49.27	44.97	0	0	0
2300	5.17	9994	49.50	45.16	0	0	0
2400	5.25	10515	49.72	45.34	0	0	0
2500	5.34	11044	49.94	45.53	0	0	0
2600	5.46	11584	50.15	45.70	0	0	0
2700	5.60	12200	50.36	45.85	0	0	0
2800	5.76	12705	50.56	46.03	0	0	0
2900	5.95	13290	50.77	46.19	0	0	0
3000	6.16	13895	50.97	46.34	0	0	0

M.P. °K
ΔH_m CAL./GFW.

B.P. °K
ΔH_v CAL./GFW.

S.P. °K
ΔH_s CAL./GFW.

T.P. °K
ΔH_t CAL./GFW.

T.P. °K
ΔH_t CAL./GFW.

$T_c =$ °K
$P_c =$ ATM.

SULFUR

SULFUR   S   Rhombic Solid from 298° to 368.6°, Monoclinic Solid from 368.6° to 392°, Liquid from 392° to 717.75°, Ideal Diatomic Gas from 717.75° to 3000°.

Gfw	32.066 GRAMS
$(H°_{298.15} - H°_0) =$	1,053 CAL./GFW.
M.P.	392.° °K
ΔH_m	337.° CAL./GFW.
B.P.	717.75 °K
ΔH_v	2,300.° CAL./GFW.
S.P.	°K
ΔH_s	CAL./GFW.
T.P.	368.6 °K
ΔH_t	90.° CAL./GFW.
T.P.	°K
ΔH_f	CAL./GFW.
$T_c =$	°K
$P_c =$	ATM.

T TEMPERATURE °K	$C°_p$ HEAT CAPACITY CAL./DEG./GFW.	$H°_T - H°_{298.15}$ HEAT CONTENT CAL./GFW.	$S°_T$ ENTROPY CAL./DEG./GFW.	$-(F°_T - H°_{298.15})/T$ FREE ENERGY FUNCTION CAL./DEG./GFW.	FORMATION FROM REFERENCE STATE		
					HEAT $\Delta H°_f$ CAL./GFW.	FREE ENERGY $\Delta F°_f$ CAL./GFW.	$LOG_{10} K_p$
298	5.40	0	7.62	7.62			
300	5.41	9	7.65	7.62			
400	7.54	1033	10.45	7.87			
500	8.84	1948	12.48	8.59			
600	8.34	2798	14.03	9.37			
700	9.00	3655	15.35	10.13			
800	4.37	17530	31.37	9.46			
900	4.40	17970	31.88	11.92			
1000	4.42	18400	32.34	13.94			
1100	4.44	18850	32.76	15.63			
1200	4.45	19290	33.15	17.08			
1300	4.46	19740	33.51	18.33			
1400	4.47	20190	33.84	19.42			
1500	4.48	20630	34.15	20.40			
1600	4.49	21080	34.44	21.27			
1700	4.50	21530	34.71	22.05			
1800	4.51	21980	34.97	22.76			
1900	4.51	22430	35.21	23.41			
2000	4.52	22880	35.44	24.00			
2100	4.52	23330	35.67	24.57			
2200	4.53	23790	35.88	25.07			
2300	4.53	24240	36.08	25.55			
2400	4.53	24690	36.27	25.99			
2500	4.54	25150	36.46	26.40			
2600	4.54	25600	36.63	26.79			
2700	4.55	26050	36.80	27.16			
2800	4.55	26510	36.97	27.51			
2900	4.55	26960	37.13	27.84			
3000	4.56	27420	37.28	28.14			

SULFUR

SULFUR S_2
IDEAL DIATOMIC GAS

Reference State for Calculating ΔH_f°, ΔF_f°, and $\log_{10} K_p$: Rhombic Solid from 298° to 368.6°, Monoclinic Solid from 368.6° to 392°, Liquid from 392° to 717.75°, Ideal Diatomic Gas from 717.75° to 3000°.

Gfw 64.132 GRAMS
$(H^\circ_{298.15} - H^\circ_0) = 2,141$ CAL./GFW.

T TEMPERATURE °K	C_p° HEAT CAPACITY CAL./DEG./GFW.	$H^\circ_T - H^\circ_{298.15}$ HEAT CONTENT CAL./GFW.	S°_T ENTROPY CAL./DEG./GFW.	$-(F^\circ_T - H^\circ_{298.15})$ FREE ENERGY FUNCTION CAL./DEG./GFW.	FORMATION FROM REFERENCE STATE		
					HEAT ΔH_f° CAL./GFW.	FREE ENERGY ΔF_f° CAL./GFW.	$\log_{10} K_p$
298	7.76	0	54.51	54.51	30840	19130	-14.023
300	7.77	13	54.55	54.51	30835	19060	-13.886
400	8.14	811	56.85	54.83	29585	15205	-8.308
500	8.39	1639	58.69	55.42	28583	11718	-5.122
600	8.54	2485	60.23	56.09	27729	8427	-3.069
700	8.65	3347	61.56	56.78	26877	5275	-1.647
800	8.73	4219	62.73	57.46	0	0	0
900	8.79	5095	63.76	58.10	0	0	0
1000	8.84	5969	64.68	58.72	0	0	0
1100	8.87	6857	65.52	59.29	0	0	0
1200	8.90	7747	66.30	59.85	0	0	0
1300	8.92	8636	67.01	60.37	0	0	0
1400	8.94	9535	67.68	60.87	0	0	0
1500	8.96	10429	68.30	61.35	0	0	0
1600	8.98	11326	68.88	61.81	0	0	0
1700	8.99	12224	69.42	62.23	0	0	0
1800	9.00	13124	69.94	62.65	0	0	0
1900	9.01	14024	70.43	63.05	0	0	0
2000	9.02	14926	70.89	63.43	0	0	0
2100	9.03	15828	71.33	63.80	0	0	0
2200	9.04	16732	71.75	64.15	0	0	0
2300	9.05	17637	72.16	64.50	0	0	0
2400	9.06	18543	72.54	64.82	0	0	0
2500	9.07	19450	72.91	65.13	0	0	0
2600	9.08	20358	73.27	65.44	0	0	0
2700	9.09	21266	73.61	65.74	0	0	0
2800	9.10	22175	73.94	66.03	0	0	0
2900	9.11	23086	74.26	66.30	0	0	0
3000	9.12	23997	74.57	66.58	0	0	0

M.P. °K
ΔH_m CAL./GFW.
B.P. °K
ΔH_v CAL./GFW.
S.P. °K
ΔH_s CAL./GFW.
T.P. °K
ΔH_t CAL./GFW.
T.P. °K
ΔH_t CAL./GFW.
T_c = °K
P_c = ATM.

SULFUR

SULFUR S_8

IDEAL OCTATOMIC GAS

Reference State for Calculating $\Delta H_f^°$, $\Delta F_f^°$, and $Log_{10} K_p$: Rhombic Solid from 298° to 368.6°, Monoclinic Solid from 368.6° to 392°, Liquid from 392° to 717.75°, Ideal Diatomic Gas from 717.75° to 1000°.

Gfw 256.528 GRAMS

$(H°_{298.15} - H°_0) = 7,547$ CAL./GFW.

T TEMPERATURE °K	$C_p°$ HEAT CAPACITY CAL./DEG./GFW.	$H°_T - H°_{298.15}$ HEAT CONTENT CAL./GFW.	$S°_T$ ENTROPY CAL./DEG./GFW.	$-(F°_T - H°_{298.15})$ FREE ENERGY FUNCTION CAL./DEG./GFW.	FORMATION FROM REFERENCE STATE		
					HEAT $\Delta H_f°$ CAL./GFW.	FREE ENERGY $\Delta F_f°$ CAL./GFW.	$LOG_{10} K_p$
298	37.17	0	102.76	102.76	24350	11880	- 8.708
300	37.25	68	102.99	102.77	24346	11809	- 8.603
400	39.73	3933	114.08	104.25	20019	7827	- 4.276
500	41.05	7983	123.13	107.17	16749	5104	- 2.231
600	41.82	12123	130.66	110.46	14089	3037	- 1.106
700	42.30	16330	137.15	113.83	11440	1395	- .435
800	42.62	20530	142.82	117.10	- 95310	- 8798	2.403
900	42.84	24850	147.85	120.24	- 94560	1911	- .464
1000	43.01	29140	152.38	123.24	- 93710	12630	- 2.750

M.P. °K
ΔH_m CAL./GFW.

B.P. °K
ΔH_v CAL./GFW.

S.P. °K
ΔH_s CAL./GFW.

T.P. °K
ΔH_t CAL./GFW.

T.P. °K
ΔH_t CAL./GFW.

$T_c =$ °K
$P_c =$ ATM.

SULFUR

SULFUR S

IDEAL MONATOMIC GAS

Gfw 32.066 GRAMS

$(H°_{298.15} - H°_0) = 1,591$ CAL./GFW.

Reference State for Calculating $\Delta H°_f$, $\Delta F°_f$, and $\text{Log}_{10} K_p$: Rhombic Solid from 298° to 368.6°, Monoclinic Solid from 368.6° to 392°, Liquid from 392° to 717.75°, Ideal Diatomic Gas from 717.75° to 3000°.

						FORMATION FROM REFERENCE STATE		
T TEMPERATURE °K	$C°_P$ HEAT CAPACITY CAL./DEG./GFW.	$H°_T - H°_{298.15}$ HEAT CONTENT CAL./GFW.	$S°_T$ ENTROPY CAL./DEG./GFW.	$-(F°_T-H°_{298.15})/T$ FREE ENERGY FUNCTION CAL./DEG./GFW.		HEAT $\Delta H°_f$ CAL./GFW.	FREE ENERGY $\Delta F°_f$ CAL./GFW.	LOG$_{10}$ K$_P$
298	5.66	0	40.09	40.09		56900	47218	-34.613
300	5.66	10	40.12	40.09		56901	47160	-34.358
400	5.55	571	41.74	40.32		56438	43922	-23.999
500	5.44	1121	42.96	40.72		56073	40833	-17.849
600	5.34	1660	43.95	41.19		55762	37810	-13.773
700	5.27	2190	44.76	41.64		55435	34848	-10.880
800	5.21	2713	45.46	42.07		42083	30811	-8.417
900	5.17	3232	46.07	42.48		42162	29391	-7.137
1000	5.14	3747	46.62	42.88		42267	27967	-6.112
1100	5.11	4250	47.10	43.23		42310	26536	-5.272
1200	5.09	4770	47.55	43.58		42380	25100	-4.571
1300	5.08	5278	47.95	43.89		42438	23666	-3.978
1400	5.07	5786	48.33	44.20		42496	22210	-3.467
1500	5.07	6292	48.68	44.49		42562	20767	-3.025
1600	5.06	6799	49.01	44.77		42619	19307	-2.637
1700	5.06	7305	49.31	45.02		42675	17855	-2.295
1800	5.07	7811	49.60	45.27		42731	16397	-1.990
1900	5.08	8318	49.88	45.51		42788	14915	-1.715
2000	5.09	8826	50.14	45.73		42846	13446	-1.469
2100	5.10	9335	50.39	45.95		42956	11993	-1.248
2200	5.11	9846	50.62	46.15		43018	10528	-1.045
2300	5.13	10358	50.85	46.35		43081	9047	-.859
2400	5.14	10871	51.07	46.55		43137	7561	-.688
2500	5.16	11387	51.28	46.73		43204	6087	-.532
2600	5.18	11904	51.48	46.91		43273	4594	-.386
2700	5.20	12423	51.68	47.08		43334	3097	-.250
2800	5.22	12944	51.87	47.25		43407	1614	-.125
2900	5.24	13467	52.05	47.41		43407	139	-.010
3000	5.26	13992	52.23	47.57		43472	-1378	.100

M.P. °K
ΔH_m CAL./GFW.

B.P. °K
ΔH_v CAL./GFW.

S.P. °K
ΔH_s CAL./GFW.

T.P. °K
ΔH_t CAL./GFW.

T.P. °K
ΔH_t CAL./GFW.

$T_c =$ °K
$P_c =$ ATM.

TANTALUM

TANTALUM Ta

Gfw 180.95 GRAMS

$(H°_{298.15} - H°_0) = 1,358$ CAL./GFW.

M.P. 3,270 °K
ΔH_m (7,500) CAL./GFW.

B.P. 5,700 °K
ΔH_v 180,000 CAL./GFW.

S.P. °K
ΔH_s CAL./GFW.

T.P. °K
ΔH_t CAL./GFW.

T.P. °K
ΔH_t CAL./GFW.

$T_c =$ °K
$P_c =$ ATM.

REFERENCE STATE

Solid from 298° to 3000°.

T TEMPERATURE °K	$C°_P$ HEAT CAPACITY CAL./DEG./GFW.	$H°_T - H°_{298.15}$ HEAT CONTENT CAL./GFW.	$S°_T$ ENTROPY CAL./DEG./GFW.	$-(F°_T - H°_{298.15})/T$ FREE ENERGY FUNCTION CAL./DEG./GFW.	FORMATION FROM REFERENCE STATE		
					HEAT $\Delta H°_f$ CAL./GFW.	FREE ENERGY $\Delta F°_f$ CAL./GFW.	LOG$_{10}$ K_P
298	6.08	0	9.90	9.90			
300	6.08	11	9.94	9.91			
400	6.27	629	11.71	10.14			
500	6.38	1260	13.13	10.61			
600	6.46	1900	14.30	11.14			
700	6.52	2550	15.30	11.66			
800	6.57	3210	16.17	12.16			
900	6.63	3870	16.95	12.65			
1000	6.67	4530	17.65	13.12			
1100	6.72	5200	18.29	13.57			
1200	6.76	5880	18.87	13.97			
1300	6.81	6550	19.42	14.39			
1400	6.85	7240	19.92	14.75			
1500	6.90	7930	20.40	15.12			
1600	6.94	8620	20.84	15.46			
1700	6.98	9310	21.26	15.79			
1800	7.02	10010	21.66	16.10			
1900	7.07	10720	22.04	16.40			
2000	7.11	11430	22.41	16.70			
2100	7.15	12140	22.76	16.98			
2200	7.19	12860	23.09	17.25			
2300	7.23	13580	23.41	17.51			
2400	7.28	14300	23.72	17.77			
2500	7.32	15030	24.02	18.01			
2600	7.36	15770	24.31	18.25			
2700	7.40	16510	24.58	18.47			
2800	7.44	17250	24.85	18.69			
2900	7.48	17990	25.11	18.91			
3000	7.53	18740	25.37	19.13			

THERMODYNAMIC PROPERTIES OF THE ELEMENTS

TANTALUM

TANTALUM   Ta

IDEAL MONATOMIC GAS

Gfw 180.95 GRAMS

$(H°_{298.15} - H°_0) = 1,482$ CAL./GFW.

Reference State for Calculating $\Delta H°_f$, $\Delta F°_f$, and $\text{Log}_{10} K_p$: Solid from 298° to 3000°.

T TEMPERATURE °K	$C°_P$ HEAT CAPACITY CAL./DEG./GFW.	$H°_T - H°_{298.15}$ HEAT CONTENT CAL./GFW.	$S°_T$ ENTROPY CAL./DEG./GFW.	$-(F°_T - H°_{298.15})/T$ FREE ENERGY FUNCTION CAL./DEG./GFW.	FORMATION FROM REFERENCE STATE		
					HEAT $\Delta H°_f$ CAL./GFW.	FREE ENERGY $\Delta F°_f$ CAL./GFW.	$\text{LOG}_{10} K_P$
298	4.99	0	44.24	44.24	186800	176561	-129.428
300	4.99	9	44.27	44.24	186798	176499	-128.590
400	5.08	512	45.72	44.44	186683	173079	-94.573
500	5.28	1029	46.87	44.82	186569	169699	-74.180
600	5.54	1570	47.86	45.25	186470	166334	-60.592
700	5.83	2138	48.73	45.68	186388	162987	-50.891
800	6.11	2735	49.53	46.12	186325	159637	-43.614
900	6.38	3359	50.26	46.53	186289	156310	-37.959
1000	6.62	4009	50.95	46.95	186279	152979	-33.436
1100	6.84	4683	51.59	47.34	186283	149653	-29.736
1200	7.04	5377	52.19	47.71	186297	146313	-26.649
1300	7.22	6091	52.77	48.09	186341	142986	-24.040
1400	7.38	6821	53.31	48.44	186381	139635	-21.799
1500	7.51	7565	53.82	48.78	186435	136305	-19.861
1600	7.63	8322	54.31	49.11	186502	132950	-18.159
1700	7.73	9090	54.77	49.43	186580	129613	-16.663
1800	7.82	9867	55.22	49.74	186657	126249	-15.327
1900	7.89	10653	55.64	50.04	186733	122893	-14.135
2000	7.96	11446	56.05	50.33	186816	119536	-13.061
2100	8.02	12245	56.44	50.61	186905	116177	-12.090
2200	8.08	13050	56.81	50.88	186990	112806	-11.206
2300	8.13	13861	57.17	51.15	187081	109433	-10.398
2400	8.18	14676	57.52	51.41	187176	106056	-9.657
2500	8.22	15496	57.86	51.67	187266	102666	-8.975
2600	8.26	16319	58.18	51.91	187349	99287	-8.346
2700	8.29	17147	58.49	52.14	187437	95880	-7.760
2800	8.32	17977	58.79	52.37	187527	92495	-7.219
2900	8.35	18811	59.09	52.61	187621	89079	-6.712
3000	8.38	19648	59.37	52.83	187708	85708	-6.243

M.P. °K
ΔH_m CAL./GFW.

B.P. °K
ΔH_v CAL./GFW.

S.P. °K
ΔH_s CAL./GFW.

T.P. °K
ΔH_t CAL./GFW.

T.P. °K
ΔH_t CAL./GFW.

$T_c =$ °K
$P_c =$ ATM.

TECHNETIUM

TECHNETIUM Tc   Solid from 298° to 2400°, Liquid from 2400° to 3000°.

REFERENCE STATE

Gfw 99. * GRAMS CAL./GFW.
$(H°_{298.15} - H°_0)$ =

						FORMATION FROM REFERENCE STATE		
T TEMPERATURE °K	$C°_P$ HEAT CAPACITY CAL./DEG./GFW.	$H°_T - H°_{298.15}$ HEAT CONTENT CAL./GFW.	$S°_T$ ENTROPY CAL./DEG./GFW.	$-(F°_T - H°_{298.15})$ FREE ENERGY FUNCTION CAL./DEG./GFW.		HEAT $\Delta H°_f$ CAL./GFW.	FREE ENERGY $\Delta F°_f$ CAL./GFW.	$\log_{10} K_p$
298	5.80	0	8.00	8.00				
300	5.80	10	8.04	8.01				
400	6.00	600	9.73	8.23				
500	6.20	1210	11.10	8.68				
600	6.40	1840	12.24	9.18				
700	6.60	2490	13.25	9.70				
800	6.80	3160	14.14	10.19				
900	7.00	3850	14.95	10.68				
1000	7.20	4560	15.70	11.14				
1100	7.40	5290	16.40	11.60				
1200	7.60	6040	17.05	12.02				
1300	7.80	6810	17.67	12.44				
1400	8.00	7600	18.25	12.83				
1500	8.20	8410	18.81	13.21				
1600	8.40	9240	19.35	13.58				
1700	8.60	10090	19.86	13.93				
1800	8.80	10960	20.36	14.28				
1900	9.00	11850	20.84	14.61				
2000	9.20	12760	21.31	14.93				
2100	9.40	13690	21.76	15.25				
2200	9.60	14640	22.20	15.55				
2300	9.80	15630	22.63	15.84				
2400	10.00	22120	25.35	16.14				
2500	10.00	23120	25.76	16.52				
2600	10.00	24120	26.15	16.88				
2700	10.00	25120	26.52	17.22				
2800	10.00	26120	26.89	17.57				
2900	10.00	27120	27.24	17.89				
3000	10.00	28120	27.58	18.21				

M.P. (2,400) °K
ΔH_m (5,500) CAL./GFW.

B.P. (4,900) °K
ΔH_v (138,000) CAL./GFW.

S.P. °K
ΔH_s CAL./GFW.

T.P. °K
ΔH_t CAL./GFW.

T.P. °K
ΔH_t CAL./GFW.

T_c = °K
P_c = ATM.

THERMODYNAMIC PROPERTIES OF THE ELEMENTS

TECHNETIUM

TECHNETIUM   Tc   Reference State for Calculating ΔH_f°, ΔF_f°, and $\log_{10} Kp$:

IDEAL MONATOMIC GAS   Solid from 298° to 2400°, Liquid from 2400° to 3000°.

Gfw 99.* GRAMS
$(H_{298.15}^\circ - H_0^\circ) = 1,481$ CAL./GFW.

T TEMPERATURE °K	C_p° HEAT CAPACITY CAL./DEG./GFW.	$H_T^\circ - H_{298.15}^\circ$ HEAT CONTENT CAL./GFW.	S_T° ENTROPY CAL./DEG./GFW.	$-(F^\circ - H_f^\circ)/T$ FREE ENERGY FUNCTION CAL./DEG./GFW.	HEAT ΔH_f° CAL./GFW.	FREE ENERGY ΔF_f° CAL./GFW.	LOG$_{10}$ K$_p$
298	4.97	0	43.25	43.25	155000	144490	-105.918
300	4.97	9	43.28	43.25	154999	144427	-105.223
400	5.00	507	44.71	43.45	154907	140915	-76.998
500	5.11	1012	45.84	43.82	154802	137432	-60.075
600	5.33	1532	46.79	44.24	154692	133962	-48.799
700	5.66	2081	47.63	44.66	154591	130525	-40.755
800	6.06	2667	48.41	45.08	154507	127091	-34.722
900	6.48	3294	49.15	45.49	154444	123664	-30.031
1000	6.86	3961	49.89	45.89	154401	120251	-26.283
1100	7.18	4664	50.52	46.28	154374	116842	-23.216
1200	7.42	5395	51.16	46.67	154355	113423	-20.658
1300	7.58	6146	51.76	47.04	154336	110019	-18.497
1400	7.66	6909	52.33	47.40	154309	106597	-16.641
1500	7.68	7676	52.85	47.74	154266	103206	-15.038
1600	7.65	8443	53.35	48.08	154203	99803	-13.632
1700	7.59	9206	53.81	48.40	154116	96401	-12.393
1800	7.50	9960	54.24	48.71	154000	93016	-11.293
1900	7.41	10706	54.65	49.02	153856	89617	-10.307
2000	7.30	11441	55.02	49.30	153681	86261	-9.425
2100	7.20	12166	55.38	49.59	153476	82874	-8.624
2200	7.09	12881	55.71	49.86	153241	79519	-7.899
2300	7.00	13585	56.02	50.12	152955	76158	-7.236
2400	6.91	14281	56.32	50.37	147161	72833	-6.632
2500	6.83	14967	56.60	50.62	146847	69747	-6.097
2600	6.75	15646	56.87	50.86	146526	66654	-5.602
2700	6.69	16318	57.12	51.08	146198	63578	-5.146
2800	6.63	16984	57.36	51.30	145864	60548	-4.725
2900	6.58	17644	57.59	51.51	145524	57509	-4.333
3000	6.53	18299	57.82	51.73	145179	54459	-3.967

M.P. °K
ΔH_m CAL./GFW.

B.P. °K
ΔH_v CAL./GFW.

S.P. °K
ΔH_s CAL./GFW.

T.P. °K
ΔH_t CAL./GFW.

T.P. °K
ΔH_t CAL./GFW.

T_c = °K
P_c = ATM.

*Isotope of Longest Known Half Life

TELLURIUM

TELLURIUM Te Solid from 298° to 723°, Liquid from 723° to 1260°, Ideal Diatomic Gas from 1260° to 3000°.

REFERENCE STATE

Gfw	127.61 GRAMS
$(H°_{298.15} - H°_0) = 1463$ CAL./GFW.
M.P.	723 °K
ΔH_m	4180 CAL./GFW.
B.P.	1260 °K
ΔH_v	12,100 CAL./GFW.
S.P.	°K
ΔH_s	CAL./GFW.
T.P.	°K
ΔH_t	CAL./GFW.
T.P.	°K
ΔH_t	CAL./GFW.
T_c =	°K
P_c =	ATM.

T TEMPERATURE °K	$C°_p$ HEAT CAPACITY CAL./DEG./GFW.	$H°_T - H°_{298.15}$ HEAT CONTENT CAL./GFW.	$S°_T$ ENTROPY CAL./DEG./GFW.	$-(F°_T - H°_{298.15})/T$ FREE ENERGY FUNCTION CAL./DEG./GFW.	FORMATION FROM REFERENCE STATE		
					HEAT $\Delta H°_f$ CAL./GFW.	FREE ENERGY $\Delta F°_f$ CAL./GFW.	LOG$_{10}$ K_p
298	6.15	0	11.88	11.88			
300	6.16	11	11.92	11.89			
400	6.68	653	13.76	12.13			
500	7.21	1347	15.31	12.62			
600	7.73	2094	16.67	13.18			
700	8.26	2894	17.90	13.77			
800	9.00	7960	24.86	14.91			
900	9.00	8860	25.92	16.03			
1000	9.00	9760	26.87	17.11			
1100	9.00	10660	27.72	18.03			
1200	9.00	11560	28.51	18.88			
1300	4.47	24250	38.58	19.93			
1400	4.47	24700	38.91	21.27			
1500	4.47	25145	39.22	22.46			
1600	4.47	25590	39.51	23.52			
1700	4.47	26040	39.78	24.47			
1800	4.47	26485	40.03	25.32			
1900	4.47	26930	40.28	26.11			
2000	4.47	27380	40.51	26.82			
2100	4.47	27825	40.73	27.48			
2200	4.47	28275	40.93	28.08			
2300	4.47	28720	41.13	28.65			
2400	4.47	29165	41.32	29.17			
2500	4.47	29615	41.50	29.66			
2600	4.47	30060	41.68	30.12			
2700	4.47	30510	41.85	30.55			
2800	4.47	30955	42.01	30.96			
2900	4.47	31400	42.17	31.35			
3000	4.47	31850	42.32	31.71			

THERMODYNAMIC PROPERTIES OF THE ELEMENTS

TELLURIUM

TELLURIUM Te_2
IDEAL DIATOMIC GAS

Gfw 255.22 GRAMS
$(H°_{298.15} - H°_0) = 2,379$ CAL./GFW.

Reference State for Calculating $\Delta H°_f$, $\Delta F°_f$, and $Log_{10}K_p$: Solid from 298° to 723°, Liquid from 723° to 1260°, Ideal Diatomic Gas from 1260° to 3000°.

T TEMPERATURE °K	$C°_P$ HEAT CAPACITY CAL./DEG./GFW.	$H°_T - H°_{298.15}$ HEAT CONTENT CAL./GFW.	$S°_T$ ENTROPY CAL./DEG./GFW.	$-(F°_T - H°_{298.15})/T$ FREE ENERGY FUNCTION CAL./DEG./GFW.	FORMATION FROM REFERENCE STATE		
					HEAT $\Delta H°_f$ CAL./GFW.	FREE ENERGY $\Delta F°_f$ CAL./GFW.	$LOG_{10} K_P$
298	8.72	0	64.10	64.10	39600	27572	-20.211
300	8.72	16	64.16	64.11	39594	27498	-20.033
400	8.81	895	66.68	64.45	39189	23525	-12.854
500	8.86	1775	68.65	65.10	38681	19666	-8.596
600	8.88	2665	70.27	65.83	38077	15919	-5.798
700	8.90	3555	71.64	66.57	37367	12279	-3.833
800	8.91	4445	72.83	67.28	28125	9637	-2.632
900	8.92	5335	73.88	67.96	27215	7379	-1.791
1000	8.92	6225	74.82	68.60	26305	5225	-1.142
1100	8.92	7117	75.67	69.20	25397	3144	-.624
1200	8.93	8010	76.44	69.77	24490	1186	-.216
1300	8.93	8903	77.16	70.32	0	0	0
1400	8.93	9795	77.82	70.83	0	0	0
1500	8.93	10690	78.44	71.32	0	0	0
1600	8.93	11585	79.01	71.77	0	0	0
1700	8.94	12475	79.56	72.23	0	0	0
1800	8.94	13370	80.06	72.64	0	0	0
1900	8.94	14265	80.55	73.05	0	0	0
2000	8.94	15160	81.01	73.43	0	0	0
2100	8.94	16050	81.45	73.81	0	0	0
2200	8.94	16950	81.86	74.16	0	0	0
2300	8.94	17840	82.26	74.51	0	0	0
2400	8.94	18730	82.64	74.84	0	0	0
2500	8.94	19630	83.00	75.15	0	0	0
2600	8.94	20520	83.36	75.47	0	0	0
2700	8.94	21420	83.69	75.76	0	0	0
2800	8.94	22310	84.02	76.06	0	0	0
2900	8.94	23200	84.33	76.33	0	0	0
3000	8.94	24100	84.64	76.61	0	0	0

M.P. °K
ΔH_m CAL./GFW.

B.P. °K
ΔH_v CAL./GFW.

S.P. °K
ΔH_s CAL./GFW.

T.P. °K
ΔH_t CAL./GFW.

T.P. °K
ΔH_t CAL./GFW.

T_c = °K
P_c = ATM.

TELLURIUM

TELLURIUM Te
Gfw 127.61 **GRAMS**
$(H°_{298.15} - H°_0) = 1,481$ CAL./GFW.

IDEAL MONATOMIC GAS

Reference State for Calculating $\Delta H°_f$, $\Delta F°_f$, and $\text{Log}_{10} K_p$: Solid from 298° to 723°, Liquid from 723° to 1260°, Ideal Diatomic Gas from 1260° to 3000°.

T TEMPERATURE °K	$C°_P$ HEAT CAPACITY CAL./DEG./GFW.	$H°_T - H°_{298.15}$ HEAT CONTENT CAL./GFW.	$S°_T$ ENTROPY CAL./DEG./GFW.	$-(F°_T - H°_{298.15})/T$ FREE ENERGY FUNCTION CAL./DEG./GFW.	FORMATION FROM REFERENCE STATE		
					HEAT $\Delta H°_f$ CAL./GFW.	FREE ENERGY $\Delta F°_f$ CAL./GFW.	$\text{LOG}_{10} K_p$
298	4.97	0	43.64	43.64	46500	37031	-27.145
300	4.97	9	43.67	43.64	46498	36973	-26.937
400	4.97	506	45.10	43.84	46353	33817	-18.478
500	4.97	1003	46.21	44.21	46156	30706	-13.422
600	4.97	1500	47.12	44.62	45906	27636	-10.067
700	4.98	1997	47.88	45.03	45603	24617	-7.686
800	4.99	2495	48.55	45.44	41035	22083	-6.033
900	5.01	2996	49.14	45.82	40636	19738	-4.793
1000	5.05	3499	49.67	46.18	40239	17439	-3.811
1100	5.09	4006	50.15	46.51	39846	15173	-3.014
1200	5.14	4517	50.60	46.84	39457	12949	-2.358
1300	5.20	5034	51.01	47.14	27284	11125	-1.870
1400	5.26	5557	51.40	47.44	27357	9871	-1.541
1500	5.32	6086	51.76	47.71	27441	8631	-1.257
1600	5.38	6621	52.11	47.98	27531	7371	-1.006
1700	5.44	7162	52.44	48.23	27622	6100	-.784
1800	5.50	7709	52.75	48.47	27724	4828	-.586
1900	5.55	8261	53.05	48.71	27831	3568	-.410
2000	5.60	8819	53.33	48.93	27939	2299	-.251
2100	5.65	9381	53.61	49.15	28056	1008	-.104
2200	5.70	9949	53.87	49.35	28174	294	.029
2300	5.74	10521	54.13	49.56	28301	-1599	.151
2400	5.78	11097	54.37	49.75	28432	-2888	.262
2500	5.82	11677	54.61	49.94	28562	-4213	.368
2600	5.85	12260	54.84	50.13	28700	-5516	.463
2700	5.88	12847	55.06	50.31	28837	-6830	.552
2800	5.91	13436	55.27	50.48	28981	-8147	.635
2900	5.94	14029	55.48	50.65	29129	-9470	.713
3000	5.96	14624	55.68	50.81	29274	-10806	.787

M.P. °K
ΔH_m CAL./GFW.

B.P. °K
ΔH_v CAL./GFW.

S.P. °K
ΔH_s CAL./GFW.

T.P. °K
ΔH_t CAL./GFW.

T.P. °K
ΔH_t CAL./GFW.

$T_c =$ °K
$P_c =$ ATM.

THERMODYNAMIC PROPERTIES OF THE ELEMENTS

TERBIUM

TERBIUM Tb

REFERENCE STATE Solid from 298° to 1700°, Liquid from 1700° to 2800°, Ideal Monatomic Gas from 2800° to 3000°.

Giw 158.93 GRAMS CAL./GFW.

$(H°_{298.15} - H°_0) =$

M.P. (1,700) °K
ΔH_m (3,900) CAL./GFW.

B.P. (2,800) °K
ΔH_v (70,000) CAL./GFW.

S.P. °K
ΔH_s CAL./GFW.

T.P. °K
ΔH_f CAL./GFW.

T.P. °K
ΔH_f CAL./GFW.

$T_c =$ °K
$P_c =$ ATM.

T TEMPERATURE °K	$C°_P$ HEAT CAPACITY CAL./DEG./GFW.	$H°_T - H°_{298.15}$ HEAT CONTENT CAL./GFW.	$S°_T$ ENTROPY CAL./DEG./GFW.	$-(F°_T - H°_{298.15})$ FREE ENERGY FUNCTION CAL./DEG./GFW.	FORMATION FROM REFERENCE STATE		
					HEAT $\Delta H°_f$ CAL./GFW.	FREE ENERGY $\Delta F°_f$ CAL./GFW.	LOG$_{10}$ K_P
298	6.54	0	17.46	17.46			
300	6.54	12	17.50	17.46			
400	6.72	675	19.41	17.73			
500	6.90	1360	20.92	18.20			
600	7.08	2055	22.20	18.78			
700	7.26	2770	23.30	19.35			
800	7.44	3510	24.28	19.90			
900	7.62	4260	25.17	20.44			
1000	7.80	5030	25.98	20.95			
1100	7.98	5820	26.74	21.45			
1200	8.16	6630	27.44	21.92			
1300	8.34	7450	28.10	22.37			
1400	8.52	8300	28.72	22.80			
1500	8.70	9160	29.32	23.22			
1600	8.88	10040	29.88	23.61			
1700	8.00	14830	32.72	24.00			
1800	8.00	15630	33.17	24.49			
1900	8.00	16430	33.61	24.97			
2000	8.00	17230	34.02	25.41			
2100	8.00	18030	34.41	25.83			
2200	8.00	18830	34.78	26.23			
2300	8.00	19630	35.14	26.61			
2400	8.00	20430	35.48	26.97			
2500	8.00	21230	35.80	27.31			
2600	8.00	22030	36.12	27.65			
2700	8.00	22830	36.42	27.97			
2800	6.00	93630	61.71	28.28			
2900	6.00	94230	61.91	29.42			
3000	6.00	94830	62.12	30.51			

THALLIUM

THALLIUM Tl Solid I from 298° to 507°, Solid II from 507° to 577°, Liquid
REFERENCE STATE from 577° to 1740°, Ideal Monatomic Gas from 1740° to 3000°.

Gfw 204.39 GRAMS
$(H°_{298.15} - H°_0) = 1,632$ CAL./GFW.

M.P. 577 °K
ΔH_m 1,020 CAL./GFW.

B.P. 1,740 °K
ΔH_v 38,740 CAL./GFW.

S.P. °K
ΔH_s CAL./GFW.

T.P. 507 °K
ΔH_t 90 CAL./GFW.

T.P. °K
ΔH_t CAL./GFW.

$T_c =$ °K
$P_c =$ ATM.

T TEMPERATURE °K	$C_p°$ HEAT CAPACITY CAL./DEG./GFW.	$H°_T - H°_{298.15}$ HEAT CONTENT CAL./GFW.	$S°_T$ ENTROPY CAL./DEG./GFW.	$-(F°_T - H°_{298.15})$ FREE ENERGY FUNCTION CAL./DEG./GFW.	HEAT $\Delta H°_f$ CAL./GFW.	FREE ENERGY $\Delta F°_f$ CAL./GFW.	LOG$_{10}$ K$_p$
298	6.29	0	15.35	15.35			
300	6.30	11	15.39	15.36			
400	6.64	658	17.25	15.61			
500	7.00	1340	18.77	16.09			
600	7.30	3170	22.03	15.75			
700	7.30	3900	23.15	17.58			
800	7.30	4630	24.13	18.35			
900	7.30	5360	24.99	19.04			
1000	7.30	6090	25.76	19.67			
1100	7.30	6820	26.45	20.25			
1200	7.30	7550	27.09	20.80			
1300	7.30	8280	27.67	21.31			
1400	7.30	9010	28.21	21.78			
1500	7.30	9740	28.72	22.23			
1600	7.30	10470	29.19	22.65			
1700	7.30	11200	29.63	23.05			
1800	5.27	50550	52.22	24.14			
1900	5.34	51080	52.50	25.62			
2000	5.42	51620	52.78	26.97			
2100	5.50	52160	53.04	28.21			
2200	5.58	52720	53.30	29.34			
2300	5.67	53280	53.55	30.39			
2400	5.75	53850	53.80	31.37			
2500	5.83	54430	54.03	32.26			
2600	5.91	55020	54.26	33.10			
2700	5.98	55610	54.49	33.90			
2800	6.05	56210	54.70	34.63			
2900	6.11	56820	54.92	35.33			
3000	6.17	57440	55.13	35.99			

THALLIUM

THALLIUM   **Tl**

IDEAL MONATOMIC GAS

Gfw   204.39   GRAMS

$(H°_{298.15} - H°_0) = 1,481$ CAL./GFW.

Reference State for Calculating $\Delta H°_f$, $\Delta F°_f$, and $Log_{10} Kp$:
Solid I from 298° to 507°, Solid II from 507° to 577°, Liquid
from 577° to 1740°, Ideal Monatomic Gas from 1740° to 3000°.

T TEMPERATURE °K	$C°_P$ HEAT CAPACITY CAL./DEG./GFW.	$H°_T - H°_{298.15}$ HEAT CONTENT CAL./GFW.	$S°_T$ ENTROPY CAL./DEG./GFW.	$-(F°_T-H°_{298.15})/T$ FREE ENERGY FUNCTION CAL./DEG./GFW.	HEAT $\Delta H°_f$ CAL./GFW.	FREE ENERGY $\Delta F°_f$ CAL./GFW.	$LOG_{10} K_P$
298	4.97	0		43.23	43000	34687	-25.427
300	4.97	9	43.23	43.23	42998	34637	-25.235
400	4.97	506	44.69	43.43	42848	31872	-17.415
500	4.97	1003	45.79	43.79	42663	29153	-12.743
600	4.97	1500	46.70	44.20	41330	26528	-9.663
700	4.97	1996	47.47	44.62	41096	24072	-7.516
800	4.97	2493	48.13	45.02	40863	21663	-5.918
900	4.97	2990	48.72	45.40	40630	19273	-4.680
1000	4.97	3487	49.24	45.76	40397	16917	-3.697
1100	4.98	3985	49.71	46.09	40165	14579	-2.896
1200	5.00	4484	50.15	46.42	39934	12262	-2.233
1300	5.02	4985	50.55	46.72	39705	9961	-1.674
1400	5.05	5489	50.92	47.00	39479	7685	-1.199
1500	5.09	5996	51.27	47.28	39256	5431	-.791
1600	5.14	6508	51.60	47.54	39038	3182	-.434
1700	5.20	7025	51.92	47.79	38825	932	-.119
1800	5.27	7549	52.22	48.03	0	0	0
1900	5.34	8079	52.50	48.25	0	0	0
2000	5.42	8617	52.78	48.48	0	0	0
2100	5.50	9164	53.04	48.68	0	0	0
2200	5.58	9718	53.30	48.89	0	0	0
2300	5.67	10280	53.55	49.09	0	0	0
2400	5.75	10851	53.80	49.28	0	0	0
2500	5.83	11430	54.03	49.46	0	0	0
2600	5.91	12017	54.26	49.64	0	0	0
2700	5.98	12612	54.49	49.82	0	0	0
2800	6.05	13214	54.70	49.99	0	0	0
2900	6.11	13822	54.92	50.16	0	0	0
3000	6.17	14436	55.13	50.32	0	0	0

M.P. °K

ΔH_m CAL./GFW.

B.P. °K

ΔH_v CAL./GFW.

S.P. °K

ΔH_s CAL./GFW.

T.P. °K

ΔH_t CAL./GFW.

T.P. °K

ΔH_t CAL./GFW.

$T_c =$ °K

$P_c =$ ATM.

THORIUM

THORIUM Th
Solid I from 298° to 1673°, Solid II from 1673° to 1968°, Liquid from 1968° to 3000°.

REFERENCE STATE

Gfw 232.05 GRAMS
$(H°_{298.15} - H°_0) = 1,556$ CAL./GFW.

M.P. 1,968 °K
ΔH_m (3,740) CAL./GFW.

B.P. 4,500 °K
ΔH_v 130,000 CAL./GFW.

S.P. °K
ΔH_s CAL./GFW.

T.P. 1,673 °K
ΔH_t (670) CAL./GFW.

T.P. °K
ΔH_t CAL./GFW.

$T_c =$ °K
$P_c =$ ATM.

T TEMPERATURE °K	$C°_P$ HEAT CAPACITY CAL./DEG./GFW.	$H°_T - H°_{298.15}$ HEAT CONTENT CAL./GFW.	$S°_T$ ENTROPY CAL./DEG./GFW.	$-(F°_T - H°_{298.15})/T$ FREE ENERGY FUNCTION CAL./DEG./GFW.	FORMATION FROM REFERENCE STATE		
					HEAT $\Delta H°_f$ CAL./GFW.	FREE ENERGY $\Delta F°_f$ CAL./GFW.	LOG$_{10}$ K_P
298	6.53	0	12.76	12.76			
300	6.54	12	12.80	12.76			
400	7.00	690	14.74	13.02			
500	7.45	1410	16.35	13.53			
600	7.90	2180	17.75	14.12			
700	8.36	2990	19.00	14.73			
800	8.81	3850	20.15	15.34			
900	9.27	4760	21.21	15.93			
1000	9.72	5710	22.21	16.50			
1100	10.17	6700	23.16	17.07			
1200	10.63	7740	24.07	17.62			
1300	11.08	8830	24.93	18.14			
1400	11.54	9960	25.77	18.66			
1500	11.99	11130	26.58	19.16			
1600	12.44	12360	27.37	19.65			
1700	11.00	14240	28.51	20.14			
1800	11.00	15340	29.14	20.62			
1900	11.00	16440	29.73	21.08			
2000	11.00	21280	32.20	21.56			
2100	11.00	22380	32.74	22.09			
2200	11.00	23480	33.25	22.58			
2300	11.00	24580	33.74	23.06			
2400	11.00	25680	34.20	23.50			
2500	11.00	26780	34.65	23.94			
2600	11.00	27880	35.09	24.37			
2700	11.00	28980	35.50	24.77			
2800	11.00	30080	35.90	25.16			
2900	11.00	31180	36.29	25.54			
3000	11.00	32280	36.66	25.90			

THERMODYNAMIC PROPERTIES OF THE ELEMENTS

THULIUM

THULIUM Tm Solid from 298° to 1900°, Liquid from 1900° to 2400°, Ideal Monatomic Gas from 2400° to 3000°.

REFERENCE STATE

Gfw	168.94	GRAMS
$(H°_{298.15} - H°_0)$ =		CAL./GFW.
M.P.	(1,900)	°K
ΔH_m	(4,400)	CAL./GFW.
B.P.	(2,400)	°K
ΔH_v	(51,000)	CAL./GFW.
S.P.		°K
ΔH_s		CAL./GFW.
T.P.		°K
ΔH_f		CAL./GFW.
T.P.		°K
ΔH_f		CAL./GFW.
T_c =		°K
P_c =		ATM.

T TEMPERATURE °K	$C°_P$ HEAT CAPACITY CAL./DEG./GFW.	$H°_T - H°_{298.15}$ HEAT CONTENT CAL./GFW.	$S°_T$ ENTROPY CAL./DEG./GFW.	$-(F°_T - H°_{298.15})/T$ FREE ENERGY FUNCTION CAL./DEG./GFW.	FORMATION FROM REFERENCE STATE		
					HEAT $\Delta H°_f$ CAL./GFW.	FREE ENERGY $\Delta F°_f$ CAL./GFW.	LOG$_{10}$ K_P
298	6.45	0	17.06	17.06			
300	6.45	11	17.10	17.07			
400	6.60	660	18.97	17.32			
500	6.75	1330	20.46	17.80			
600	6.90	2010	21.71	18.36			
700	7.05	2710	22.78	18.91			
800	7.20	3420	23.73	19.46			
900	7.35	4150	24.59	19.98			
1000	7.50	4890	25.37	20.48			
1100	7.65	5650	26.09	20.96			
1200	7.80	6420	26.77	21.42			
1300	7.95	7210	27.40	21.86			
1400	8.10	8010	27.99	22.27			
1500	8.25	8830	28.56	22.68			
1600	8.40	9660	29.09	23.06			
1700	8.55	10510	29.61	23.43			
1800	8.70	11370	30.10	23.79			
1900	8.00	16650	32.89	24.13			
2000	8.00	17450	33.30	24.58			
2100	8.00	18250	33.69	25.00			
2200	8.00	19050	34.07	25.42			
2300	8.00	19850	34.42	25.79			
2400	5.18	71220	55.83	26.16			
2500	5.20	71740	56.04	27.35			
2600	5.23	72260	56.24	28.45			
2700	5.25	72780	56.44	29.49			
2800	5.28	73310	56.63	30.45			
2900	5.30	73840	56.82	31.36			
3000	5.32	74370	57.00	32.21			

TIN

TIN   Sn   White Solid from 298° to 505°, Liquid from 505° to 2960°, Ideal Monatomic Gas from 2960° to 3000°.

REFERENCE STATE

Gfw 118.70 GRAMS
$(H^\circ_{298.15} - H^\circ_0) = 1,507$ CAL./GFW.

M.P. 505 °K
ΔH_m 1,720 CAL./GFW.

B.P. 2,960 °K
ΔH_v 69,400 CAL./GFW.

S.P.   °K
ΔH_s CAL./GFW.

T.P. 292 °K
ΔH_t 535 CAL./GFW.

T.P.   °K
ΔH_t CAL./GFW.

$T_c =$ °K
$P_c =$ ATM.

T TEMPERATURE °K	C°_p HEAT CAPACITY CAL./DEG./GFW.	$H^\circ_T - H^\circ_{298.15}$ HEAT CONTENT CAL./GFW.	S°_T ENTROPY CAL./DEG./GFW.	$-(F^\circ_T - H^\circ_{298.15})$ FREE ENERGY FUNCTION CAL./DEG./GFW.	FORMATION FROM REFERENCE STATE		
					HEAT ΔH°_f CAL./GFW.	FREE ENERGY ΔF°_f CAL./GFW.	LOG$_{10}$ K$_p$
298	6.30	0	12.29	12.29			
300	6.31	11	12.33	12.30			
400	6.94	680	14.25	12.55			
500	7.57	1400	15.87	13.07			
600	7.30	3850	20.59	14.18			
700	7.30	4580	21.72	15.18			
800	7.30	5310	22.69	16.06			
900	7.30	6040	23.55	16.84			
1000	7.30	6770	24.32	17.55			
1100	7.30	7500	25.02	18.21			
1200	7.30	8230	25.65	18.80			
1300	7.30	8960	26.23	19.34			
1400	7.30	9690	26.77	19.85			
1500	7.30	10420	27.28	20.34			
1600	7.30	11150	27.75	20.79			
1700	7.30	11880	28.19	21.21			
1800	7.30	12610	28.61	21.61			
1900	7.30	13340	29.00	21.98			
2000	7.30	14070	29.38	22.35			
2100	7.30	14800	29.73	22.69			
2200	7.30	15530	30.07	23.02			
2300	7.30	16260	30.40	23.34			
2400	7.30	16990	30.71	23.64			
2500	7.30	17720	31.01	23.93			
2600	7.30	18450	31.29	24.20			
2700	7.30	19180	31.57	24.47			
2800	7.30	19910	31.83	24.72			
2900	7.30	20640	32.18	25.07			
3000	6.27	90720	55.89	25.65			

THERMODYNAMIC PROPERTIES OF THE ELEMENTS

TIN

TIN	Sn	Reference State for Calculating ΔH_f°, ΔF_f°, and $\log_{10} K_p$:
Gfw	118.70 GRAMS CAL./GFW.	White Solid from 298° to 505°, Liquid from 505°
IDEAL MONATOMIC GAS		to 2960°, Ideal Monatomic Gas from 2960° to 3000°.
$(H^\circ_{298.15} - H^\circ_0) = 1,485$		

	T TEMPERATURE °K	C_p° HEAT CAPACITY CAL./DEG./GFW.	$H^\circ_T - H^\circ_{298.15}$ HEAT CONTENT CAL./GFW.	S°_T ENTROPY CAL./DEG./GFW.	$-(F^\circ_T - H^\circ_{298.15})/T$ FREE ENERGY FUNCTION CAL./DEG./GFW.	FORMATION FROM REFERENCE STATE		
						HEAT ΔH_f° CAL./GFW.	FREE ENERGY ΔF_f° CAL./GFW.	$\log_{10} K_p$
	298	5.08	0	40.24	40.24	72000	63667	-46.671
	300	5.09	9	40.28	40.25	71998	63613	-46.345
M.P. °K	400	5.47	535	41.79	40.46	71855	60839	-33.243
ΔH_m CAL./GFW.	500	6.05	1110	43.07	40.85	71710	58110	-25.401
	600	6.66	1747	44.22	41.31	69897	55719	-20.297
	700	7.18	2440	45.29	41.81	69860	53361	-16.661
B.P. °K	800	7.57	3179	46.28	42.31	69869	50997	-13.932
	900	7.80	3948	47.18	42.80	69908	48641	-11.812
ΔH_v CAL./GFW.	1000	7.91	4735	48.01	43.28	69965	46275	-10.114
	1100	7.92	5527	48.77	43.75	70027	43902	-8.723
S.P. °K	1200	7.86	6316	49.46	44.20	70086	41514	-7.561
	1300	7.75	7097	50.08	44.63	70137	39132	-6.579
ΔH_s CAL./GFW.	1400	7.62	7866	50.65	45.04	70176	36744	-5.736
	1500	7.48	8621	51.17	45.43	70201	34366	-5.007
	1600	7.34	9363	51.65	45.80	70213	31973	-4.367
	1700	7.21	10090	52.09	46.16	70210	29580	-3.802
T.P. °K	1800	7.09	10805	52.50	46.50	70195	27193	-3.301
	1900	6.97	11508	52.88	46.83	70168	24796	-2.852
ΔH_t CAL./GFW.	2000	6.87	12200	53.23	47.13	70130	22430	-2.450
	2100	6.78	12882	53.57	47.44	70082	20018	-2.083
	2200	6.69	13555	53.88	47.72	70025	17643	-1.752
T.P. °K	2300	6.62	14221	54.18	48.00	69961	15267	-1.450
	2400	6.55	14879	54.46	48.27	69889	12889	-1.173
ΔH_t CAL./GFW.	2500	6.49	15532	54.72	48.51	69812	10547	-.921
	2600	6.44	16178	54.98	48.76	69728	8134	-.683
	2700	6.39	16820	55.22	49.00	69640	5785	-.468
$T_c =$ °K	2800	6.34	17456	55.45	49.22	69546	3410	-.266
	2900	6.30	18089	55.67	49.44	69449	1328	-.100
$P_c =$ ATM.	3000	6.27	18717	55.89	49.66	0	0	0

TITANIUM

TITANIUM   Ti

Solid I from 298° to 1155°, Solid II from 1155° to 1950°, Liquid from 1950° to 3000°.

REFERENCE STATE

Gfw	47.90	GRAMS
$(H^°_{298.15} - H^°_0) =$	1,150	CAL./GFW.
M.P.	1,950	°K
ΔH_m	(3,700)	CAL./GFW.
B.P.	3,550	°K
ΔH_v	102,500	CAL./GFW.
S.P.		°K
ΔH_s		CAL./GFW.
T.P.	1,155	°K
ΔH_t	950	CAL./GFW.
T.P.		°K
ΔH_t		CAL./GFW.
$T_c =$		°K
$P_c =$		ATM.

T TEMPERATURE °K	$C^°_P$ HEAT CAPACITY CAL./DEG./GFW.	$H^°_T - H^°_{298.15}$ HEAT CONTENT CAL./GFW.	$S^°_T$ ENTROPY CAL./DEG./GFW.	$-(F^°_T - H^°_{298.15})$ FREE ENERGY FUNCTION CAL./DEG./GFW.	FORMATION FROM REFERENCE STATE HEAT $\Delta H^°_f$ CAL./GFW.	FORMATION FROM REFERENCE STATE FREE ENERGY $\Delta F^°_f$ CAL./GFW.	LOG$_{10}$ K$_P$
298	5.98	0	7.33	7.33			
300	5.98	11	7.37	7.34			
400	6.36	629	9.15	7.58			
500	6.62	1280	10.60	8.04			
600	6.84	1950	11.82	8.57			
700	7.02	2640	12.89	9.12			
800	7.18	3355	13.84	9.65			
900	7.33	4080	14.70	10.17			
1000	7.47	4820	15.48	10.66			
1100	7.60	5575	16.19	11.13			
1200	7.72	7290	17.68	11.61			
1300	7.84	8070	18.30	12.10			
1400	7.95	8860	18.89	12.57			
1500	8.06	9660	19.44	13.00			
1600	8.16	10470	19.96	13.42			
1700	8.26	11290	20.46	13.82			
1800	8.36	12130	20.94	14.21			
1900	8.46	12970	21.39	14.57			
2000	8.00	17500	23.71	14.96			
2100	8.00	18300	24.10	15.39			
2200	8.00	19100	24.47	15.79			
2300	8.00	19900	24.83	16.18			
2400	8.00	20700	25.17	16.55			
2500	8.00	21500	25.50	16.90			
2600	8.00	22300	25.81	17.24			
2700	8.00	23100	26.11	17.56			
2800	8.00	23900	26.40	17.87			
2900	8.00	24700	26.68	18.17			
3000	8.00	25500	26.96	18.46			

THERMODYNAMIC PROPERTIES OF THE ELEMENTS

TITANIUM

TITANIUM   **Ti**

IDEAL MONATOMIC GAS

Gfw   47.90   GRAMS

$(H°_{298.15} - H°_0) = 1,802$ CAL./GFW.

Reference State for Calculating $\Delta H°_f$, $\Delta F°_f$, and $Log_{10} Kp$: Solid I from 298° to 1155°, Solid II from 1155° to 1950°, Liquid from 1950° to 3000°.

T TEMPERATURE °K	$C°_p$ HEAT CAPACITY CAL./DEG./GFW.	$H°_T - H°_{298.15}$ HEAT CONTENT CAL./GFW.	$S°_T$ ENTROPY CAL./DEG./GFW.	$-(F°_T - H°_{298.15})/T$ FREE ENERGY FUNCTION CAL./DEG./GFW.	FORMATION FROM REFERENCE STATE		
					HEAT $\Delta H°_f$ CAL./GFW.	FREE ENERGY $\Delta F°_f$ CAL./GFW.	$LOG_{10} K_p$
298	5.84	0	43.07	43.07	112600	101944	-74.730
300	5.83	11	43.10	43.07	112600	101881	-74.226
400	5.52	577	44.74	43.30	112548	98312	-53.719
500	5.34	1119	45.95	43.72	112439	94764	-41.424
600	5.24	1648	46.91	44.17	112298	91244	-33.238
700	5.17	2168	47.71	44.62	112128	87754	-27.400
800	5.13	2683	48.40	45.05	111928	84280	-23.026
900	5.10	3194	49.00	45.46	111714	80844	-19.632
1000	5.10	3704	49.54	45.84	111484	77424	-16.922
1100	5.11	4214	50.03	46.20	111239	74015	-14.706
1200	5.13	4726	50.47	46.54	110036	70688	-12.875
1300	5.18	5241	50.88	46.85	109771	67417	-11.334
1400	5.24	5762	51.27	47.16	109502	64170	-10.018
1500	5.31	6289	51.63	47.44	109229	60944	-8.880
1600	5.40	6825	51.98	47.72	108955	57723	-7.884
1700	5.50	7370	52.31	47.98	108680	54535	-7.011
1800	5.62	7926	52.63	48.23	108396	51354	-6.234
1900	5.74	8494	52.93	48.46	108124	48198	-5.543
2000	5.86	9074	53.23	48.70	104174	45134	-4.931
2100	5.99	9666	53.52	48.92	103966	42184	-4.390
2200	6.13	10273	53.80	49.14	103773	39247	-3.898
2300	6.27	10893	54.08	49.35	103593	36318	-3.450
2400	6.41	11527	54.35	49.55	103427	33395	-3.040
2500	6.55	12175	54.61	49.74	103275	30500	-2.666
2600	6.70	12837	54.87	49.94	103137	27581	-2.318
2700	6.84	13514	55.13	50.13	103014	24662	-1.995
2800	6.99	14206	55.38	50.31	102906	21772	-1.698
2900	7.13	14912	55.63	50.49	102812	18857	-1.421
3000	7.28	15633	55.87	50.66	102733	16003	-1.165

M.P. °K
ΔH_m CAL./GFW.

B.P. °K
ΔH_v CAL./GFW.

S.P. °K
ΔH_s CAL./GFW.

T.P. °K
ΔH_t CAL./GFW.

T.P. °K
ΔH_t CAL./GFW.

$T_c =$ °K
$P_c =$ ATM.

TUNGSTEN

Solid from 298° to 3000°.

TUNGSTEN	W							FORMATION FROM REFERENCE STATE		
Gfw	183.86	GRAMS	T TEMPERATURE °K	C°p HEAT CAPACITY CAL./DEG./GFW.	H°T − H°298.15 HEAT CONTENT CAL./GFW.	S°T ENTROPY CAL./DEG./GFW.	−(F°T − H°298.15)/T FREE ENERGY FUNCTION CAL./DEG./GFW.	HEAT ΔH°f CAL./GFW.	FREE ENERGY ΔF°f CAL./GFW.	LOG₁₀ Kp
REFERENCE STATE										
$(H°_{298.15} - H°_0) =$	1,216	CAL./GFW.	298	5.92	0	8.04	8.04			
			300	5.92	10	8.08	8.05			
M.P.	3,650	°K	400	6.00	606	9.79	8.28			
ΔHm	(8,420)	CAL./GFW.	500	6.09	1211	11.14	8.72			
			600	6.17	1824	12.25	9.21			
			700	6.25	2445	13.21	9.72			
B.P.	5,800	°K	800	6.34	3075	14.05	10.21			
ΔHv	191,000	CAL./GFW.	900	6.42	3710	14.80	10.68			
			1000	6.50	4360	15.48	11.12			
			1100	6.58	5010	16.11	11.56			
			1200	6.67	5680	16.68	11.95			
S.P.		°K	1300	6.75	6350	17.22	12.34			
ΔHs		CAL./GFW.	1400	6.83	7030	17.72	12.70			
			1500	6.91	7710	18.20	13.06			
			1600	7.00	8410	18.65	13.40			
			1700	7.08	9110	19.07	13.72			
T.P.		°K	1800	7.16	9820	19.48	14.03			
ΔHt		CAL./GFW.	1900	7.24	10540	19.87	14.33			
			2000	7.33	11270	20.24	14.61			
			2100	7.41	12010	20.60	14.89			
			2200	7.49	12760	20.95	15.15			
T.P.		°K	2300	7.58	13510	21.29	15.42			
ΔHt		CAL./GFW.	2400	7.66	14270	21.61	15.67			
			2500	7.74	15040	21.92	15.91			
			2600	7.82	15820	22.23	16.15			
			2700	7.91	16600	22.53	16.39			
Tc =		°K	2800	7.99	17400	22.81	16.60			
Pc =		ATM.	2900	8.07	18200	23.10	16.83			
			3000	8.15	19010	23.37	17.04			

TUNGSTEN

TUNGSTEN W
Gfw 183.86 **GRAMS**
IDEAL MONATOMIC GAS
$(H°_{298.15} - H°_0) = 1,486$ CAL./GFW.

Reference State for Calculating $\Delta H°_f$, $\Delta F°_f$, and $\log_{10} K_p$: Solid from 298° to 3000°.

T TEMPERATURE °K	$C°_p$ HEAT CAPACITY CAL./DEG./GFW.	$H°_T - H°_{298.15}$ HEAT CONTENT CAL./GFW.	$S°_T$ ENTROPY CAL./DEG./GFW.	$-(F°_T-H°_{298.15})/T$ FREE ENERGY FUNCTION CAL./DEG./GFW.	FORMATION FROM REFERENCE STATE		
					HEAT $\Delta H°_f$ CAL./GFW.	FREE ENERGY $\Delta F°_f$ CAL./GFW.	$\log_{10} K_p$
298	5.09	0	41.55	41.55	200000	190009	-139.286
300	5.10	9	41.58	41.55	199999	189949	-138.389
400	5.54	538	43.10	41.76	199932	186608	-101.966
500	6.30	1128	44.41	42.16	199917	183282	-80.118
600	7.25	1804	45.64	42.64	199980	179946	-65.550
700	8.22	2578	46.83	43.16	200133	176599	-55.141
800	9.03	3442	47.99	43.69	200367	173215	-47.324
900	9.58	4375	49.08	44.23	200665	169813	-41.239
1000	9.86	5349	50.11	44.77	200989	166359	-36.361
1100	9.90	6338	51.05	45.29	201328	162894	-32.367
1200	9.79	7324	51.91	45.81	201644	159368	-29.027
1300	9.57	8292	52.69	46.32	201942	155831	-26.199
1400	9.30	9236	53.39	46.80	202206	152268	-23.772
1500	9.01	10151	54.02	47.26	202441	148711	-21.668
1600	8.72	11038	54.59	47.70	202628	145124	-19.822
1700	8.45	11896	55.11	48.12	202786	141518	-18.193
1800	8.21	12729	55.59	48.52	202909	137911	-16.743
1900	7.99	13538	56.02	48.91	202998	134313	-15.448
2000	7.80	14327	56.43	49.27	203057	130677	-14.279
2100	7.63	15099	56.81	49.62	203089	127048	-13.221
2200	7.50	15855	57.16	49.96	203095	123433	-12.261
2300	7.38	16598	57.49	50.28	203088	119828	-11.386
2400	7.29	17332	57.80	50.58	203062	116206	-10.581
2500	7.23	18058	58.10	50.88	203018	112568	9.840
2600	7.18	18778	58.38	51.16	202958	108968	9.159
2700	7.15	19494	58.65	51.43	202894	105388	8.528
2800	7.13	20208	58.91	51.70	202808	101728	7.939
2900	7.14	20921	59.16	51.95	202721	98147	7.396
3000	7.15	21636	59.40	52.19	202626	94536	6.886

M.P. °K
ΔH_m CAL./GFW.
B.P. °K
ΔH_v CAL./GFW.
S.P. °K
ΔH_s CAL./GFW.
T.P. °K
ΔH_t CAL./GFW.
T.P. °K
ΔH_t CAL./GFW.
$T_c =$ °K
$P_c =$ ATM.

URANIUM

URANIUM U
Gfw 238.07 GRAMS

Solid I from 298° to 941°, Solid II from 941° to 1047°,
Solid III from 1047° to 1406°, Liquid from 1406° to 3000°.

REFERENCE STATE
$(H°_{298.15} - H°_0) = 1,559$ CAL./GFW.

M.P.	1,406	°K
ΔH_m	3,700	CAL./GFW.
B.P.	4,200	°K
ΔH_v	101,000	CAL./GFW.
S.P.		°K
ΔH_s		CAL./GFW.
T.P.	941	°K
ΔH_t	674	CAL./GFW.
T.P.	1,047	°K
ΔH_t	1,083	CAL./GFW.
$T_c =$		°K
$P_c =$		ATM.

T TEMPERATURE °K	$C°_P$ HEAT CAPACITY CAL./DEG./GFW.	$H°_T - H°_{298.15}$ HEAT CONTENT CAL./GFW.	$S°_T$ ENTROPY CAL./DEG./GFW.	$-(F°_T - H°_{298.15})$ FREE ENERGY FUNCTION CAL./DEG./GFW.	FORMATION FROM REFERENCE STATE		
					HEAT $\Delta H°_f$ CAL./GFW.	FREE ENERGY $\Delta F°_f$ CAL./GFW.	$LOG_{10} K_P$
298	6.64	0	12.03	12.03			
300	6.65	12	12.07	12.03			
400	7.08	700	14.04	12.29			
500	7.61	1430	15.67	12.81			
600	8.25	2220	17.12	13.42			
700	8.97	3080	18.44	14.04			
800	9.88	4020	19.70	14.68			
900	11.12	5070	20.93	15.30			
1000	10.15	6810	22.77	15.96			
1100	9.15	8860	24.72	16.67			
1200	9.15	9770	25.52	17.38			
1300	9.15	10690	26.25	18.03			
1400	9.15	11600	26.93	18.65			
1500	9.15	16215	30.17	19.36			
1600	9.15	17130	30.78	20.08			
1700	9.15	18045	31.33	20.72			
1800	9.15	18960	31.86	21.33			
1900	9.15	19875	32.35	21.89			
2000	9.15	20790	32.82	22.43			
2100	9.15	21705	33.27	22.94			
2200	9.15	22620	33.69	23.41			
2300	9.15	23535	34.10	23.87			
2400	9.15	24450	34.49	24.31			
2500	9.15	25365	34.86	24.72			
2600	9.15	26280	35.22	25.12			
2700	9.15	27195	35.57	25.50			
2800	9.15	28110	35.90	25.87			
2900	9.15	29025	36.22	26.22			
3000	9.15	29940	36.53	26.55			

THERMODYNAMIC PROPERTIES OF THE ELEMENTS

URANIUM

URANIUM   U

IDEAL MONATOMIC GAS

Reference State for Calculating ΔH_f°, ΔF_f°, and $\log_{10} Kp$:
Solid I from 298° to 941°, Solid II from 941° to 1047°,
Solid III from 1047° to 1406°, Liquid from 1406° to 3000°.

Gfw 238.07   GRAMS

$(H^\circ_{298.15} - H^\circ_0) = 1,553$ CAL./GFW.

TEMPERATURE °K	C_p° HEAT CAPACITY CAL./DEG./GFW.	$H^\circ_T - H^\circ_{298.15}$ HEAT CONTENT CAL./GFW.	S°_T ENTROPY CAL./DEG./GFW.	$-(F^\circ_T - H^\circ_{298.15})/T$ FREE ENERGY FUNCTION CAL./DEG./GFW.	FORMATION FROM REFERENCE STATE		
					HEAT ΔH_f° CAL./GFW.	FREE ENERGY ΔF_f° CAL./GFW.	$\log_{10} K_p$
298	5.66	0	47.73	47.73	117168	106515	-78.080
300	5.67	10	47.76	47.73	117158	106451	-77.555
400	5.72	582	49.40	47.95	117042	102898	-56.225
500	5.66	1151	50.67	48.37	116881	99381	-43.442
600	5.59	1714	51.70	48.85	116654	95906	-34.936
700	5.54	2270	52.56	49.32	116350	92466	-28.871
800	5.53	2823	53.30	49.78	115963	89083	-24.338
900	5.58	3378	53.95	50.20	115468	85750	-20.824
1000	5.65	3939	54.54	50.61	114289	82519	-18.036
1100	5.75	4509	55.08	50.99	112809	79413	-15.779
1200	5.87	5090	55.59	51.35	112480	76396	-13.914
1300	5.99	5683	56.06	51.69	112153	73400	-12.340
1400	6.11	6288	56.51	52.02	111848	70436	-10.996
1500	6.22	6905	56.94	52.34	107850	67695	-9.863
1600	6.33	7533	57.34	52.64	107563	65067	-8.887
1700	6.42	8170	57.73	52.93	107285	62405	-8.022
1800	6.49	8816	58.10	53.21	107016	59784	-7.258
1900	6.56	9469	58.45	53.47	106754	57164	-6.575
2000	6.61	10127	58.79	53.73	106497	54557	-5.961
2100	6.64	10790	59.11	53.98	106245	51981	-5.409
2200	6.67	11455	59.42	54.22	105995	49389	-4.906
2300	6.68	12123	59.72	54.45	105748	46822	-4.449
2400	6.69	12792	60.00	54.67	105502	44278	-4.031
2500	6.69	13461	60.28	54.90	105256	41706	-3.645
2600	6.69	14130	60.53	55.10	105010	39204	-3.295
2700	6.68	14798	60.79	55.31	104763	36669	-2.967
2800	6.66	15465	61.03	55.51	104515	34151	-2.665
2900	6.65	16130	61.27	55.71	104265	31620	-2.382
3000	6.63	16794	61.49	55.90	104014	29134	-2.122

M.P.   °K
ΔH_m CAL./GFW.

B.P.   °K
ΔH_v CAL./GFW.

S.P.   °K
ΔH_s CAL./GFW.

T.P.   °K
ΔH_t CAL./GFW.

T.P.   °K
ΔH_t CAL./GFW.

$T_c =$ °K
$P_c =$ ATM.

VANADIUM

VANADIUM V Solid from 298° to 2190°, Liquid from 2190° to 3000°.

REFERENCE STATE

Gfw	50.95	GRAMS
$(H°_{298.15} - H°_0)$ =	1,122	CAL./GFW.
M.P.	2,190	°K
ΔH_m	(4,200)	CAL./GFW.
B.P.	3,650	°K
ΔH_v	109,600	CAL./GFW.
S.P.		°K
ΔH_s		CAL./GFW.
T.P.		°K
ΔH_t		CAL./GFW.
T.P.		°K
ΔH_t		CAL./GFW.
T_c =		°K
P_c =		ATM.

T TEMPERATURE °K	$C°_P$ HEAT CAPACITY CAL./DEG./GFW.	$H°_T - H°_{298.15}$ HEAT CONTENT CAL./GFW.	$S°_T$ ENTROPY CAL./DEG./GFW.	$-(F°_T - H°_{298.15})/T$ FREE ENERGY FUNCTION CAL./DEG./GFW.	FORMATION FROM REFERENCE STATE		
					HEAT $\Delta H°_f$ CAL./GFW.	FREE ENERGY $\Delta F°_f$ CAL./GFW.	$LOG_{10} K_P$
298	5.91	0	7.01	7.01			
300	5.92	11	7.05	7.02			
400	6.23	620	8.78	7.23			
500	6.43	1254	10.19	7.69			
600	6.58	1910	11.38	8.20			
700	6.70	2570	12.41	8.74			
800	6.85	3250	13.31	9.25			
900	7.05	3940	14.13	9.76			
1000	7.30	4660	14.89	10.23			
1100	7.55	5400	15.59	10.69			
1200	7.83	6170	16.26	11.12			
1300	8.12	6970	16.90	11.54			
1400	8.43	7800	17.52	11.95			
1500	8.74	8650	18.11	12.35			
1600	9.05	9550	18.68	12.72			
1700	9.35	10470	19.24	13.09			
1800	9.67	11430	19.79	13.44			
1900	9.97	12410	20.32	13.79			
2000	10.25	13410	20.84	14.14			
2100	10.50	14450	21.34	14.46			
2200	9.50	19700	23.75	14.80			
2400	9.50	20650	24.17	15.20			
2400	9.50	21600	24.57	15.57			
2500	9.50	22550	24.96	15.94			
2600	9.50	23500	25.33	16.30			
2700	9.50	24450	25.69	16.64			
2800	9.50	25400	26.04	16.97			
2900	9.50	26350	26.37	17.29			
3000	9.50	27300	26.69	17.59			

THERMODYNAMIC PROPERTIES OF THE ELEMENTS

VANADIUM

VANADIUM	V		Reference State for Calculating ΔH_f°, ΔF_f°, and $\log_{10} K_p$:
Gfw	50.95	GRAMS	Solid from 298° to 2190°, Liquid from 2190° to 3000°.
IDEAL MONATOMIC GAS			
$(H^\circ_{298.15} - H^\circ_0) =$	1,890	CAL./GFW.	

						FORMATION FROM REFERENCE STATE		
T TEMPERATURE °K	C_p° HEAT CAPACITY CAL./DEG./GFW.	$H_T^\circ - H_{298.15}^\circ$ HEAT CONTENT CAL./GFW.	S_T° ENTROPY CAL./DEG./GFW.	$-(F_T^\circ - H_{298.15}^\circ)/T$ FREE ENERGY FUNCTION CAL./DEG./GFW.		HEAT ΔH_f° CAL./GFW.	FREE ENERGY ΔF_f° CAL./GFW.	$\log_{10} K_p$
298	6.22	0	43.55	43.55		122750	111856	-81.996
300	6.21	12	43.58	43.54		122751	111792	-81.447
400	5.89	615	45.32	43.79		122745	108129	-59.083
500	5.78	1197	46.62	44.23		122693	104478	-45.670
600	5.80	1775	47.68	44.73		122615	100835	-36.732
700	5.88	2359	48.58	45.21		122539	97220	-30.355
800	5.95	2951	49.36	45.68		122451	93611	-25.575
900	6.00	3548	50.07	46.13		122358	90012	-21.859
1000	6.03	4150	50.70	46.55		122240	86430	-18.891
1100	6.04	4754	51.28	46.96		122104	82845	-16.461
1200	6.03	5357	51.80	47.34		121937	79289	-14.441
1300	6.00	5959	52.28	47.70		121739	75745	-12.735
1400	5.97	6558	52.73	48.05		121508	72214	-11.274
1500	5.94	7154	53.14	48.38		121254	68709	-10.011
1600	5.91	7746	53.52	48.68		120946	65202	-8.905
1700	5.89	8336	53.88	48.98		120616	61728	-7.935
1800	5.87	8924	54.22	49.27		120244	58270	-7.074
1900	5.85	9510	54.53	49.53		119850	54851	-6.308
2000	5.85	10095	54.83	49.79		119435	51455	-5.622
2100	5.85	10679	55.12	50.04		118979	48041	-4.999
2200	5.86	11264	55.39	50.27		114314	44706	-4.441
2300	5.88	11851	55.65	50.50		113951	41547	-3.947
2400	5.90	12440	55.90	50.72		113590	38398	-3.496
2500	5.94	13032	56.14	50.93		113232	35282	-3.084
2600	5.98	13628a	56.38	51.14		112878	32148	-2.702
2700	6.04	14229	56.60	51.33		112529	29072	-2.353
2800	6.10	14836	56.82	51.53		112186	26002	-2.029
2900	6.17	15449	57.04	51.72		111849	22906	-1.726
3000	6.24	16070	57.25	51.90		111520	19840	-1.445

M.P.		°K
ΔH_m		CAL./GFW.
B.P.		°K
ΔH_v		CAL./GFW.
S.P.		°K
ΔH_s		CAL./GFW.
T.P.		°K
ΔH_t		CAL./GFW.
T.P.		°K
ΔH_t		CAL./GFW.
$T_c =$		°K
$P_c =$		ATM.

XENON

XENON Xe

REFERENCE STATE

Gfw	131.30	GRAMS
$(H°_{298.15} - H°_0) =$	1,481	CAL./GFW.
M.P.	161.3	°K
ΔH_m	549.	CAL./GFW.
B.P.	165.04	°K
ΔH_v	3,021.	CAL./GFW.
S.P.		°K
ΔH_s		CAL./GFW.
T.P.		°K
ΔH_f		CAL./GFW.
T.P.		°K
ΔH_f		CAL./GFW.
$T_c =$	256.57	°K
$P_c =$	58.0	ATM.

Ideal Monatomic Gas from 298° to 3000°.

T TEMPERATURE °K	$C°_p$ HEAT CAPACITY CAL./DEG./GFW.	$H°_T - H°_{298.15}$ HEAT CONTENT CAL./GFW.	$S°_T$ ENTROPY CAL./DEG./GFW.	$-(F°_T - H°_{298.15})$ FREE ENERGY FUNCTION CAL./DEG./GFW.	FORMATION FROM REFERENCE STATE		
					HEAT $\Delta H°_f$ CAL./GFW.	FREE ENERGY $\Delta F°_f$ CAL./GFW.	$LOG_{10} K_p$
298	4.97	0	40.53	40.53			
300	4.97	9	40.56	40.53			
400	4.97	506	41.99	40.73			
500	4.97	1003	43.10	41.10			
600	4.97	1500	44.00	41.50			
700	4.97	1996	44.77	41.92			
800	4.97	2493	45.43	42.32			
900	4.97	2990	46.02	42.70			
1000	4.97	3487	46.54	43.06			
1100	4.97	3984	47.07	43.40			
1200	4.97	4480	47.45	43.72			
1300	4.97	4977	47.85	44.03			
1400	4.97	5474	48.21	44.30			
1500	4.97	5971	48.56	44.58			
1600	4.97	6468	48.88	44.84			
1700	4.97	6964	49.18	45.09			
1800	4.97	7461	49.46	45.32			
1900	4.97	7958	49.73	45.55			
2000	4.97	8455	49.99	45.77			
2100	4.97	8952	50.23	45.97			
2200	4.97	9448	50.46	46.17			
2300	4.97	9945	50.68	46.36			
2400	4.97	10442	50.89	46.54			
2500	4.97	10939	51.09	46.72			
2600	4.97	11436	51.29	46.90			
2700	4.97	11932	51.48	47.07			
2800	4.97	12429	51.66	47.23			
2900	4.97	12926	51.83	47.38			
3000	4.97	13423	52.00	47.53			

YTTERBIUM

YTTERBIUM Yb

REFERENCE STATE: Solid I from 298° to 1071°, Solid II from 1071° to 1097°, Liquid from 1097° to 1800°, Ideal Monatomic Gas from 1800° to 3000°.

Gfw	173.04	GRAMS
$(H°_{298.15} - H°_0) =$		CAL./GFW.
M.P.	1,097	°K
ΔH_m	(2,200)	CAL./GFW.
B.P.	1,800	°K
ΔH_v	37,100	CAL./GFW.
S.P.		°K
ΔH_s		CAL./GFW.
T.P.	1,071	°K
ΔH_t	(300)	CAL./GFW.
T.P.		°K
ΔH_t		CAL./GFW.
$T_c =$		°K
$P_c =$		ATM.

T TEMPERATURE °K	$C°_P$ HEAT CAPACITY CAL./DEG./GFW.	$H°_T - H°_{298.15}$ HEAT CONTENT CAL./GFW.	$S°_T$ ENTROPY CAL./DEG./GFW.	$-(F°_T - H°_{298.15})/T$ FREE ENERGY FUNCTION CAL./DEG./GFW.	FORMATION FROM REFERENCE STATE		
					HEAT $\Delta H°_f$ CAL./GFW.	FREE ENERGY $\Delta F°_f$ CAL./GFW.	LOG$_{10}$ K$_P$
298	6.00	0	15.04	15.00			
300	6.00	11	15.04	15.01			
400	6.20	620	16.79	15.24			
500	6.40	1250	18.19	15.69			
600	6.60	1900	19.38	16.22			
700	6.80	2570	20.41	16.74			
800	7.00	3260	21.33	17.26			
900	7.20	3970	22.17	17.76			
1000	7.40	4700	22.94	18.24			
1100	7.50	7950	25.94	18.72			
1200	7.50	8700	26.59	19.34			
1300	7.50	9450	27.19	19.93			
1400	7.50	10200	27.75	20.47			
1500	7.50	10950	28.27	20.97			
1600	7.50	11700	28.75	21.44			
1700	7.50	12450	29.20	21.88			
1800	7.50	13200	29.64	22.31			
1900	4.97	50860	50.55	23.79			
2000	4.97	51360	50.81	25.13			
2100	4.98	51850	51.05	26.36			
2200	4.98	52350	51.28	27.49			
2300	4.99	52850	51.50	28.53			
2400	5.00	53350	51.72	29.50			
2500	5.01	53850	51.92	30.38			
2600	5.03	54350	52.12	31.22			
2700	5.05	54860	52.31	32.00			
2800	5.08	55360	52.49	32.72			
2900	5.12	55870	52.67	33.41			
3000	5.16	56390	52.85	34.06			

YTTERBIUM

YTTERBIUM Yb
Gfw 173.04 GRAMS
$(H°_{298.15} - H°_0) = 1,481$ CAL./GFW.

Reference State for Calculating $\Delta H°_f$, $\Delta F°_f$, and $Log_{10} K_p$:
Solid I from 298° to 1071°, Solid II from 1071° to 1097°, Liquid from 1097° to 1800°, Ideal Monatomic Gas from 1800° to 3000°.

T TEMPERATURE °K	$C°_p$ HEAT CAPACITY CAL./DEG./GFW.	$H°_T - H°_{298.15}$ HEAT CONTENT CAL./GFW.	$S°_T$ ENTROPY CAL./DEG./GFW.	$-(F°_T - H°_{298.15})/T$ FREE ENERGY FUNCTION CAL./DEG./GFW.	FORMATION FROM REFERENCE STATE		
					HEAT $\Delta H°_f$ CAL./GFW.	FREE ENERGY $\Delta F°_f$ CAL./GFW.	$LOG_{10} K_p$
298	4.97	0	41.35	41.35	42900	35044	-25.689
300	4.97	9	41.38	41.35	42898	34996	-25.496
400	4.97	506	42.81	41.55	42786	32378	-17.691
500	4.97	1003	43.92	41.92	42653	29788	-13.021
600	4.97	1500	44.83	42.33	42500	27230	-9.919
700	4.97	1996	45.59	42.74	42326	24700	-7.712
800	4.97	2493	46.26	43.15	42133	22189	-6.062
900	4.97	2990	46.84	43.52	41920	19717	-4.788
1000	4.97	3487	47.36	43.88	41687	17267	-3.774
1100	4.97	3984	47.84	44.22	38934	14844	-2.949
1200	4.97	4480	48.27	44.54	38680	12664	-2.306
1300	4.97	4977	48.67	44.85	38427	10503	-1.765
1400	4.97	5474	49.04	45.13	38174	8368	-1.306
1500	4.97	5971	49.38	45.40	37921	6256	-.911
1600	4.97	6468	49.70	45.66	37668	4148	-.566
1700	4.97	6965	50.00	45.91	37415	2055	-.264
1800	4.97	7461	50.28	46.14	37161	9	-.001
1900	4.97	7958	50.55	46.37	0	0	0
2000	4.97	8456	50.81	46.59	0	0	0
2100	4.98	8953	51.05	46.79	0	0	0
2200	4.98	9451	51.28	46.99	0	0	0
2300	4.99	9950	51.50	47.18	0	0	0
2400	5.00	10449	51.72	47.37	0	0	0
2500	5.01	10950	51.92	47.54	0	0	0
2600	5.03	11452	52.12	47.72	0	0	0
2700	5.05	11956	52.31	47.89	0	0	0
2800	5.08	12463	52.49	48.04	0	0	0
2900	5.12	12973	52.67	48.20	0	0	0
3000	5.16	13486	52.85	48.36	0	0	0

M.P. °K
ΔH_m CAL./GFW.

B.P. °K
ΔH_v CAL./GFW.

S.P. °K
ΔH_s CAL./GFW.

T.P. °K
ΔH_t CAL./GFW.

T.P. °K
ΔH_t CAL./GFW.

$T_c =$ °K
$P_c =$ ATM.

THERMODYNAMIC PROPERTIES OF THE ELEMENTS

YTTRIUM

YTTRIUM Y
REFERENCE STATE

Gfw 88.92 GRAMS
$(H^°_{298.15} - H^°_0) =$ CAL./GFW.

Solid from 298° to 1773°, Liquid from 1773° to 3000°.

T TEMPERATURE °K	$C^°_P$ HEAT CAPACITY CAL./DEG./GFW.	$H^°_T - H^°_{298.15}$ HEAT CONTENT CAL./GFW.	$S^°_T$ ENTROPY CAL./DEG./GFW.	$-(F^°_T - H^°_{298.15})$ FREE ENERGY FUNCTION CAL./DEG./GFW.	FORMATION FROM REFERENCE STATE		
					HEAT $\Delta H^°_f$ CAL./GFW.	FREE ENERGY $\Delta F^°_f$ CAL./GFW.	LOG$_{10}$ K$_P$
298	6.01	0	11.00	11.00			
300	6.01	11	11.04	11.01			
400	6.11	617	12.78	11.24			
500	6.21	1233	14.15	11.69			
600	6.31	1859	15.29	12.20			
700	6.41	2495	16.27	12.71			
800	6.52	3141	17.14	13.22			
900	6.62	3798	17.91	13.69			
1000	6.72	4465	18.61	14.15			
1100	6.82	5142	19.26	14.59			
1200	6.92	5829	19.86	15.01			
1300	7.03	6527	20.41	15.39			
1400	7.13	7235	20.94	15.78			
1500	7.23	7953	21.43	16.13			
1600	7.33	8681	21.90	16.48			
1700	7.43	9419	22.35	16.81			
1800	8.00	14280	25.10	17.17			
1900	8.00	15080	25.53	17.60			
2000	8.00	15880	25.94	18.00			
2100	8.00	16680	26.33	18.39			
2200	8.00	17480	26.70	18.76			
2300	8.00	18280	27.06	19.12			
2400	8.00	19080	27.40	19.45			
2500	8.00	19880	27.72	19.77			
2600	8.00	20680	28.04	20.09			
2700	8.00	21480	28.34	20.39			
2800	8.00	22280	28.63	20.68			
2900	8.00	23080	28.91	20.96			
3000	8.00	23880	29.18	21.22			

M.P. (1,773) °K
ΔH_m (4,100) CAL./GFW.

B.P. (3,500) °K
ΔH_v (94,000) CAL./GFW.

S.P. °K
ΔH_s CAL./GFW.

T.P. °K
ΔH_t CAL./GFW.

T.P. °K
ΔH_t CAL./GFW.

$T_c =$ °K
$P_c =$ ATM.

YTTRIUM

YTTRIUM   Y
IDEAL MONATOMIC GAS

Reference State for Calculating ΔH_f°, ΔF_f°, and $\mathrm{Log}_{10} K_p$:
Solid from 298° to 1773°, Liquid from 1773° to 3000°.

G_{fw} 88.92   GRAMS
$(H^\circ_{298.15} - H^\circ_0) = 1{,}639$ CAL./GFW.

T TEMPERATURE °K	C_p° HEAT CAPACITY CAL./DEG./GFW.	$H^\circ_T - H^\circ_{298.15}$ HEAT CONTENT CAL./GFW.	S_T° ENTROPY CAL./DEG./GFW.	$-(F^\circ_T - H^\circ_{298.15})$ FREE ENERGY FUNCTION CAL./DEG./GFW.	FORMATION FROM REFERENCE STATE		
					HEAT ΔH_f° CAL./GFW.	FREE ENERGY ΔF_f° CAL./GFW.	$\mathrm{LOG}_{10} K_p$
298	6.18	0	42.87	42.87	102000	92497	-67.804
300	6.18	11	42.91	42.88	102000	92439	-67.347
400	6.04	625	44.67	43.11	102008	89252	-48.769
500	5.83	1218	46.00	43.57	101985	86060	-37.619
600	5.64	1791	47.04	44.06	101932	82882	-30.192
700	5.49	2347	47.90	44.55	101852	79711	-24.888
800	5.39	2891	48.63	45.02	101750	76558	-20.916
900	5.31	3425	49.26	45.46	101627	73412	-17.828
1000	5.25	3953	49.81	45.86	101488	70288	-15.362
1100	5.20	4476	50.31	46.25	101334	67179	-13.348
1200	5.17	4994	50.76	46.60	101165	64085	-11.672
1300	5.14	5510	51.17	46.94	100983	60995	-10.255
1400	5.13	6023	51.56	47.26	100788	57920	-9.042
1500	5.12	6535	51.91	47.56	100582	54862	-7.993
1600	5.12	7047	52.24	47.84	100366	51822	-7.078
1700	5.13	7559	52.55	48.11	100140	48800	-6.273
1800	5.15	8074	52.84	48.36	95794	45862	-5.568
1900	5.19	8590	53.12	48.60	95510	43089	-4.956
2000	5.24	9111	53.39	48.84	95231	40331	-4.406
2100	5.30	9638	53.65	49.07	94958	37586	-3.911
2200	5.38	10172	53.90	49.28	94692	34852	-3.462
2300	5.47	10714	54.14	49.49	94434	32150	-3.054
2400	5.58	11267	54.37	49.68	94187	29459	-2.682
2500	5.71	11831	54.60	49.87	93951	26751	-2.338
2600	5.86	12410	54.83	50.06	93730	24076	-2.023
2700	6.02	13004	55.05	50.24	93524	21407	-1.732
2800	6.20	13614	55.27	50.41	93334	18742	-1.462
2900	6.39	14243	55.50	50.59	93163	16052	-1.209
3000	6.59	14892	55.72	50.76	93012	13392	-.975

M.P. °K ΔH_m CAL./GFW.

B.P. °K ΔH_v CAL./GFW.

S.P. °K ΔH_s CAL./GFW.

T.P. °K ΔH_t CAL./GFW.

T.P. °K ΔH_t CAL./GFW.

T_c =   °K

P_c =   ATM.

THERMODYNAMIC PROPERTIES OF THE ELEMENTS

ZINC

ZINC   **Zn**   Solid from 298° to 692.7°, Liquid from 692.7° to 1181°, Ideal Monatomic Gas from 1181° to 3000°.

REFERENCE STATE

Gfw	65.38 GRAMS
$(H°_{298.15} - H°_0)$ =	1,349 CAL./GFW.
M.P.	692.7 °K
ΔH_m	1,765 CAL./GFW.
B.P.	1,181 °K
ΔH_v	27,560 CAL./GFW.
S.P.	°K
ΔH_s	CAL./GFW.
T.P.	°K
ΔH_t	CAL./GFW.
T.P.	°K
ΔH_t	CAL./GFW.
T_c =	°K
P_c =	ATM.

T TEMPERATURE °K	$C°_p$ HEAT CAPACITY CAL./DEG./GFW.	$H°_T - H°_{298.15}$ HEAT CONTENT CAL./GFW.	$S°_T$ ENTROPY CAL./DEG./GFW.	$-(F°_T - H°_{298.15})$ FREE ENERGY FUNCTION CAL./DEG./GFW.	FORMATION FROM REFERENCE STATE		
					HEAT $\Delta H°_f$ CAL./GFW.	FREE ENERGY $\Delta F°_f$ CAL./GFW.	LOG$_{10}$ K$_p$
298	6.07	0	9.95	9.95			
300	6.07	11	9.99	9.96			
400	6.31	630	11.76	10.19			
500	6.55	1270	13.20	10.66			
600	6.79	1940	14.41	11.18			
700	7.50	4400	18.03	11.75			
800	7.50	5150	19.03	12.60			
900	7.50	5900	19.92	13.37			
1000	7.50	6650	20.71	14.06			
1100	7.50	7400	21.42	14.70			
1200	4.97	35660	45.37	15.66			
1300	4.97	36157	45.77	17.96			
1400	4.97	36654	46.14	19.96			
1500	4.97	37150	46.48	21.72			
1600	4.97	37647	46.80	23.28			
1700	4.97	38144	47.10	24.67			
1800	4.97	38641	47.38	25.92			
1900	4.97	39138	47.65	27.06			
2000	4.97	39634	47.91	28.10			
2100	4.97	40131	48.15	29.04			
2200	4.97	40628	48.38	29.92			
2300	4.97	41125	48.60	30.72			
2400	4.97	41622	48.81	31.47			
2500	4.97	42118	49.02	32.18			
2600	4.97	42615	49.21	32.82			
2700	4.97	43112	49.40	33.44			
2800	4.97	43609	49.58	34.01			
2900	4.97	44106	49.75	34.55			
3000	4.97	44602	49.92	35.06			

ZINC

ZINC   **Zn**   Reference State for Calculating ΔH_f°, ΔF_f°, and $\log_{10} K_p$:
Solid from 298° to 692.7°, Liquid from 692.7° to 1181°, Ideal Monatomic Gas from 1181° to 3000°.

IDEAL MONATOMIC GAS

Gfw 65.38 GRAMS
$(H^\circ_{298.15} - H^\circ_0) = 1,481$ CAL./GFW.

	T TEMPERATURE °K	C_p° HEAT CAPACITY CAL./DEG./GFW.	$H^\circ_T - H^\circ_{298.15}$ HEAT CONTENT CAL./GFW.	S°_T ENTROPY CAL./DEG./GFW.	$-(F^\circ_T - H^\circ_{298.15})/T$ FREE ENERGY FUNCTION CAL./DEG./GFW.	FORMATION FROM REFERENCE STATE		
						HEAT ΔH_f° CAL./GFW.	FREE ENERGY ΔF_f° CAL./GFW.	$\log_{10} K_p$
	298	4.97	0	38.45	38.45	31180	22682	−16.627
	300	4.97	9	38.48	38.45	31178	22631	−16.488
	400	4.97	506	39.91	38.65	31056	19796	−10.816
	500	4.97	1003	41.02	39.02	30913	17003	−7.432
	600	4.97	1500	41.93	39.43	30740	14228	−5.182
	700	4.97	1996	42.69	39.84	28776	11514	−3.595
B.P.	800	4.97	2493	43.35	40.24	28523	9067	−2.477
	900	4.97	2990	43.94	40.62	28270	6652	−1.615
ΔH_v	1000	4.97	3487	44.46	40.98	28017	4267	−.932
	1100	4.97	3984	44.94	41.32	27764	1892	−.375
S.P.	1200	4.97	4480	45.37	41.64	0	0	0
	1300	4.97	4977	45.77	41.95	0	0	0
	1400	4.97	5474	46.14	42.23	0	0	0
ΔH_s	1500	4.97	5971	46.48	42.50	0	0	0
	1600	4.97	6468	46.80	42.76	0	0	0
T.P.	1700	4.97	6964	47.10	43.01	0	0	0
	1800	4.97	7461	47.38	43.24	0	0	0
ΔH_t	1900	4.97	7958	47.65	43.47	0	0	0
	2000	4.97	8455	47.91	43.69	0	0	0
	2100	4.97	8952	48.15	43.89	0	0	0
	2200	4.97	9448	48.38	44.09	0	0	0
	2300	4.97	9945	48.60	44.28	0	0	0
T.P.	2400	4.97	10442	48.81	44.46	0	0	0
	2500	4.97	10939	49.02	44.65	0	0	0
ΔH_t	2600	4.97	11436	49.21	44.82	0	0	0
	2700	4.97	11932	49.40	44.99	0	0	0
	2800	4.97	12429	49.58	45.15	0	0	0
$T_c =$ °K	2900	4.97	12926	49.75	45.30	0	0	0
$P_c =$ ATM.	3000	4.97	13423	49.92	45.45	0	0	0

THERMODYNAMIC PROPERTIES OF THE ELEMENTS

ZIRCONIUM

ZIRCONIUM   Zr   Gfw 91.22 GRAMS

$(H°_{298.15} - H°_0) = 1,313$ CAL./GFW.

Solid I from 298° to 1143°, Solid II from 1143° to 2125°, Liquid from 2125° to 3000°.

M.P. 2,125 °K ΔH_m (4,000) CAL./GFW.

B.P. 4,650 °K ΔH_v 139,000 CAL./GFW.

S.P. 1,143 °K ΔH_s 1,040 CAL./GFW.

T.P.   °K ΔH_t CAL./GFW.

T.P.   °K ΔH_t CAL./GFW.

T_c =   °K P_c =   ATM.

T TEMPERATURE °K	$C°_P$ HEAT CAPACITY CAL./DEG./GFW.	$H°_T - H°_{298.15}$ HEAT CONTENT CAL./GFW.	$S°_T$ ENTROPY CAL./DEG./GFW.	$-(F°_T - H°_{298.15})$ FREE ENERGY FUNCTION CAL./DEG./GFW.	FORMATION FROM REFERENCE STATE		
					HEAT $\Delta H°_f$ CAL./GFW.	FREE ENERGY $\Delta F°_f$ CAL./GFW.	$LOG_{10} K_P$
298	6.01	0	9.29	9.29			
300	6.01	11	9.33	9.30			
400	6.36	630	11.11	9.54			
500	6.63	1280	12.56	10.00			
600	6.88	1960	13.79	10.53			
700	7.12	2660	14.87	11.07			
800	7.34	3380	15.84	11.62			
900	7.56	4125	16.71	12.13			
1000	7.79	4890	17.52	12.63			
1100	8.01	5680	18.28	13.12			
1200	6.79	7450	19.86	13.66			
1300	6.95	8140	20.41	14.15			
1400	7.11	8840	20.93	14.62			
1500	7.27	9560	21.42	15.05			
1600	7.43	10300	21.90	15.47			
1700	7.59	11050	22.35	15.85			
1800	7.75	11820	22.79	16.23			
1900	7.91	12600	23.21	16.58			
2000	8.07	13400	23.62	16.92			
2100	8.23	14220	24.02	17.25			
2200	8.00	19000	26.27	17.64			
2300	8.00	19800	26.62	18.02			
2400	8.00	20600	26.96	18.38			
2500	8.00	21400	27.29	18.73			
2600	8.00	22200	27.60	19.07			
2700	8.00	23000	27.91	19.40			
2800	8.00	23800	28.20	19.70			
2900	8.00	24600	28.48	20.00			
3000	8.00	25400	28.75	20.29			

ZIRCONIUM

ZIRCONIUM   Zr

IDEAL MONATOMIC GAS

G*w* 91.22 GRAMS

$(H°_{298.15} - H°_0) = 1,629$ CAL./GFW.

Reference State for Calculating $\Delta H°_f$, $\Delta F°_f$, and $\text{Log}_{10} K_p$: Solid I from 298° to 1143°, Solid II from 1143° to 2125°, Liquid from 2125° to 3000°.

T TEMPERATURE °K	$C°_p$ HEAT CAPACITY CAL./DEG./GFW.	$H°_T - H°_{298.15}$ HEAT CONTENT CAL./GFW.	$S°_T$ ENTROPY CAL./DEG./GFW.	$-(F°_T - H°_{298.15})$ FREE ENERGY FUNCTION CAL./DEG./GFW.	FORMATION FROM REFERENCE STATE			FREE ENERGY FROM REFERENCE STATE	
					HEAT $\Delta H°_f$ CAL./GFW.	FREE ENERGY $\Delta F°_f$ CAL./GFW.	$\text{LOG}_{10} K_p$	
298	6.37	0	43.32	43.32	146000	135853	-99.587	
300	6.38	12	43.36	43.32	146001	135792	-98.932	
400	6.61	664	45.23	43.57	146034	132386	-72.338	
500	6.59	1326	46.71	44.06	146046	128971	-56.377	
600	6.46	1979	47.90	44.61	146019	125553	-45.732	
700	6.32	2618	48.88	45.14	145958	122151	-38.140	
800	6.20	3243	49.72	45.67	145863	118759	-32.446	
900	6.13	3859	50.44	46.16	145734	115377	-28.019	
1000	6.12	4472	51.09	46.62	145582	112012	-24.482	
1100	6.16	5085	51.67	47.05	145405	108676	-21.593	
1200	6.23	5704	52.21	47.46	144254	105434	-19.203	
1300	6.32	6331	52.71	47.84	144191	102201	-17.183	
1400	6.43	6968	53.19	48.22	144128	98964	-15.450	
1500	6.54	7617	53.63	48.56	144057	95742	-13.950	
1600	6.66	8277	54.06	48.89	143977	92521	-12.637	
1700	6.76	8948	54.47	49.21	143898	89294	-11.479	
1800	6.87	9629	54.86	49.52	143809	86083	-10.451	
1900	6.96	10320	55.23	49.80	143720	82882	-9.533	
2000	7.05	11020	55.59	50.08	143620	79680	-8.706	
2100	7.13	11730	55.94	50.36	143510	76478	-7.959	
2200	7.20	12450	56.27	50.62	139450	73450	-7.296	
2300	7.27	13170	56.59	50.87	139370	70439	-6.693	
2400	7.34	13900	56.90	51.11	139300	67444	-6.141	
2500	7.41	14640	57.20	51.35	139240	64465	-5.635	
2600	7.48	15380	57.49	51.58	139180	61466	-5.166	
2700	7.55	16140	57.78	51.81	139140	58491	-4.734	
2800	7.61	16890	58.05	52.02	139090	55510	-4.332	
2900	7.68	17660	58.32	52.24	139060	52524	-3.958	
3000	7.75	18430	58.58	52.44	139030	49540	-3.608	

M.P. °K
ΔH_m CAL./GFW.

B.P. °K
ΔH_v CAL./GFW.

S.P. °K
ΔH_s CAL./GFW.

T.P. °K
ΔH_t CAL./GFW.

T.P. °K
ΔH_t CAL./GFW.

$T_c =$ °K
$P_c =$ ATM.

Bibliography

(1) Adams, G. B., Johnston, H. L., Kerr, E. C., *J. Am. Chem. Soc.* **74**, 4784 (1952).
(2) Adenstedt, H. K., *Trans. Am. Soc. Metals* **44**, 949 (1952).
(3) Adenstedt, H. K., Pequignot, J. R., Raymer, J. M., *Ibid.*, **44**, 990 (1952).
(4) Ahmann, D. H., *Iowa State College J. Sci.* **27**, 120 (1953).
(5) Albertson, W., *Phys. Rev.* **52**, 644 (1937).
(6) Altshuller, A. P., *J. Chem. Phys.* **22**, 1947 (1954).
(7) Anderson, C. T., *J. Am. Chem. Soc.* **52**, 2296 (1930).
(8) *Ibid.*, p. 2301.
(9) *Ibid.*, p. 2712.
(10) *Ibid.*, p. 2720.
(11) *Ibid.*, **58**, 564 (1936).
(12) *Ibid.*, **59**, 488 (1937).
(13) *Ibid.*, p. 1036.
(14) Anderson, J. S., *J. Chem. Soc.* **1943**, 141.
(15) Armstrong, G. T., *J. Research Natl. Bur. Standards* **53**, 263 (1954).
(16) Armstrong, L. D., Grayson-Smith, H., *Can. J. Research* **27A**, 9 (1949).
(17) *Ibid.*, **28A**, 51 (1950).
(18) Aston, J. G., Moessen, G. W., *J. Chem. Phys.* **21**, 948 (1953).
(19) Bacher, R. F., Goudsmit, S., "Atomic Energy States," McGraw-Hill, New York, 1932.
(20) Barrow, R. F., Dodsworth, P. G., Downie, A. R., Jeffries, E. A. N. S., Pugh, A. C. P., Smith, F. J., Swinstead, J. M., *Trans. Faraday Soc.* **51**, 1354 (1955).
(21) Barrow, R. F., Rowlinson, H. C., *Proc. Roy. Soc. (London)* **A224**, 374 (1954).
(22) Baughan, E. C., *Quart. Revs. (London)* **7**, 103 (1953).
(23) Baur, E., Brunner, R., *Helv. Chim. Acta* **17**, 958 (1934).
(24) Beale, A. F., private communication, Research Dept., Dowell, Inc., Tulsa, Okla.
(25) Beale, A. F., Univ. Microfilms, Publ. No. **5009**, 70 pp.; *Dissertation Abstr.* **13**, 295 (1953).
(26) Beamer, W. H., Maxwell, C. R., *J. Chem. Phys.* **17**, 1293 (1949).
(27) Blocher, J. M., Campbell, I. E., *J. Am. Chem. Soc.* **71**, 4040 (1949).
(28) Bloom, D. S., Putnam, J. W., Grant, N. J., *J. Metals* **4**, *Trans.* 626 (1952).
(29) Booth, G. L., Hoare, F. E., Murphy, B. T., *Proc. Phys. Soc. (London)* **68B**, 830–2 (1955).
(30) Borelius, G., Paulson, K. A., *Arkiv. Mat., Astron. Fysik* **A33**, No. 7, 16 pp. (1946).
(31) Braune, H., Moller, O., *Z. Naturforsch.* **9a**, 210 (1954).
(32) Braune, H., Peter, S., Neveling, U., *Ibid.*, **6a**, 32 (1951).
(33) Brewer, L., private communication, Dept. of Chemistry, University of California, Berkeley, Calif.
(34) Brewer, L., quoted in article by Fickett, W., Cowan, R. D., *J. Chem. Phys.* **23**, 1349 (1955).
(35) Brewer, L., Paper 3, National Nuclear Energy Series **IV-19B**, L. L. Quill, ed., McGraw-Hill, New York, 1950.
(36) Brewer, L., Bromley, L. A., Gilles, P. W., Lofgren, N. F., *Ibid.*, Paper 6, Appendix 4.
(37) Brewer, L., Kane, J. S., *J. Phys. Chem.* **59**, 105 (1955).
(38) Brewer, L., Porter, R. F., *J. Chem. Phys.* **21**, 2012 (1953).
(39) Brewer, L., Searcy, A. W., *J. Am. Chem. Soc.* **73**, 5308 (1951).
(40) Brix, P., *Z. Physik* **126**, 431 (1949).
(41) Brix, P., Herzberg, G., *Can. J. Phys.* **32**, 110 (1954).
(42) Bronson, H. L., MacHattie, L. F., *Can. J. Research* **16A**, 177 (1938).
(43) Brooks, L. S., *J. Am. Chem. Soc.* **74**, 227 (1952).
(44) *Ibid.*, **77**, 3211 (1955).
(45) Brown, A., Zemansky, M. W., Boorse, H. A., *Proc. NBS Semicentennial Symposium Low-Temp. Phys Natl. Bur. Standards* **519**, 99–101 (1952); *Phys. Rev.* **92**, 52 (1953).
(46) Burns, J. F., *J. Chem. Phys.* **23**, 1347 (1955).
(47) Busey, R. H., Giauque, W. F., *J. Am. Chem. Soc.* **74**, 3157 (1952).
(48) *Ibid.*, **75**, 806 (1953).

(49) Buzzard, R. W., Liss, R. B., Fickle, D. P., *J. Research Natl. Bur. Standards* **50,** 209 (1953).
(50) Cady, G. H., Hildebrand, J. H., *J. Am. Chem. Soc.* **52,** 3839 (1930).
(51) Carpenter, L. G., Mair, W. N., *Proc. Phys. Soc. (London)* **64B,** 57 (1951).
(52) Centnerszwer, M., *Z. physik. Chem.* **85,** 99 (1913).
(53) Chiotti, P., *J. Electrochem. Soc.* **101,** 567 (1954).
(54) Chiotti, P., U. S. Atomic Energy Comm. **A.E.C.D.-3072,** June 5, 1950.
(55) Chupka, W. A., Inghram, M. G., *J. Phys. Chem.* **59,** 100 (1955).
(56) Chupka, W. A., Inghram, M. G., *Mem. soc. roy. sci. Liège* **15,** 373 (1955).
(57) Clark, A. M., Din, F., Robb, J., Michels, A., Wassenaar, T., Zwietering, T., *Physica* **17,** 876 (1951).
(58) Clement, J. R., Quinnell, E. H., *Phys. Rev.* **92,** 258 (1953).
(59) Clusius, K., *Z. physik. Chem.* **4B,** 1 (1929).
(60) *Ibid.,* **31B,** 459 (1936).
(61) Clusius, K., Frank, A., *Z. Elektrochem.* **49,** 308 (1943).
(62) Clusius, K., Losa, C. G., *Z. Naturforsch.* **10A,** 939 (1955).
(63) Clusius, K., Konnertz, F., *Ibid.,* **4a,** 117 (1949).
(64) Clusius, K., Kruis, A., Konnertz, F., *Ann. Physik* **33,** 642 (1938).
(65) Clusius, K., Riccoboni, L., *Z. physik. Chem.* **B38,** 81 (1937).
(66) Clusius, K., Schachinger, L., *Z. angew. Phys.* **4,** 442 (1952).
(67) Clusius, K., Schachinger, L., *Z. Naturforsch.* **2a,** 90 (1947).
(68) *Ibid.,* **7a,** 185 (1952).
(69) Clusius, K., Stern, H., *Z. angew. Phys.* **6,** 194 (1954).
(70) Clusius, K., Vaughen, J. V., *J. Am. Chem. Soc.* **52,** 4686 (1930).
(71) Clusius, K., Wiegand, K., *Z. physik. Chem.* **B42,** 111 (1939).
(72) Coleman, F. F., Egerton, A., *Trans. Roy. Soc. (London)* **234A,** 177 (1935).
(73) Coughlin, J. P., *U. S. Bur. Mines, Bull.* **542,** 21 (1954).
(74) Coughlin, J. P., King, E. G., *J. Am. Chem. Soc.* **72,** 2262 (1950).
(75) Craig, R. S., Krier, C. A., Coffer, L. W., Bates, E. A., Wallace, W. E., *Ibid.,* **76,** 238 (1954); (rev. according to private communication from W. E. Wallace, July 8, 1955).
(76) Cristescu, S., Simon, F., *Z. physik. Chem.* **16B,** 143 (1932).
(77) Cueilleron, J., *Ann. chim.* **19,** 459 (1944).
(78) Daane, A. H., U. S. Atomic Energy Commission **A.E.C.D.-3209,** Dec. 14, 1950.
(79) Dahl, A. I., Cleaves, H. E., *J. Research Natl. Bur. Standards* **43,** 513 (1949).
(80) Dancy, T. E., *J. Iron Steel Inst. (London)* **167,** 160 (1951).
(81) Darken, L. S., Smith, R. P., *Ind. Eng. Chem.* **43,** 1815 (1951).
(82) Dauphinee, T. M., Martin, D. L., Preston-Thomas, H., *Proc. Roy. Soc. (London)* **233A,** 214 (1955).
(83) DeSorbo, W., *Acta Met.* **1,** 503 (1953).
(84) DeSorbo, W., *J. Chem. Phys.* **21,** 1144 (1953).
(85) DeSorbo, W., Tyler, W. W., *Ibid.,* **21,** 1660 (1953).
(86) Doolan, J. J., Partington, J. R., *Trans. Faraday Soc.* **20,** 342 (1929).
(87) Douglas, A. E., *J. Phys. Chem.* **59,** 109 (1955).
(88) Douglas, P. E., *Proc. Phys. Soc. (London)* **67,** 783 (1954).
(89) Douglas, T. B., Ball, A. F., Ginnings, D. C., *J. Research Natl. Bur. Standards* **46,** 334 (1951).
(90) Douglas, T. B., Dever, J. L., *J. Am. Chem. Soc.* **76,** 4824 (1954).
(91) Duwez, P., *J. Appl. Phys.* **22,** 1174 (1951).
(92) Duyckaerts, G., *Physica* **6,** 817 (1939).
(93) Eastman, E. D., Cubicciotti, D. D., Thurmond, C. D., Paper 2, National Nuclear Energy Series **IV-19B,** L. L. Quill, ed., McGraw-Hill, New York, 1950.
(94) Eastman, E. D., McGavock, W. C., *J. Am. Chem. Soc.* **59,** 145 (1937).
(95) Edwards, J. W., Johnston, H. L., Blackburn, P. E., *Ibid.,* **73,** 172 (1951).
(96) *Ibid.,* p. 4727.
(97) *Ibid.,* **74,** 1539 (1952).
(98) Edwards, J. W., Johnston, H. L., Ditmars, W. E., *Ibid.,* **73,** 4729 (1951).
(99) *Ibid.,* **75,** 2467 (1953).
(100) Egerton, A. C. G., *Proc. Roy. Soc. (London)* **A103,** 469 (1923).
(101) Eighteenth Conference of International Union of Pure and Applied Chemistry, *Chem. Eng. News* **33,** 5578 (1955).
(102) Elson, R. G., Smith, H. G., Wilhelm, J. O., *Can. J. Research* **18A,** 83 (1940).
(103) Epstein, M. B., Pitzer, K. S., Rossini, F. D., *J. Research Natl. Bur. Standards* **42,** 379 (1949).

(104) Estermann, I., Weertman, J. R., *J. Chem. Phys.* **20**, 972 (1952).
(105) Evans, W. H., Jacobson, R., Munson, T. R., Wagman, D. D., *J. Research Natl. Bur. Standards* **55**, 83 (1955).
(106) Evans, W. H., Munson, T. R., Wagman, D. D., *Ibid.*, **55**, 147 (1955).
(107) Evans, W. H., Wagman, D. D., *Ibid.*, **49**, 141 (1952).
(108) Farkas, L., *Z. Physik* **70**, 737 (1931).
(109) Farr, T. D., Tennessee Valley Authority, Chemical Engineering Report No. 8, U. S. Govt. Printing Office, Washington, D. C., 1950.
(110) Fast, J. D., *J. Appl. Phys.* **23**, 350 (1952).
(111) Fischer, J., *Z. anorg. Chem.* **219**, 1, 367 (1934).
(112) Fischer, J., Bingle, J., *J. Am. Chem. Soc.* **77**, 6511 (1955).
(113) Fishbeck, K., Eich, H., *Ber.* **71**, 520 (1938).
(114) Fiske, M. D., *Phys. Rev.* **61**, 513 (1942).
(115) Foster, K. W., U. S. Atomic Energy Commission **MLM-901**, Mound Laboratory, Miamisburg, Ohio, July 7, 1953.
(116) Fuchtbauer, C., Bartels, H., *Z. Physik* **4**, 337 (1921).
(117) Furukawa, G. T., McCoskey, R. E., *Natl. Advisory Comm. Aeronaut., Tech. Note* **2969**, 1953.
(118) Gaydon, A. G., "Dissociation Energies and Spectra of Diatomic Molecules," Chapman & Hall, London, 1953.
(119) Geballe, T. H., Giauque, W. F., *J. Am. Chem. Soc.* **74**, 2368 (1952).
(120) Giauque, W. F., *Ibid.*, **53**, 507 (1931).
(121) Giauque, W. F., Clayton, J., *Ibid.*, **55**, 4875 (1933).
(122) Giauque, W. F., Johnston, H. L., *Ibid.*, **51**, 2300 (1929).
(123) Giauque, W. F., Meads, P. F., *Ibid.*, **63**, 1897 (1941).
(124) Giauque, W. F., Powell, T. M., *Ibid.*, **61**, 1970 (1939).
(125) Gibson, G. E., dissertation, Breslau, 1911, reported in Landolt-Bornstein Tabellen, vol. II, p. 1338, Julius Springer, Berlin, 1923.
(126) Ginnings, D. C., Corruccini, R. J., *J. Research Natl. Bur. Standards* **39**, 309 (1947).
(127) Ginnings, D. C., Douglas, T. B., Ball, A. F., *J. Am. Chem. Soc.* **73**, 1236 (1951).
(128) Glockler, G., *J. Chem. Phys.* **22**, 159 (1954).
(129) Gordon, A. R., *Ibid.*, **5**, 350 (1937).
(130) Greenwood, H. C., *Proc. Roy. Soc. (London)* **A82**, 396 (1909).
(131) *Ibid.*, **A83**, 483 (1910).
(132) Greenwood, H. C., *Z. physik. Chem.* **76**, 484 (1911).
(133) Greiner, E. S., *J. Metals* **5**, 1044 (1952).
(134) Griffel, M., Skochdopole, R. E., *J. Am. Chem. Soc.* **75**, 5250 (1953).
(135) Griffel, M., Skochdopole, R. E., Spedding, F. H., *Phys. Rev.* **93**, 657 (1954).
(136) Griffel, M., Skochdopole, R. E., Spedding, F. H., private communication by R. E. Skochdopole.
(137) Grube, G., Knabe, R., *Z. Electrochem.* **42**, 793 (1936).
(138) Gulbransen, E. A., Andrew, K. F., *J. Electrochem. Soc.* **97**, 383 (1952).
(139) *Ibid.*, **99**, 402 (1952).
(140) Guthrie, Jr., G. B., Scott, D. W., Waddington, G., *J. Am. Chem. Soc.* **76**, 1488 (1954).
(141) Hackspill, M. L., *Ann. chim. et phys.* [Ser. 8] **28**, 613 (1913).
(142) Hall, L. D., *J. Am. Chem. Soc.* **73**, 757 (1951).
(143) Hansen, M., Kamen, E. L., Kessler, D. J., McPherson, D. J., *J. Metals* **3**, 881 (1951).
(144) Hansen, M., Kessler, H. D., McPherson, D. J., *Trans. Am. Soc. Metals* **44**, 518 (1952).
(145) Harteck, P., *Z. physik. Chem.* **134**, 1 (1928).
(146) Hartmann, H., Schneider, R., *Z. anorg. Chem.* **180**, 275 (1929).
(147) Hassion, F. X., Thurmond, G. D., Trumbore, F. A., *J. Phys. Chem.* **59**, 1076 (1955).
(148) Hendrie, J. M., *J. Chem Phys.* **22**, 1503 (1954).
(149) Henning, F., Otto, J., *Physik. Z.* **37**, 633 (1936).
(150) Henning, O., Wensel, H., *Ann. Physik* **17**, 620 (1933).
(151) Hersh, H. N., *J. Am. Chem. Soc.* **75**, 1529 (1953).
(152) Herzberg, G., "Molecular Spectra and Molecular Structure. I. Spectra of Diatomic Molecules," 2nd Ed., Van Nostrand, New York, 1950.
(153) Herzberg, G., "Molecular Spectra and Molecular Structure. II. Infrared and Raman Spectra of Polyatomic Molecules," Van Nostrand, New York, 1945.
(154) Hicks, Jr., J. F. G., *J. Am. Chem. Soc.* **60**, 1000 (1938).
(155) Hill, R. W., Parkinson, D. H., *Phil. Mag.* **43**, 309 (1952).
(156) Hill, R. W., Smith, P. L., *Ibid.*, **44**, 636 (1953).
(157) Hoch, M., private communication, Ohio State University, Columbus, Ohio.

(158) Hoch, M., Blackburn, P. E., Dingledy, D. P., Johnston, H. L., *J. Phys. Chem.* **59,** 97 (1955).
(159) Hoge, H. J., *J. Research Natl. Bur. Standards* **44,** 321–45 (1950).
(160) Holden, R. B., Speiser, R., Johnston, H. L., *J. Am. Chem. Soc.* **70,** 3897 (1948).
(161) Honig, R. E., *J. Chem. Phys.* **22,** 1610 (1954).
(162) Hu, J. H., White, D., Johnston, H. L., *J. Am. Chem. Soc.* **75,** 5642 (1953).
(163) Huff, V. N., Gordon, S., Morrell, V. E., *Natl. Advisory Comm. Aeronaut.*, Rept. No. **1037,** 1951.
(164) Ingold, C. K., *J. Chem. Soc.* **121,** 2419 (1922).
(165) Jaeger, F. M., Rosenbohm, E., *Rec. trav. chim.* **51,** 1 (1932).
(166) Jaeger, F. M., Rosenbohm, E., Fonteyne, R., *Ibid.*, **55,** 615 (1936).
(167) Jaeger, F. M., Veenstra, W. A., *Ibid.*, **53,** 677 (1934).
(168) Johnston, H. L., Hersh, H. N., Kerr, E. C., *J. Am. Chem. Soc.* **73,** 1112 (1951).
(169) Johnston, H. L., Marshall, A. L., *Ibid.*, **62,** 1382 (1940).
(170) Jones, H. A., Langmuir, I., Mackay, G. M. J., *Phys. Rev.* **30,** 201 (1927).
(171) Jones, W. M., Gordon, J., Long, E. A., *J. Chem. Phys.* **20,** 695 (1952).
(172) Jovanovic, S. L., *Bull. soc. chim. Belgrade* **12,** 51 (1947).
(173) Katz, T. J., Margrave, J. L., *J. Chem. Phys.* **23,** 983 (1955).
(174) Keesom, W. H., "Helium," Elsevier, Amsterdam, 1942.
(175) Keesom, W. H., Desirant, M., *Physica* **8,** 273 (1941).
(176) Keesom, W. H., Ende, J. N. van den, *Proc. Acad. Sci. Amsterdam* **33,** 243 (1930); *Ibid.*, **34,** 210 (1931).
(177) Kelley, K. K., *J. Am. Chem. Soc.* **61,** 203 (1939).
(178) Kelley, K. K., *J. Chem. Phys.* **8,** 316 (1940).
(179) Kelley, K. K., *U. S. Bur. Mines, Bull.* **371,** (1934).
(180) *Ibid.*, **383** (1935).
(181) *Ibid.*, **384** (1935).
(182) *Ibid.*, **393** (1936).
(183) *Ibid.*, **406** (1937).
(184) *Ibid.*, **407** (1937).
(185) *Ibid.*, **476** (1949).
(186) *Ibid.*, **477** (1950).
(187) Kelley, K. K., private communication, Pacific Experimental Station, Bureau of Mines, University of California, Berkeley, Calif.
(188) Killian, T. J., *Phys. Rev.* **27,** 578 (1926).
(189) Kilpatrick, J. E., Prosen, E. J., Pitzer, K. S., Rossini, F. D., *J. Research Natl. Bur. Standards* **36,** 559 (1946).
(190) Klinkenberg, P. F. A., *Physica* **12,** 33 (1946).
(191) *Ibid.*, **14,** 269–84 (1948).
(192) *Ibid.*, **21,** 53 (1955).
(193) Kobe, K. A., Lynn, Jr., R. E., *Chem. Revs.* **52,** 117 (1953).
(194) Kohlmeyer, E. J., Spandau, H., *Z. anorg. Chem.* **253,** 37–40 (1945).
(195) Kok, J. A., Keesom, W. H., *Physica* **3,** 1035–45 (1936).
(196) Kolsky, H. G., Gilles, P. W., *J. Chem. Phys.* **22,** 232 (1954).
(197) *Ibid.*, **24,** 828 (1956).
(198) Korber, F., Oelsen, W., *Mitt. Kaiser-Wilhelm-Inst. Eisenforsch. Dusseldorf* **18,** 109 (1936).
(199) Kornev, I. V., Golubkin, V. N., *Doklady Akad. Nauk S.S.S.R.* **99,** 565 (1954).
(200) Kothen, C. W., dissertation, Ohio State University, 1952.
(201) Kothen, C. W., Johnston, H. L., *J. Am. Chem. Soc.* **75,** 3101 (1953).
(202) Kotov, E. I., *Vestnik Akad. Nauk Kazakh. S.S.R.* **6,** No. 1 (46), 37–51 (1949).
(203) Krauss, F., Warncke, H., *Z. Metallkunde* **46,** 61–9 (1955).
(204) Kroner, A., *Ann. Physik* [Ser. 4] **40,** 438 (1913).
(205) Kubaschewski, O., *Z. Metallkunde* **41,** 445 (1950).
(206) Kubaschewski, O., Brizgys, P., Huchler, O., Jauch, R., Reinartz, K., *Z. Elektrochem.* **54,** 275 (1950).
(207) Kubaschewski, O., Evans, E. L., "Metallurgical Thermochemistry," Academic Press, New York, 1951.
(208) Landolt-Bornstein Tabellen, Sechste Auflage, Band **I,** Teil I, Springer-Verlag, Berlin, 1950.
(209) Langmuir, D. B., Malter, L., *Phys. Rev.* **55,** 743 (1949).
(210) Latimer, W. M., "The Oxidation States of the Elements and Their Potentials in Aqueous Solutions," Prentice-Hall, New York, 1938.

(211) Leitgebel, W. von, *Z. anorg. Chem.* **202**, 305 (1931).
(212) Lewis, G. N., Gibson, G. E., *J. Am. Chem. Soc.* **39**, 2554 (1917).
(213) Litton, F. B., *J. Electrochem. Soc.* **98**, 488 (1951).
(214) Luft, N. W., *Monatsh.* **86**, 474 (1955).
(215) Lyashenko, V. S., *Izvest. Akad. Nauk S.S.S.R., Otdel. Khim. Nauk* **1951**, 242.
(216) Lyman, T., "Metals Handbook," American Society for Metals, Cleveland, Ohio, 1948.
(217) Lyubimov, A. P., Granovskaya, A. A., *Zhur. Fiz. Khim.* **27**, 473 (1953).
(218) MacRae, D., Van Voorhis, C. C., *J. Am. Chem. Soc.* **43**, 547 (1921).
(219) Magnus, A., *Ann. Physik* **70**, 303 (1923).
(220) Magnus, A., Holzman, H., *Ann. Physik* [Ser. 5] **3**, 585 (1929).
(221) Makansi, M. M., Madsen, W., Selke, W. A., Bonilla, C. F., *J. Phys. Chem.* **60**, 128 (1956).
(222) Margrave, J. L., *J. Chem. Educ.* **32**, 520 (1955).
(223) Marshall, A. L., Dornte, R. W., Norton, F. J., *J. Am. Chem. Soc.* **59**, 116 (1937).
(224) Maxwell, C. R., *J. Chem. Phys.* **17**, 1288 (1949).
(225) Mayer, J., Mayer, M., "Statistical Thermodynamics," Wiley, New York, 1940.
(226a) Maykuth, D. J., Ogden, H. R., Jaffe, R. I., *J. Metals* **5**, 225 (1953).
(226b) McCabe, C. L., Birchenall, C. E., *J. Metals* **5**, *AIME Trans.* **197**, 707 (1953).
(227) McDonald, R. A., unpublished measurements at The Dow Chemical Co., June 27, 1955.
(228) McDonald, R. A., Stull, D. R., *J. Am. Chem. Soc.* **77**, 5293 (1955).
(229) McQuillan, A. D., *Proc. Roy. Soc. (London)* **A204**, 309 (1950).
(230) Meads, P. F., Forsythe, W. R., Giauque, W. F., *J. Am. Chem. Soc.* **63**, 1902 (1941).
(231) Meggers, W. F., *Rev. Mod. Phys.* **14**, 96 (1942).
(232) Meggers, W. F., *J. Research Natl. Bur. Standards* **47**, 7 (1951).
(233) Meggers, W. F., Humphreys, C. J., *J. Research Natl. Bur. Standards* **28**, 463 (1942).
(234) Meggers, W. F., Murphy, R. J., *Ibid.*, **48**, 334 (1952).
(235) Meyer, A. J. P., Taglang, P., *Compt. rend.* **231**, 612 (1950).
(236) Michels, A., Wassenaar, T., *Physica* **16**, 253 (1950).
(237) Michels, A., Wassenaar, T., Zwietering, T. N., *Ibid.*, **18**, 63 (1952).
(238) Moessen, G. W., Aston, J. G., Ascah, R. G., *J. Chem. Phys.* **22**, 2096 (1954).
(239) Molnar, J. P., Hitchcock, W. J., *J. Opt. Soc. Amer.* **30**, 523 (1940).
(240) Monval, P. M., *Bull. soc. chim.* [Ser. 4] **39**, 1349 (1926).
(241) Moore, C. E., *Natl. Bur. Standards (U. S.), Circ.* **467** (1949).
(242) Morris, L. D., Scholes, S. R., *J. Am. Ceram. Soc.* **18**, 359 (1935).
(243) Myers, C. E., M.S. thesis, Purdue University, 1953.
(244) Naylor, B. F., *J. Chem. Phys.* **13**, 329 (1945).
(245) Neel, L., *Compt. rend.* **207**, 1384 (1938).
(246) Nernst, W., Schwers, F., *Sitzber. preuss. Akad. Wiss. Physik.-math. Kl.*, 1914, p. 355.
(247) Nesmeyanov, A. N., Iofa, B. Z., *Doklady Akad. Nauk S.S.S.R.* **98**, 993 (1954).
(248) Neumann, K., Lichtberger, E., *Z. physik. Chem.* **A184**, 89 (1939).
(249) Niwa, K., Sibata, Z., *J. Fac. Sci. Hokkaido Univ. Ser. III*, **3**, 53 (1940).
(250) *Ibid.*, p. 75.
(251) O'Donnell, T. A., *Australian J. Sci., Ser. A* **8**, 485 (1955).
(252) *Ibid.*, pp. 493–500.
(253) Oelsen, W., *Arch. Eisenhüttenw.* **26**, 519 (1955).
(254) Oelsen, W., Oelsen, O., Thiel, D., *Z. Metallkunde* **46**, 555 (1955).
(255) Oriani, R. A., Jones, T. S., *Rev. Sci. Instr.* **25**, 248 (1954).
(256) Parkinson, D. H., Quarrington, J. E., *Proc. Phys. Soc. (London)* **68A**, 762 (1955).
(257) Parkinson, D. H., Simon, F. E., Spedding, F. H., *Proc. Roy. Soc. (London)* **207A**, 137 (1951.)
(258) Pearlman, N., Keesom, P. H., *Phys. Rev.* **88**, 398 (1952).
(259) Persoz, B., *Compt. rend.* **208**, 1632 (1939).
(260) Pickard, G. L., Simon, F. E., *Proc. Phys. Soc. (London)* **61**, 1 (1948).
(261) Pilling, N. B., *Phys. Rev.* **18**, 362 (1921).
(262) Preuner, G., Brockmöller, J., *Z. physik. Chem.* **81**, 129 (1912).
(263) Priselkov, I. A., Nesmeianov, A. N., *Doklady Akad. Nauk S.S.S.R* **95**, No. 6, 1207 (1954)
(264) Rauh, E. G., Thorn, R. J., *J. Chem. Phys.* **22**, 1414 (1954).
(265) Redfield, T. A., Hill, J. H., U. S. Atomic Energy Commission **ORNL** 1087, Sept. 24, 1951.
(266) Reimann, A. L., Grant, C. K., *Phil. Mag.* **22**, 34 (1936).
(267) Richards, A. W., private communication, Imperial Smelting Corp., Ltd., Avonmouth, England.

(268) Richardson, D., "Proc. Fifth Summer Conference on Spectroscopy and Applications," p. 64, Wiley, New York, 1938.
(269) Rodebush, W. H., Dixon, A. L., *J. Am. Chem. Soc.* **47**, 1036 (1925).
(270) Rodebush, W. H., Dixon, A. L., *Phys. Rev.* **6**, 851 (1925).
(271) Rossini, F. D., "Chemical Thermodynamics," Wiley, New York, 1950.
(272) Rossini, F. D., Gucker, F. T., Johnston, H. L., Pauling, L., Vinal, G. W., *J. Am. Chem. Soc.* **74**, 2699 (1952).
(273) Rossini, F. D., Pitzer, K. S., Arnett, R. L., Braun, R. M., Pimentel, G. C., "Selected Values of Physical and Thermodynamic Properties of Hydrocarbons and Related Compounds," Am. Petroleum Inst. Research Project **44**, Carnegie Press, Pittsburgh, Pa., 1953.
(274) Rossini, F. D., Wagman, D. D., Evans, E. H., Levine, S., Jaffe, I., *Natl. Bur. Standards (U. S.), Circ.* **500** (1952).
(275) Roth, W. A., Meyer, I., Zeumer, H., *Z. anorg. Chem.* **214**, 309 (1933).
(276) Roth, W. L., DeWitt, T. W., Smith, A. J., *J. Am. Chem. Soc.* **69**, 2881 (1947).
(277) Rudberg, E., *Phys. Rev.* **46**, 763 (1934).
(278) Rudberg, E., Lempert, J., *J. Chem. Phys.* **3**, 627 (1935).
(279) Ruff, O., Bergdahl, G., *Z. anorg. Chem.* **106**, 76 (1919).
(280) Ruff, O., Hartmann, H., *Ibid.*, **133**, 29 (1924).
(281) Ruff, O., Johannsen, O., *Ber.* **38**, 3601 (1905).
(282) Ruff, O., Keilig, F., *Z. anorg. Chem.* **88**, 410 (1914).
(283) Ruff, O., Konschak, M., *Z. Elektrochem.* **32**, 515–54 (1926).
(284) Russell, H. N., *J. Opt. Soc. Amer.* **40**, 550 (1950).
(285) Schneider, A., Esch, U., *Z. Elektrochem.* **45**, 888 (1939).
(286) Schneider, A., Schupp, K., *Ibid.*, **50**, 163 (1944).
(287) Schofield, T. H., Bacon, A. E., *J. Inst. Metals* **82**, 167 (1953).
(288) Schuman, R., Garrett, A. B., *J. Am. Chem. Soc.* **66**, 442 (1944).
(289) Schuurmans, P., *Physica* **11**, 419 (1946).
(290) Scott, D. H., *Phil. Mag.* **47**, 32 (1924).
(291) Searcy, A. W., *J. Am. Chem. Soc.* **74**, 4789 (1952).
(292) Searcy, A. W., Freeman, R. D., *Ibid.*, **76**, 5229 (1954).
(293) Searcy, A. W., Freeman, R. D., *J. Chem. Phys.* **23**, 88–90 (1955).
(294) Searcy, A. W., Freeman, R. D., Michel, M. C., *J. Am. Chem. Soc.* **76**, 4050 (1954).
(295) "Selected Values of Chemical Thermodynamic Properties," National Bureau of Standards, Washington, D. C.
(296) Selincourt, M. de, *Proc. Phys. Soc. (London)* **52**, 348 (1940).
(297) Serebrennikov, N. N., Gel'd, P. V., *Doklady Akad. Nauk S.S.S.R.* **87**, 1021 (1952).
(298) Shenstone, A. G., *Proc. Roy. Soc. (London)* **A219**, 419 (1953).
(299) Sherwood, E. M., Rosenbaum, D. M., Blocher, Jr., J. M., Campbell, I. E., *J. Electrochem. Soc.* **102**, 650 (1955).
(300) Shomate, C. H., *J. Chem. Phys.* **13**, 326 (1945).
(301) Shomate, C. H., *J. Phys. Chem.* **58**, 368 (1954).
(302) Simon, F., Lange, F., *Z. Physik.* **15**, 312–21 (1923).
(303) Simon, F., Ruhemann, F., *Z. physik. Chem.* **129**, 321 (1927).
(304) Simon, F. E., Swenson, C. A., *Nature* **165**, 829 (1950).
(305) Simon, F., Zeidler, W., *Z. physik Chem.* **123**, 383 (1926).
(306) Sims, C. T., Craighead, C. M., Jaffee, R. I., *J. Metals* **7**, *AIME Trans.* **203**, 168 (1955).
(307) Skinner, G. B., dissertation, Ohio State University, 1951.
(308) Skinner, G. B., Edwards, J. W., Johnston, H. L., *J. Am. Chem. Soc.* **73**, 174 (1951).
(309) Skinner, G. B., Johnston, H. L., *Ibid.*, **73**, 4599 (1951).
(310) Skochdopole, R. E., Griffel, M., Spedding, F. H., *J. Chem. Phys.* **23**, 2258 (1955).
(311) Slansky, C. M., Coulter, L. V., *J. Am. Chem. Soc.* **61**, 564 (1939).
(312) Smith, Jr., W. T., Oliver, G. D., Cobble, J. W., *Ibid.*, **75**, 5785 (1953).
(313) Spedding, F. H., private communication, Iowa State College, Ames, Iowa.
(314) Spedding, F. H., Daane, A. H., *J. Metals* **6**, 504 (1954).
(315) Spedding, F. H., Miller, C. F., U. S. Atomic Energy Commission **I.S.C. 167**, July 25, 1951.
(316) Speiser, R., Johnston, H. L., *J. Am. Chem. Soc.* **75**, 1469 (1953).
(317) Speiser, R., Johnston, H. L., Blackburn, P. E., *Ibid.*, **72**, 4142 (1950).
(318) Stephenson, C. C., private communication, Dept. of Chemistry, Massachusetts Institute of Technology, Cambridge, Mass.
(319) Stimson, H. F., *Am. J. Phys.* **23**, 614 (1955).
(320) Stock, A., Gibson, G. E., Stamm, E., *Ber.* **45**, 3527 (1912).

(321) St. Pierre, G., Chipman, J., *J. Am. Chem. Soc.* **76,** 4787 (1954).
(322) Stull, D. R., *Ind. Eng. Chem.* **39,** 517 (1947).
(323) Sykes, C., Wilkinson, H., *Proc. Phys. Soc. (London)* **50,** 834 (1938).
(324) Taylor, J. B., Langmuir, I., *Phys. Rev.* **51,** 753 (1937).
(325) Thomson, G. W., Garelis, E., "Sodium—Its Manufacture, Properties and Uses," ACS Monograph, chap. 3, M. Sittig, ed.; to be published in 1956.
(326) Thorn, R. J., Winslow, G. H., *J. Chem. Phys.* **23,** 1369 (1955).
(327) Todd, S. S., *J. Am. Chem. Soc.* **72,** 2914 (1950).
(328) Toennies, J. P., Greene, E. F., *J. Chem. Phys.* **23,** 1356 (1955).
(329) Tomlin, D. H., *Proc. Phys. Soc. (London)* **67B,** 787 (1954).
(330) Trees, R. E., Harvey, M. M., *J. Research Natl. Bur. Standards* **49,** 397 (1952).
(331) Udy, M. C., Boulger, F. W., *J. Metals* **6,** 207 (1954).
(332) Umino, S., *Sciences Repts. Tôhoku Imp. Univ., First Ser.* **15,** 597 (1926).
(333) Valentiner, S., *Z. anorg. u. allgem. Chem.* **277,** 201 (1954).
(334) Van Den Berg, G. J., Klinkenberg, P. F. A., Van Den Bosch, J. C., *Physica* **18,** 221 (1952).
(335) Van Den Bosch, J. C., Van Den Berg, G. J., *Ibid.*, **15,** 329 (1949).
(336) Van Dusen, M. S., Dahl, A. I., *J. Research Natl. Bur. Standards* **39,** 291 (1947).
(337) Vetter, F. A., Kubaschewski, O., *Z. Elektrochem.* **57,** 243 (1953).
(338) Vines, R. F., "The Platinum Metals and Their Alloys," International Nickel Co., New York, 1941.
(339) Wagman, D. D., Kilpatrick, J. E., Taylor, W. J., Pitzer, K. S., Rossini, F. D., *J. Research Natl. Bur. Standards* **34,** 143 (1945).
(340) Wallace, W. E., Craig, R. S., Krier, C. A., U. S. Atomic Energy Commission **NYO 6325,** April 6, 1955.
(341) Wartenberg, H. von, *Z. Elektrochem.* **19,** 482 (1913).
(342) Weertman, J., Burk, D., Goldman, J. E., Am. Phys. Soc. Meeting, Jan. 31 to Feb. 2, 1952; private communication by J. E. Goldman.
(343) White, D., Friedman, A. S., Johnston, H. L., *J. Am. Chem. Soc.* **72,** 3565 (1950).
(344) *Ibid.*, **73,** 5713 (1951).
(345) Wichers, E., *Ibid.*, **76,** 2033 (1954).
(346) Wilhelm, R. M., *Natl. Bur. Standards (U. S.), Bull.* **13,** 655 (1917).
(347) Wittig, F. E., *Z. Elektrochem.* **54,** 288 (1950).
(348) Wittig, F. E., *Z. Metallkunde* **43,** 158 (1952).
(349) Woolley, H. W., *J. Research Natl. Bur. Standards* **40,** 163 (1948).
(350) Woolley, H. W., Scott, R. B., Brickwedde, F. G., *Ibid.*, **41,** 379 (1948).
(351) Yost, D. M., Russell, Jr., H., "Systematic Inorganic Chemistry," Prentice-Hall, New York, 1944.
(352) Zwikker, C., *Physica* **6,** 361 (1926).

Index

Element	Discussion	Reference State Table	Element	Discussion	Reference State Table
Actinium	10	36	Neodymium	23	132
Aluminum	10	37	Neon	23	134
Antimony	10	39	Nickel	24	135
Argon	11	43	Niobium	24	137
Arsenic	11	44	Nitrogen	24	139
Astatine	11	48	Osmium	24	141
Barium	11	51	Oxygen	25	143
Beryllium	12	53	Palladium	25	145
Bismuth	12	55	Phosphorus	25	147
Boron	12	58	Platinum	26	152
Bromine	13	60	Polonium	26	154
Cadmium	13	63	Potassium	27	157
Calcium	13	65	Praseodymium	27	160
Carbon	14	67	Promethium	27	161
Cerium	14	71	Protactinium	27	162
Cesium	15	72	Radium	27	163
Chlorine	15	75	Radon	27	165
Chromium	15	77	Rhenium	28	166
Cobalt	16	79	Rhodium	28	168
Copper	16	81	Rubidium	28	170
Dysprosium	16	83	Ruthenium	28	173
Erbium	17	84	Samarium	29	175
Europium	17	85	Scandium	29	177
Fluorine	17	87	Selenium	29	179
Francium	17	89	Silicon	29	183
Gadolinium	17	91	Silver	30	185
Gallium	17	93	Sodium	30	187
Germanium	18	95	Strontium	30	190
Gold	18	97	Sulfur	31	192
Hafnium	18	99	Tantalum	31	196
Helium	19	101	Technetium	32	198
Holmium	19	102	Tellurium	32	200
Hydrogen	19	103	Terbium	32	203
Indium	19	105	Thallium	32	204
Iodine	20	107	Thorium	32	206
Iridium	20	110	Thulium	33	207
Iron	20	112	Tin	33	208
Krypton	21	114	Titanium	33	210
Lanthanum	21	115	Tungsten	34	212
Lead	21	117	Uranium	34	214
Lithium	21	119	Vanadium	34	216
Lutetium	22	122	Xenon	34	218
Magnesium	22	124	Ytterbium	35	219
Manganese	22	126	Yttrium	35	221
Mercury	22	128	Zinc	35	223
Molybdenum	23	130	Zirconium	35	225